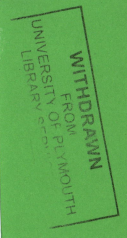

GREENING
A BROWN
LAND

THE AUSTRALIAN
SEARCH FOR
SUSTAINABLE
LAND USE

For Julie and Roseann;
and our children Ilka, Evan and Ceira,
and Joanna, Belinda and Alice,
who will live through another generation
of the search for sustainable land use.

GREENING A BROWN LAND

THE AUSTRALIAN
SEARCH FOR
SUSTAINABLE
LAND USE

NEIL BARR
Sustainable Development Unit,
Victorian Department of Agriculture

JOHN CARY
School of Agriculture and Forestry,
University of Melbourne

M

First published 1992 by
MACMILLAN EDUCATION AUSTRALIA PTY LTD
107 Moray Street, South Melbourne 3205
6 Clarke Street, Crows Nest 2065

Associated companies and representatives
throughout the world

National Library of Australia
cataloguing in publication data

Barr, N. F. (Neil Francis).
    Greening a brown land: an Australian search for sustainable land use.

    Bibliography.
    Includes index.
    ISBN 0 7329 1423 X.
    ISBN 0 7329 1422 1 (pbk.).

    1. Land use, Rural – Australia. 2. Land use, Rural – Australia –
History. 3. Sustainable agriculture – Australia. 4. Sustainable
agriculture – Australia – History. I. Cary, J. W. (John William). II.
Title.

333.76150994

Set in Janson
by Superskill Graphics, Singapore
Cover illustration: Arthur Boyd *Wheatfield, Berwick* 1948
                Courtesy Joseph Brown Collection

Contents

The smell of sex
Natural farming
Poisons in our food

Part Three: A green vision

Acknowledgments

There are many people whose advice and work we have benefited from when writing this book. We particularly wish to thank Bruce Davidson, Gerald Halloran, Roger Wilkinson, Craig Ewers, Tony Ellington, David Coventry, Meredith Mitchell, Alan Wilson, John Avery, Michael Read, Ian Barrass, John Griffiths, Ken McDougal, Andrew Campbell, Bob Edgar, Kate Irving, Joseph Woelfel and Alistair Watson.

The following art galleries and libraries have kindly given permission to reproduce paintings and drawings: City of Hamilton Art Gallery, National Library of Australia, National Gallery of Victoria, Mitchell Library, State Library of New South Wales, La Trobe Library, State Library of Victoria. We gratefully acknowledge the use of photographs kindly supplied by the La Trobe Collection, State Library of Victoria (p. 21); Andrew Campbell (pp. 26, 89, 90, 94, 102, 103, 167); the Department of Agriculture, Victoria (pp. 27, 28, 33, 36, 37, 82, 120, 124, 150, 186, 189, 226, 244, 252, 265); Department of Conservation and Environment, Victoria (p. 58); A. Ellington (p. 139); G. Steed (p. 140); John Avery (p. 159); Bill Washington (p. 190); and Rural Water Commission (p. 275). Thanks are also due to Angus and Robertson for permission to quote from *Been There Before* by A. B. Paterson.

Some of the work concerning current perceptions of land use was supported by research grants from the National Soil Conservation Program and the Land and Water Research and Development Corporation.

Introduction

In a flight across the ancient land of Australia, from east to west or north to south, the pervading colour is a parched brown or red ochre. The settled seaboard is a green fringe. The other striking feature is the flatness of the country. Australia is the flattest, driest and oldest inhabited continent. The vast flatness, predominating dryness and age of the continent are inextricably linked. Australia is a dry country because of its orientation in the middle latitudes and because it has no really high mountains. Australia is flat because it is the most geologically stable and oldest of the continents.

The flatness and dryness of the continent bear pervasively on our history and patterns of land use. The ancient, weathered, and often easily erodible, soils are much older and much less fertile than soils in most other places. Australia is thirty times larger than that small European country, the United Kingdom, yet the United Kingdom has about one-quarter of Australia's sheep population, half of its cattle population and nearly the same cereal production. Human settlement has been a process of learning to live within the delicately balanced capacity of the dry infertile soils. As in any learning, there have been failures and successes. There has been a recurring temptation to intensify land use beyond soil capacity.

An even stronger theme throughout the history of European agriculture in Australia has been the longing for water. Much of the failure of our farming methods can be traced back to an inability to comprehend or to adapt to the land of droughts and flooding rains. Today, we might worry whether our agriculture is sustainable into the next century and beyond. The early settlers worried whether their farms would be sustainable through summer until the autumn break. They applied English farming methods, which had been developed in a land of cool wet summers, to a land where summers were dry, hot and long. Farming methods had to be relearned in the brown land. The time for learning has been brief and the experience forced. This book explores that learning. In part it is a history of our agriculture. It is also a study of adaptation, which continues today. It is the story of a 200-year search for sustainable farming systems.

The idea for this book came while we were on a hill in Australia's green margin. Being an Australian hill it was called a mountain: Mount Greenock. We had only recently completed interviewing farmers as part of a study of tree planting. We were investigating why farmers did not plant trees on their farms. The implicit assumptions were simple and comfortable. Previous generations of farmers had cleared the landscape of trees.

Successive governments had given active support to this destruction. Now governments had more enlightened and sensitive policies. Farmers were being encouraged to replant at least some of the trees to control such problems as soil salting. Some of the farmers we interviewed had a more complex understanding of the issues behind tree planting. One farmer wondered why his neighbours were keen for him to plant trees on his hill. There had never been trees on the hill. It had always been bare. His 'hill' was an extinct volcano — grassy, rocky and treeless. Had it always been so?

We sought insight from the journal of the first European to record a passage through that part of the country. Thomas Mitchell passed close by the farmer's hill, but not close enough to record any useful description of its tree cover. Mitchell had climbed nearby Mount Greenock, a similarly sleeping volcano, a day's march away. He had sketched the vista from the summit and reproduced it in his journal. The hills were mostly bare. The sketch was the best evidence available of where trees had grown. We too climbed Mitchell's hill. The scene that lay at our feet was remarkably similar to that which Mitchell had sketched over 150 years earlier. The forests and isolated stands of trees were still there. Very little seemed to have changed.

A few weeks later a report was released, which mapped the loss of trees across Australia since white settlement. Surprisingly, it was claimed that in the small shire in which Mount Greenock stood, 60 per cent of the trees had been cleared since white settlement.[1] Our evidence suggested that, at least in some places, little had changed since Mitchell passed in 1836. The idea of a book was born — a book to explore the history of the land, to describe its current state and the successes and failures in its management.

It became clear to us that much of the public debate over sustainable land use was conducted from a position of simple certainty. Simple certainties had pushed the debate to the forefront of the political discussion. Behind the simple certainties there was no agreement about even whether there was a problem. For example, soil scientists have assessed that over half the land in agricultural use in Australia is in need of treatment because it has been affected by land degradation of one form or another.[2] Another strongly held view is that such pessimism doesn't stand the test of commonsense analysis. If the aggregate productivity of land has not declined, can we say that land degradation is such a serious problem? Would agricultural production have more than doubled in Australia over the past forty years, almost exclusively on land that has been farmed for over forty years, if soils were steadily deteriorating?[3]

Disagreement such as this is inevitable in any debate over sustainable land use. Sustainability, like goodness, is an abstract idea with which few people disagree. When you come to grapple more closely with sustainability it is a slippery notion. Ideas such as wealth and poverty are relatively simple because, in the end, they are part of the same thing — relative positions along the length of a piece of string. We measure wealth or poverty against a scale that reflects our access to the goods and facilities provided by society. Sustainability doesn't have readily measured properties; and we require more than one piece of string because sustainability is expressed in a multiplicity of forms and values. These forms often cannot be laid end to end for a simple additive sum, providing an answer that is obvious to everyone. Instead, we argue about the relative importance of each piece of string in framing a final solution.

The journey to understanding the debate over sustainable land use is best started in the more concrete foundations of the past, though the quest must be finished in the more abstract future. So the past is a good place to begin and provides an important theme in this book. It is the hindsight on which to build a balanced perspective in looking to the future. The community view of the environment is a shifting tide. In the previous 200 years, and in the years preceding those, different settlers have viewed the land with different expectations and requirements. Land management problems have changed with different generations. Many of today's perceived problems have long been present. Some of today's perceived problems will be viewed differently with future hindsight. Time will provide solutions to some land conservation problems and will give a sense of perspective to others. Some environmental problems will remain as challenges; some will remain as serious threats to our future well-being.

We explore the history of the use of Australia's agricultural land with the premise that, more often than not, people *at the time* perceived what they were doing to be sustainable, at least at a level that was relevant to them at the time. With hindsight, it often turned out that, at some other level, what was done was unsustainable. Scientifically developed solutions to problems of unsustainable land use often subsequently presented a new set of problems. This is not simply to deride the original solutions as wrong, but to acknowledge that at a given time we have neither all the facts nor perfect knowledge. It is to acknowledge that humankind is fallible and the solution of problems to which the answers are not obvious is a characteristically human quest.

We have taken the same approach in examining the way that landholders perceive and use their land today. There is a wide range of perceptions and behaviours regarding care of the land on contemporary Australian farms. The predominating, publicly espoused view of caring for the land is often different from the perception of the problem on average farms and different from the reality of current day-to-day activity carried out on many farms. We try to explain how landholders manage their land today and the likelihood of management practices changing. Behaviour that, from a distance, may seem irrational or short-sighted may have its own underlying rationality. The management of farms and the use of land is a series of complex trade-offs. If we only focus on a single purpose we can lose the full understanding of what drives a farmer and directs decisions about land use.

The theme that pervades our account is the 200-year struggle to *green* the brown land, from the early attempts to recreate England in the land of exiles, through the dreams of a sturdy yeomanry and of turning the coastal streams inland, to the search for new crop and pasture species better suited to our climate. Three attempts to grapple with the dry summer stand out: the use and abuse of the grasslands, long fallow and the development of improved cropping practices now threatened by new forms of land deterioration, and irrigation on the parched plains. These form the three sections of the book.

Part One

Grasslands and forests

1

The early grasslands

The first Europeans to explore south-eastern Australia did not struggle through large areas of dense forest. Except for Strzelecki, the journals of the explorers record that the forest was open. In 1824 Hamilton Hume and William Hovell travelled down the northern edge of the Great Dividing Range, a route now traversed by the Hume Highway. In their journal they described some of the hills as well timbered, and others as scantily timbered or bare. Almost without exception the country to the west was grassy plains or open timber.[1] They described the plains to the north of Melbourne as wanting nothing but timber. In 1836 Major Thomas Mitchell passed farther west through the Loddon Valley, over the Central Highlands and across the Western District before returning through the lower Goulburn Valley. In his journal he described large areas of open forest and treeless plain. Mitchell could drive his cumbersome drays and livestock through these areas with little difficulty. South of the Great Dividing Range where he crossed the volcanic plains there were very few trees, mainly majestic red gums growing along the water courses. Kangaroo, poa and plains grasses covered wide areas.[2] In 1838 Joseph Hawdon passed down the Goulburn and Murray rivers to Adelaide, exploring a possible droving route for cattle. He described the lower Campaspe and Loddon river valleys as 'immense plains, here and there intersected by belts of trees'.[3] Angus McMillan discovered Gippsland to be wide open forests on the ranges and open grassy plains and alluvial river valleys.[4] The only major exception was Count Strzelecki's struggle with the dense forests of southern Gippsland.

Later migrants were surprised at how the open countryside resembled the ordered fields and copses of England. The cultured Quaker gold digger William Howitt wrote evocative letters home while he traipsed between Victorian goldfields. In these letters he expressed wonder at the park-like appearance of the land:

> We advanced over open plains, bounded on our right by downs, green, flowing in their outlines, as free from trees as the downs in England . . . The clear soft

swells rising and falling like the downs or the green hills of Derbyshire, in many places perfectly clear of wood, in others only thinly sprinkled, and the edges of forest showing themselves round at a distance, had a most civilised look . . . You would have said it was a cultivated sheep farming country, like Wiltshire; but it was as nature had left it.[5]

Howitt lacked the insight into the affairs of the native people that his son Alfred was to show in later years. The scenery was not as nature had left it! What the explorers were describing was not a natural forest, but a legacy of Australia's first farmers.

The Aboriginal heritage

Australia had been farmed for thousands of years. The Aborigines did not till the soil and they did not fence in the animals. Instead the Aboriginal people used fire to control game and increase the productive capacity of the land. Early Europeans, such as Abel Tasman, Joseph Banks, James Cook and Arthur Phillip, commented on the Aborigines' use of fire. The land the first Europeans explored had been skilfully managed and shaped by continuous and creative use of fire. Aboriginal firestick farming was reported in Queensland, in central and northern New South Wales, in south-eastern Australia and in south-western Australia.[6] Burning the bush encouraged lush young growth from the grasses. This attracted the kangaroo, wallaby and other game, which the Aborigines hunted. Burning stimulated the production of manna by the manna gum.[7] Burning cleared tracks. Burning herded game. The early sea-borne explorers recorded observations of Aborigines' constant use of fire. Tasman saw Aborigines trying to set fire to the rainforest while it was raining.[8] Cook saw fires all along the eastern seaboard. He noted the Aborigines' efficiency at firing the country:

I have observed that when they went from our tents on the Endeavour River we could trace them by the fire which they kindled in their way; and we imagined that the fires were intended some way for the taking the kanguroo [sic] . . . They produce fire with great facility and spread it in a wonderful manner . . . from the smallest spark they increase it with great speed and dexterity. We have often seen one of them run along the shore, to all appearances with nothing in his hand, who stooping down for a moment, at the distance of every fifty or hundred yards, left fire behind him, as we could see first by the smoke, and then by the flame . . . We had the curiosity to examine one of these planters of fire as he set off, and we saw him wrap up a small spark in dry grass, which, when he had run a little way, having been fanned by the air that his motion produced, began to blaze; he then laid it down in a place convenient for his purpose, enclosing a spark of it in another quantity of grass, and to continue his course.[9]

It was the same inland. Hume and Hovell's journals make constant references to the Aborigines' use of fire. Many times they described the land all around as all lit up by the natives' grass fires.[10] They fed their horses by 'leap-frogging' from one area of fresh unburnt grass to the next. At one stage the group was compelled to stop and wait for the ground ahead to cool sufficiently to allow the passage of the horses.

Fire was a crucial part of Aboriginal life. Major Mitchell described the importance of fire in his journals of his inland travels:

> Fire, grass, kangaroos, and human inhabitants, seem all dependent on each other for existence in Australia; for any one of these being wanting, the others could no longer continue. Fire is necessary to burn the grass, and form the open forests, in which we find the large forest-kangaroo . . . the omission of the annual periodic burning of the grass and the young saplings, has already pro-duced in the open forests near Sydney thick forests of young trees, where, formerly, a man might gallop without impediment, and see whole miles before him. Kangaroos are no longer to be seen there; the grass is choked by underwood; neither are there natives to burn grass . . .[11]

Archaeological research suggests that the work of countless generations of these farmers changed the landscape.[12] The Aborigines helped to create and maintain eucalypt-dominated forests. Analysis of pollen grains trapped in the mud sediments of Lake George near Canberra tells an interesting story. In the wetter areas of Australia the Aborigines arrived to a land covered with a diverse forest of casuarina, native cypress, pine, eucalypt and rainforest species.[13] A dramatic change in the vegetation of the Lake George catchment occurred about 130 000 years ago. Grasses and euca-lypts took over the catchment. The sudden increase in the amount of charcoal in the sediments at the same time is strong evidence that fire caused the decline of forest diversity. Rainforest trees, casuarinas, cypress and pine were not able to compete with the fire-resistant eucalypts in the new fire regime. Were the Aboriginal people the catalyst for the dramatic appearance of fire? The 130 000-year time scale does not yet correlate with other records of Aboriginal habitation, but the same pattern is repeated in pollen sediments in northern Queensland. The changes there date well within the currently accepted physical evidence of human occupation of the continent.

The forests created by the Aborigines were open on the hillsides and plains and denser along watercourses. The Aborigines managed these open forests with regular firing to prevent the eucalypts taking over completely in dense regrowth. The floors of the open forests had deep spongy soil and were colonised by deep rooted grasses. The rocky hilltops often had shallow soil and were treeless. They were colonised by shallow rooted grasses. The opening of the forests increased the marsupial stocking rate.

It is attractive and comforting to assume that the land's first human inhabitants shared our much more recent conservation ethic.[14] This view is our own modern equivalent of the 'noble savage' ideal. The evidence suggests we should be careful in projecting our own ideology onto an ancient culture. The Aborigines reshaped the land within the limits of their technology. Their fires decreased tree cover and reduced the genetic diversity of the forest. They changed the soil structure with the constant regime of fire, increasing erosion rates.[15] They introduced their own feral animal, the dingo. The dingo is implicated in the explosion of bush fly numbers. The Aborigines may have played a part in the extinction of large marsupials such as the *Diprotodon* and the *Phascolonus*.[16] In short, the Aboriginal farming system did not conserve the landscape of Australia. It created a new landscape, which was more productive than the landscape they found. Although they did not conserve what their ancestors found, the Aborigines created a sustainable agricultural system that lasted for tens of thousands of years, a system dependent on continued burning. In the end it was their grasslands that made their society unsustainable. The landscape they produced was particularly inviting to a new race of colonisers, the Europeans.

In the final analysis, the main difference between the white settlers and the Aborigines may well be the former's greatly enhanced capacity for destruction.[17] What would the Aborigines have done with greater access to European technology? In north-eastern Victoria the tomahawk preceded the first explorers. Presumably traded between tribes, it travelled faster than white settlement. As Hume and Hovell blazed the first European trail south from the Murray they noted in their journal the attacks made on trees with the new tool.[18] To the Aborigine, the tomahawk was a better tool to remove witchetty grubs from trees. That the tool caused greater damage to the tree seemed of little importance to a culture that had created its livelihood by burning down the forest. What would have happened after 60 000 years of the tomahawk? Possibly the country's first inhabitants would have created even more grasslands.

Invasion of the sheep

Major Mitchell was the first European to explore western Victoria. His journey into the heartland of Victoria was a turning point for both the Victorian Aborigines and the landscape they created. His transitory passing over the plains and hill country of western Victoria left an imprint on both the country and its people, white and black. Mitchell believed this was a land ripe for development, and unlike other Europeans making secret journeys, he publicised his discoveries. His enthusiastic chronicles of the lightly treed country created by the Aborigines left a deep impression in the minds of the European settlers. It was a land seen as ready for the taking, and many raced to fill the grasslands with sheep.

The aspiring pastoralists saw great opportunities. The local entrepreneurs had realised that the way to wealth in the new colony was through sheep. Australia was short of labour and far from significant markets. The climate was very variable. John Macarthur was one of the first to understand the importance of these limitations:

A petty population, established at so vast a distance from other civilised parts of the globe, could have no prospect of ultimately succeeding unless by raising as an export some raw material, which would be produced with little labour, be in considerable demand and be capable of bearing the expense of a long sea voyage; that only by the production of such a commodity, whatever might be the natural fertility of the country, could it hope to escape the alternations of abundance and scarcity, even of bread.[19]

Macarthur gained his place on Australia's former two-dollar note both because he was the first to realise wool production and the Spanish Merino were the answer to this riddle and because he had the sense or good fortune to choose a wife who was able to manage and develop their Merino flock while he attended to his many bitter personal squabbles. The Merino sheep was better adapted to heat than British breeds. It was able to walk farther to water. Most importantly, it produced fine wool, which was a valuable non-perishable commodity.

The government of the day tightly controlled the release of land for grazing and tried to encourage the development of crop farming to ensure the colony's food security. The entrepreneurs ignored the government's policies and land controls. Instead they followed the Macarthurs' business plan and Mitchell's tracks out across the land of the Aborigines. Following Mitchell was simple. His heavy carts had left a deep and lasting impression on the soft ground they traversed. The dray ruts were a double signpost to the settlers. They directly pointed the way to the grasslands. In the following decade settlers followed Mitchell's 'line' in their search for a run in Australia Felix. But the ruts were also a signpost to the fragility of the soil over which the squatters rushed. The soil was at risk to overstocking and subsequent compaction. Despite the physical signs, only the message of new lands was heeded.

Destruction of the native grasslands

The first squatters were enthusiastic at the naturally clear land and open forests. The land was covered with fine grass waiting for sheep. (See box: *Squatters' Impressions of the Aboriginal Grasslands*.)[20] There was no need to clear the land, an expense they couldn't afford.

Squatters' impressions of the Aboriginal grasslands

'This country possessed . . . the inestimable advantage of being ready for immediate use without the outlay of a sixpence. This absence of preliminary outlay will be particularly noticed by those who remember that many lands in Tasmania cost the early colonist, with prison labour, as high as 40 pounds an acre to clear, before they could be thoroughly fitted for the plough. In the category of prime country ready to hand . . . must be much of the district now known as the Riverina . . .'

Edward Curr describing the Riverina grasslands in the 1840s.

'After crossing the Wannon River, we made a new route almost east; and we met with no kind of obstruction . . . We reached Geelong on the fifth day after leaving the Glenelg. I may remark that during this journey we did not meet with any natives; the country was desolate and uninhabited, and was covered with rich kangaroo grass three and four feet high.'

Captain Foster Fyans, describing land between present-day Casterton and Geelong in 1839.

'The country we passed through today consisted of open forest, well grassed, the timber consisting of red and white gum, box, he-and she-oak, and occasionally wattle . . . the grass was up to our stirrup irons.'

Angus McMillan, first European to describe the Bairnsdale district.

'We were delighted with the country, and pronounced it first rate, either for sheep or cattle, or for agriculture . . . the natives had burnt all the grass in Gippsland late in summer. Heavy rains must have fallen before we reached there, in the month of March. The whole country was very green. It had the appearance of young corn fields; the young grass about six inches high, and in places very thick.'

William Brodribb describing the plains of Central Gippsland in 1840.

'When I first arrived . . . I cannot express the joy I felt at seeing such splendid country before me . . . the whole of the Wannon had been swept by a bushfire in December, and there had been a heavy fire in January (which has happened, less or more, for the last 13 years), and the grasses were about four inches high, of that lovely dark green; the sheep had no trouble to fill their bellies; all was edible; nothing had trodden the grasses before them. I could neither think nor sleep for admiring this new world to me who was fond of sheep.'

John Robertson describing Victoria's Western District in 1840.

'Kangaroo grass two feet high, and as thick as it could stand. Good hay could be made, and in any quantity. The trees not more than six to the acre, and those small she-oak and wattle . . . most of the high hills were covered with grass to the summit, and not a tree, although the land was as good as land could be. The whole appeared like land laid out in farms for some hundred years back, and every tree trans-planted.'

John Batman describing the Corio district in 1835.

In the pioneering society of last century, Aborigines were viewed as a primitive and unskilled people who lived off the land without making any effort to improve either their lot or the land. The image of the impotence of the native was popular in the towns. Many years later in 1885 the popular correspondent of the Melbourne *Argus*, 'Vagabond', wrote for his Melbourne audience on the subject: 'Aboriginal races do nothing to assist nature. The white man alone coaxes and cultivates mother earth and gives unto it the sweat of his brow.'[21]

This popular portrayal was the basis of the European justification for the displacement of the Aboriginal race from their grassland home. If the Aborigine did nothing to improve nature, then the race must give way to those who will. The impact of the Aborigines on their environment was noticed by the more observant farmers of the day. Two years before Vagabond's dismissal of the achievements of the Aboriginal race, the retired Echuca district squatter Edward Curr had published a more charitable and perceptive account of the Aboriginal legacy:

> Mere hunters, who absolutely cultivated nothing — the spear, the net and the tomahawk — could have produced no appreciable effect on the natural products of a large continent. Nor did they; but there was another instrument in the hands of these savages which must be credited with results which it would be difficult to overestimate. I refer to the firestick; for the blackfellow was constantly setting fire to the grass and trees, both accidentally and systematically for hunting purposes. Living principally on wild roots and animals, he tilled his land with fire and cultivated his pastures with fire; and we shall not, perhaps, be far from the truth if we conclude that every part of New Holland was swept over by a fierce fire, on an average, once every five years. That constant and extensive conflagrations could have occurred without something more than temporary consequences seems impossible, and I am disposed to attribute to them many important features of nature here: for instance the baked, calcined, indurated condition of the ground common to so many parts of the continent, the remarkable absence of mould which should have resulted from the accumulation of decayed vegetation, the comparative unproductiveness of our soils, the character of our vegetation and its scantiness, the retention within bounds of our insect life.[22]

The European invasion destroyed the Aboriginal way of lie. Rangeland skirmishes, poisoning and disease all took a terrible toll. With the grassland people gone, the squatters had to learn to emulate the Aborigines. The native grasses quickly became dry and rank if left unburnt. The sheep did not do well on the dry stalks. Burning stimulated fresh, lush growth. The discovery of fire was a liberation to Edward Curr; he regularly burnt portions of his run to produce feed that was 'particularly wholesome and fattening' for his sheep.[23] Burning the rank growth in autumn each year became a major activity on many squatters' runs. In the Wimmera the use of fire was well integrated into grazing practice, as the squatter J. C. Hamilton recalled:

> Sheep were all shepherded in those days, and provision was made for a place to fly for safety in case of fire. This was done by burning the grass in patches when it was half green. These patches were all over the run, and when a shepherd saw a fire coming he could drive his flock there, and remain in safety until the fire passed . . . We were in the habit of burning all rubbishy country in autumn. I,

myself, made a practice of setting aside all station work in March, and, taking five or six men and a supply of water, we burned the country in comparative safety.[24]

In western Victoria, William Moodie observed that his neighbour was sanguine about the effect of fire, which deprived him of his woolshed, but improved the feed for his sheep and lambs such that 'they would make full compensation in one year'.[25]

Burning was not the complete answer to sustaining the native pastures. Further challenges loomed. The native grasses were adapted to the Australian cycle of drought and plenty, and to the eating habits and the soft paws of the marsupials. The palatable grasses of the hillsides and flats were clumpy and deep rooted. These native perennial grasses were adapted to the mix of good and bad seasons typical of much of Australia. Their clumpiness protected their root systems from desiccation during dry weather. The perennials held their place in the dry seasons and grew prolifically in the good seasons. The marsupials were not aggressive eaters. They nibbled the leaves at the base of the native perennial grasses, leaving the seed head to the parrots and finches and leaving the roots to restore the grass when the rain came again. Seeds spread simply by growing where they fell.

Sheep were alien to this ecosystem. They could not travel far to water, so their movement was concentrated around watering holes and streams. Their solid hooves compacted the soil. They were selective grazers, eating out the grasses, which they preferred. They were much more aggressive grazers. They ate the whole of the plant, bottom leaves and seed head. When feed was short they grubbed at the protected roots under the clumps of grass. In many areas the more palatable native perennial grasses were not able to compete against other species in this new environment. They were over grazed, their seeds were not given a chance to germinate and they were not able to exploit the compacted soil. To make matters worse, squatters often grazed too many sheep, lulled into false optimism by the lush growth of good seasons after burning.

The grassland environment came to resemble the shallow soiled environment of the stony hilltops. Deep rooted summer growing perennials were replaced by the shallow rooted native annual grasses and short winter growing perennials. Many of these annual grasses had wiry or burr-like seeds. Corkscrew grass produces corkscrew shaped seeds. Its seeds use changes in moisture to contract and expand, burrowing into soil, or flesh. In 1855 William Howitt complained about these seeds:

When these become ripe they are like so many needles: and it is a point to get the sheep washed and clipped before these seeds are ripe, or they fill the wool and ruin the fingers of all who attempt to clean, spin or weave it. They seem furnished with little barbs or scales, which continually push them forwards on the least motion, so that they are forced into the sheep's skin in the thousands, and even penetrating to the lungs . . . I have seen skins shown to me after the sheep were killed, regularly bristled inside with the points of those vegetable needles . . . the seeds penetrated our trousers in all directions like so many pins.[26]

These grasses quickly colonised the compacted slopes and flats by taking a ride on the sheep's back, much to the discomfort of the sheep. Native annual grasses or introduced weeds replaced the palatable perennial grasses. The new grasses were not as productive as the original native grass pastures and the carrying capacity of the land declined. A handful of squatters described the decline of pastures in their memoirs. (See box: *The Decline of the Native Grasses*.)[27] The squatter John Robertson saw the dreaded silver grass invading his Western District property, just as it had done in Tasmania. In reality, the silver grass was merely colonising bare patches where native grasses had been destroyed. It is not considered a weed today, only a sign of poor pasture management.[28]

The decline of the native grasses

'Over pasturing certainly seems to do considerable injury, and the old residents will still smile to hear the new-comers extol the pastoral richness of the newly discovered county: "Wait, wait", they will say, "till they have been as long and as heavily pastured as the old country".

> P. Cunningham describing the settled areas of New South Wales in 1827.

'. . . the grasses originally grew in tussocks, standing from two to twenty feet apart, depending on the circumstances. It bore no resemblance to a sward . . . Gradually, as the tussocks got fed down by sheep and cattle, they stooled out; and the seed got trampled into the ground around them, and in the absence of bushfires grew, so that presently a sward more or less resulted, such as we see at present [1883] . . . throughout the continent the most nutritious grasses were originally the most common; but in consequence of constant overstocking and scourging the pastures, these, where not eradicated, have very much decreased, their places taken by inferior sorts and weeds.'

> Edward Curr describing changes on the Riverina Plains

'In 1846 all the country round here, [called] the West Wimmera was covered with kangaroo grass, splendid summer feed for stock of all kinds. It was at its best during January, February and March, and remained good up to May, but it lost its colour after that and gave way to a finer grass — herbs such as yams, etc . . . When we were children at Ozenkadnook we used to play hide and seek in the long grass near the homestead, and I have known a flock of sheep to be hidden by the grass, and only be discovered by its waving as they made their way through it . . . The country was like this for some years after 1846, until destroyed.'

> J. C. Hamilton describing changes in the Wimmera

'The few sheep at first made little impression on the face of the
country for the first three or four years . . . [Then] many of our
herbaceous plants began to disappear from the pasture land; the silk
grass began to show itself in the edge of the bush track, and in
patches here and there on the hill. The patches have grown larger
every year; herbaceous plants and grasses give way for the silk grass
and the little annuals, beneath which are annual peas, and die in our
deep clay soil with a few hot days in spring, and nothing returns to
their place until late in the winter following.'

  John Robertson describing the Western District in 1853

'The indigenous kangaroo grass had not then been displaced by its
numerous weedy successors, but was in full vigour and luxuriance,
and was the most fattening grass we had. When in full growth and
ripe, it was more like a field of corn than grass, with dark brown seeds
waving in the wind and its rich green leaves at the bottom. Its
disappearance is to be attributed chiefly to the strong partiality the
sheep had for it.'

  Alfred Joyce describing grasses in the Maryborough district

Other early settlers did not appear to be concerned about pasture
decline. Thomas Chirnside of the Ararat district, writing to Governor La
Trobe at the same time as John Robertson, noted that settlers generally
believed stocking had improved the land. Writing seventy-five years later,
agricultural scientists were warning farmers with native pastures against
the same mistakes:

In [drought] years, the better types of native grasses are eaten or killed out
owing to their slow growth and seeding habits. These grasses are replaced by
such plants as barley grasses, the useless soft brome, barren fescue, thistles of
various kinds, capeweed and plants of low grazing value . . . Even in normal
times many stock owners carried stock in such large numbers that the good
grasses had no chance to seed, with the results that the better grasses are
replaced by . . . herbage of lower grazing value than the original pasture.[29]

What do we make of this? Was pasture decline widespread, or did the
majority not notice it? If pasture decline was widespread, it is possible that
few squatters survived on their holdings long enough to observe it. Financial problems were far more pressing matters for most squatters than slow
and subtle changes in pasture composition. In the last few years of the
1830s the squatting rush was frenetic. Many of the more recent entrepre-
neurial excesses of the 1980s were similarly in evidence: extensive borrow-
ing against assets rapidly increasing in value well beyond intrinsic or
productive worth. This was a period of conspicuous consumption excesses,
funded, temporarily, by speculative gain and high wool prices. Similarly,
much of this excess was brought about by foreign (in this case British)
capital pouring into Australia faster than it could be absorbed in legitimate

business, even taking into account the squatting expansion. In 1839 allotments purchased originally for £150 changed hands at £10 244.[30] Such
booms come to an end and many of the squatters without large capital
backing went out of business; others took their place buying stations in
some cases for little more than a year's income.[31]

The earliest squatters had little security of tenure; they had, in the
words of Governor Bourke, 'an uncertain and insecure occupation'.[32] Many
of the squatters were out to make a quick fortune and then return to the
comfort of a wealthy life in England or Scotland. The inevitable depression of the early 1840s saw all but the most solidly established squatters
forced off their runs by a combination of low prices and high interest rates.
Most of Robertson's squatting neighbours were forced to sell when wool
prices fell and commitments could not be met. The value of Robertson's
holdings declined to one-seventh of what he had paid. He was only able to
survive because of his exceptional and significant savings. Throughout the
squatting period runs changed hands at regular intervals. Many who held
runs for long periods were absentee business men who spent their time in
Melbourne rather than living on their runs. Very few had the luxury of
observing changes over a sufficiently long time span.

The decline of the perennial native grasses changed the water balance
on the grasslands. The original vegetation, such as kangaroo grass, was
adapted to respond quickly to the occasional heavy rainfall of summer. The
originally uncompacted and spongy soil was able to store large amounts of
water. Where the native grass declined and the soil was compacted the land
was no longer protected from the occasional but heavy summer rains.
Runoff on the bare compacted soil increased dramatically. Erosion developed around water courses and waterholes where there was greater stock
traffic, and spread from there.

In New South Wales, Cunningham was aware of the possible damage
by stock to the new pasture regime:

> It is astonishing to see how quickly and how luxuriously the new grasses push up
> after a burning, if a shower of rain should happen to follow them. When
> judiciously accomplished, this certainly produces most beneficial effects . . . You
> must caution, however, not to put cattle or horses upon the burnt ground too
> early as they . . . materially injure the pasture.[33]

When annual grasses and winter growing perennials replaced the summer growing perennial grasses, the soil was at the mercy of heavy summer
rainstorms in areas where livestock concentrated. On many clay based
Australian soils the pounding of heavy rain brings the smaller soil clay
particles to the surface where they crust together. The crust reduces the
speed with which soil can absorb water. The compaction of soil by hoofed
animals reduces the capacity of the soil to hold water. Reduced infiltration
and reduced water storage capacity increases the speed and quantity of
runoff immediately after rainfall. The force of flowing water increases
dramatically with small increases in speed and the fast runoff carves gullies
in previously stable creek beds. On his run near the Murray River Edward

Curr noticed that, by trampling of stock, 'the ground has been hardened and drainage increased'.[34] John Robertson made some, since often quoted, observations about soil erosion on his western Victorian run:

> The long deep rooted grasses that held our clay together have died out; the ground is now exposed to the sun . . . the clay is left perfectly bare in summer. The strong clay cracks, the winter rain washes out the clay; now mostly every little gully has a deep rut; when the rain falls it runs off the hard ground, rushes down these ruts, runs into the larger creeks, and is carrying earth, trees and all before it. Over Wannon country is now as difficult a ride as if it were fenced. Ruts seven, eight, and ten feet deep, and as wide, are found for miles, where two years ago it was covered with tussocky grass like a land marsh. When I first came here, I knew of but two landslips, both of which I went to see; now there are hundreds found within the last three years.[35]

Once a gully has carved its way through a paddock, water flowing from the paddock into the gully will fall over a vertical drop to reach the gully floor. In heavy rain, sides of gullies become waterfalls, which create new side gullies. In time, side gullies will grow out radially from the main gully. Large amounts of soil and subsoil are lost as the sides of the gully are undercut, fall and are swept away. Much of the fertility of soils is found in the fine particles near the surface, which are washed away. Where the runoff is able to flow over large areas of exposed soil without being concentrated into channels, sheet erosion can be the result.

On the upper hillsides the soil was not as compacted. Rainfall seeped into the soil profile in greater quantities than before. In some situations this led to changes in the local or regional watertables, causing salting. Whatever water falls as rain must be used by plants, be lost as evaporation, drain away in runoff, drain to the watertable or increase soil moisture.[36] The erosion and salt problems that John Robertson observed were inevitable results of his inadvertent tampering with this water balance. Before European settlement, native grass and leaf litter covered the soils. The trees and perennial grasses were adapted to use water whenever it was available. Felling trees removed the leaf litter. Heavy grazing by stock and rabbits decimated the perennial grasses, leaving annual grasses, which died off in summer. Less vegetation meant less plant water use. Rainfall water had to be disposed as drainage, evaporation, the watertable or wetter soil. Increased drainage often meant erosion. Some of the unused water was seeping down until it reached an impervious layer of rock or clay, where it formed a 'watertable'. As the watertable rose, it dissolved the salts that had collected in the soil over thousands of years. When the watertable approached the surface of the soil it brought the salinity with it.[37]

The changes in the water balance were a mystery to the few squatters who noticed the phenomenon. Robertson was one of the few early squatters to describe this transformation. He started squatting with 1000 sheep; after five years he had 7300 sheep. Stocking pressure increased considerably; after fourteen years of settlement he estimated that his 11 810 acres were fully stocked with 8000 to 10 000 sheep, but the run was still unfenced. The grazing pressure along the creeks and water courses would have been many times higher than his average of a little under one sheep to the acre. After ten years of grazing, land slips became much more frequent. Robertson described saline seepage and associated soil erosion appearing thirteen years after he let his sheep loose on the plains.[38]

> A rather strange thing is going on now. One day all the creeks and little watercourses were covered with a large tussocky grass, with other grasses and plants, to the middle of every watercourse but the Glenelg and Wannon, and in many places of these rivers; now that the only soil is getting trodden hard with stock, springs of salt water are bursting out in every hollow or watercourse, as it trickles down the watercourse in summer, the strong tussocky grasses die before it, with all others.[39]

The canny Robertson retired to his native Scotland a year later. His observations of the limits of his land were exceptional. He is often quoted in accounts of soil degradation because few other squatters recorded concerns about erosion or salting. It is difficult to tell whether it was Robertson's property, his powers of observation derived from having been an assistant to the Tasmanian Botanist, or his zeal for correspondence that placed him apart. His peers seem to have been preoccupied with neighbouring squatters, prices or blacks.

Creating paddocks and forests

Initially the squatters' runs did not have fences. Only natural landmarks showed the borders between the land of competing squatters. Although these boundaries were recorded on lease agreements with the Crown, there was ample room for re-interpretation, and re-interpretation was common. Squatters had endless squabbles. They hired shepherds to watch their flocks and spent their own time watching their neighbours' flocks to catch boundary cribbing or spy an opportunity. The Crown Lands Commissioner in each district had the job of mediating disputes, and it was in every squatter's interest to be on the good side of the inspector.

There were no fences because, before 1847, squatters had no security of tenure. The Crown could terminate their leases with little warning, and there was no compensation for any improvements the squatter might have made. A squatter who fenced took a major risk. A change in the whim of a government inspector might see the squatter lose his lease and capital improvement without appeal. The tenancy licences prohibited many improvements. The lack of fences had one major advantage. Squatters could burn the grasslands without the risk of destroying fences.

As time passed fences became more attractive. With the arrival of the gold rushes squatters had the greatest difficulty in recruiting shepherds. As the first generation of shepherds succumbed to syphilitic madness, few other men were willing to accept the low pay in exchange for the lonely work of watching the sheep. Without fences, improving stock with better breeding was a waste of time. One neighbour's ill-bred ram could threaten any gains made by running a well-bred ram. Lack of fences hindered attempts to control stock disease. Catarrh, scab and pleuro-pneumonia caused great losses in many flocks. Sheep scab was caused by a small mite that fed on dead skin. In order to breed, it buried into the sheep's skin, causing incessant irritation to the sheep. A scabby sheep was a useless sheep for a wool producer. The scratching by the sheep pulled out wool, which was nearly worthless, and created horrible scabs. The mite was very contagious, able to jump from one sheep to another, spreading rapidly between flocks when they mixed. An outbreak of scab was disastrous for a squatter. Alfred Joyce was distraught at the appearance of scab among his flocks:

> . . . scab has broken out in every flock on the Norwood station. The discovery at first filled me with such utter dismay that the only alternative appeared to be whether to sell the place and sheep for what they would fetch, or remain and await certain ruin, for the disease being now so general, it seemed almost impossible to effect a cure, the quantities of material (arsenic or sublimate) required in dressing, in the whole colony, not being sufficient to dress one fourth of them. The scarcity of it may be imagined when it rose from 9 pence to 5 shillings per pound.[40]

No one wanted scabby sheep, so squatters began to search for cures. Mercury and lard ointment, lime and water, dilute sulfuric acid dip were all tried, with marginal success and great expense.[41] Quite by chance a Warrnambool squatter discovered that a dip of dilute acid and tobacco did the trick. A new industry of tobacco growing sprang up to provide for the squatters' dips. The tobacco cure was later superseded by a cheaper lime and sulfur mixture. By laborious hand dipping squatters could eradicate the mite from their flocks, but without fences there was no way of stopping new infections. Neighbours could not be relied on to control their infections. For drovers taking flocks over the countryside it was often easier to let infected sheep go feral, infecting local flocks.

The first fencing began in the late 1840s. Fencing accelerated when the early Selection Acts allowed squatters to secure their land by 'dummying' and 'peacocking'. As fencing spread, fire became a contentious issue. A grass fire could destroy fences in a matter of minutes. Replacement was costly. In his memoirs the Wimmera squatter Hamilton complained of the damage these fires would cause:

> The boundary of Bringalbert and Ozenkadnook would measure seventy miles, and the subdivision in to paddocks another seventy miles. These fences cost us 15 pounds per mile, and in two years had to be topped up . . . which cost nearly ten pounds a mile. This was our position in the sixties and seventies — inflammable property to the amount of upwards of 3000 pounds, and we could never tell when a fire would sweep down on us and burn both fences and sheep.[42]

Other squatters balanced the advantages of fences against the pasture gains from a good burn-off and decided in favour of the pasture. This led to many furtive and occasional open acts of grassland arson and later negotiation over compensation:

We found out by dearly bought experience that it was desirable to have some burnt country for our ewes and lambs, so there were a good many unaccountable fires! In later years when neighbours became more numerous it was not so easy to get all you would wish done in this line. While a good many of the new folk did not like the fire when it was burning they were not averse to the result. They said it was a dangerous place to live . . . [One] neighbour, who had land in a very rough place, would frequently meet me to assess the damage after some of these 'unaccountable' fires. I generally gave him fencing wire and the right to split posts on my land. I rode out one morning to find him scraping live coals off a log he thought dangerous for another outbreak, and without raising his head, when I got within speaking distance he said 'Three coils [of wire] will be enough this time'.[43]

The old ways died and fencing became the norm. The advantages were soon realised. The combination of medication, fences and an Act of Parliament enabled graziers to totally eliminate scab from the country's flocks by 1875. With properties fenced, graziers no longer had to find shepherds. The graziers also reaped the benefits of better breeding once stray 'mongrel' rams could be kept from the flocks. But fences had unintended consequences. A well-bred flock was more valuable than an expendable ill-bred mob. Graziers had a greater incentive to retain their stock when the dry seasons came, thus damaging pastures and risking erosion. As fencing took over, fire was no longer used to rejuvenate the native pastures. In some areas the results were unexpected. Vigorous woodland regrowth appeared. Modern farmers know that the easiest way to encourage regrowth is to fence an area around an old eucalypt tree and leave it to seed. Regrowth in the open forests of the Aborigines had been controlled by a combination of regular burning and grazing by the small kangaroo rats that lived in the clumpy grasslands. The sheep had decimated the environment of the small marsupials, but the squatters had unintentionally controlled the regrowth by maintaining the burning. When the regular fires stopped, the seeds of the forest were now free to recolonise.[44] The Wimmera squatter J. C. Hamilton saw parts of his own lease revert to forest, destroying the understorey grasses:

> The country when we took it up was lightly covered with red and white gum, yellow box, sheoak and honeysuckle. There were pines on the sandbanks, belts of stringybark, but only a few bulloaks here and there. The country remained open until brush fences were started, and the use of wholesale fire given up. This gave the timber a chance of going ahead as it liked. Something favourable to the honeysuckle started it first on the light sandy soil, and it became a dense scrub. This went on for some time, and then it stopped . . . Then the bulloak sprang up everywhere, taking possession of the best of the country. Where the seed came from is a mystery. Red gum also went ahead, but it does not poison the ground as much as white box, which, when it grows thickly, is the greatest enemy to vegetation.[45]

The early pastoralists were unwittingly undoing the work of their Aboriginal predecessors.

Weeds invade the pastures

To some European eyes the Australian landscape was alien, forbidding and unwelcoming. After emigrating across the world it was natural to long for home. Holidays to the homeland were rarely possible. It took a six-month boat journey to visit the 'old country'. The alternative was to try to recreate Europe in Australia. The classic homesteads of the Western District of Victoria are nearly all surrounded by green English gardens, modelled on the stately homes of the English aristocracy.

The less wealthy followed the cottage-garden style of the English yeoman farmer. The visiting observer, William Howitt, was struck by the fervour of the recreation of England:

> . . . the English stamp and English character are on all their settlements. They are English houses, English enclosures that you see; English farms, English gardens, English cattle and horses, English fowls about the yard, English flowers and plants carefully cultivated. You see great bushes of furze, even by the rudest settlers' cottages. There are hedges of sweet briar about their gardens, bushes of holly . . . There are hawthorns and young oaks in the shrubberies . . . England reproduces herself in new lands.[46]

Many of the plants of the English-style gardens ran wild. William Howitt also observed this:

The Wimmera mansion 'Woodlands', a grand Italianate homestead, was surrounded by a magnificent garden of exotic trees and shrubs.

In the glades of the woods and even in the cultivated fields I observed enormous thickets of sweetbriars, in some cases covering whole acres. The sweetbriar, the furze and the thistle — the first two introduced for fences and the third coming over with seed corn — have propagated in this climate to such an extent, that not only the thistle, but the sweetbriar and furze are beginning to be regarded as real nuisances. The birds disperse the seeds of these plants everywhere, the winds assisting to disseminate the thistles.[47]

The Scotch thistle was a particular menace. Being a biennial plant, the thistle could be controlled in two years if all standing thistles were cut down before they seeded. Large stations were forced to employ many labourers to control the thistles. If everyone in the district cut their thistles for two or three years, the weed was contained. But just as with scab, not everyone was cooperative. Some were lazy, some parsimonious and some complacent. Some graziers believed the thistle provided useful green fodder in dry seasons and were reluctant to eliminate it from their land. Where it was eliminated from pasture, the thistle still colonised the roadsides and reserves. The government of the colony of Victoria acted to overcome the lack of cooperation between landholders. In 1856 it legislated to force Victorian landowners to cut thistles each December before flowering ended. Thistle inspectors were appointed with the task of enforcing this edict.[48]

The struggle with thistles was the first skirmish in an invasion by exotic grasses and weeds. Foreign seeds came into the country accidentally in ship ballast, packaging, mattresses, crop seed and fodder. Other seeds came in because someone decided the country would be better with them established — Patterson's curse, blackberry, St John's wort and boneseed to name some of the most well known. By the turn of the century over 300 exotic plants had become established in Victoria. Two-thirds of these were of no benefit. Many were nuisance weeds.

The need for food was an obvious reason to introduce new plants. Many settlers decided to plant blackberries. We can see the pattern in the spread of the blackberry in northern Tasmania. In 1843 a Launceston resident, Phillip Oakden, imported a blackberry bush in a pot from England. He generously gave cuttings to friends who wanted them. Jacob Broadway took six cuttings by ship to Forth. Three of these went to a Mr Fenton who longed for the luscious fruit. Broadway struck his cuttings by the riverside. Fenton lovingly tended his in his garden. Three years later the vines were out of control in both locations, and worse was to come. The birds spread their seeds far and wide. New thickets developed in unpredictable places.[49] This same story was repeated in communities across southern Australia. As late as the 1890s the Victorian government botanist was encouraging blackberry planting. Today the blackberry is a major pest, choking stream banks and invading paddock gullies.

Sometimes new plants were introduced for their medicinal properties. Around 1880 an elderly German lady living in the Victorian township of Bright imported some seeds of St John's wort. The woman was a traditional midwife and St John's wort was a traditional drug for inducing human abortion. She planted the seeds in her garden. This was one

delivery she must have bitterly regretted. Once established the wort con-
solidated into a dense thicket, smothering all other growth. She took to
uprooting the plants and throwing them over the fence. Travelling stock
helped it spread to the racetrack. The locals called wort 'racecourse weed'.[50]
Horses came from local and outlying valleys to race at the track, and took
the seed home with them. Soon the wort seeds were mixing with chaff, and
the invasion could not be stopped. Within a few years the wort was
growing on properties throughout the north-eastern mountain valleys. In
1902 there were 2500 hectares of wort-infested land in the mid Ovens
Valley.[51] The degraded native pastures were no competition for this vigorous weed.

Cattle could not eat the wort. It was poisonous and it induced photosen-
sitivity in exposed skin, irritating cattle. The cattle rubbed the irritation
incessantly, causing infection and often death. The wort could grow in
shallow rocky soil or in gullies of running water. It could survive the dry
summers or a covering of snow and it spread into the mountain hillsides.
Cutting the weed back, burning and spreading dry salt proved the only
affordable methods of control. According to local opinion, a farmer's
worth was judged by the quality of his stud stock and the absence of St
John's wort from his pastures.

Entomologists sought to find natural enemies to control the weed.
About 1900 it was noticed that the native climber dodder had attacked the
wort and killed some patches. Dodder itself was a pest and farmers were
advised to be cautious in its use.[52] A few native insects fed on the wort
plant, but rarely did any damage. In other countries the phytophagous
beetle attacked the plant, but it was considered 'a dangerous experiment
indeed to introduce this beetle to Australia'.[53] In 1929 the first of five
attempts was made to import and establish natural predators. Of seven
species introduced, only one had any impact: *Chrysomela gemellata* from
southern France.[54] Initially, it defoliated the wort plants as swarms migrated through wort thickets. But, unless the farmer quickly planted a
vigorous grass pasture, the wort reappeared. Planting the new pasture was
often a waste of time because of another imported pest, the rabbit.

The *Chrysomela* could not survive in the native bushland, so it had no
impact on the weed in the bush. A farmer could control St John's wort on
his own property, only to have it reinvade from neighbouring bush. In the
narrow mountain valleys almost every property had a bush boundary.
Today, in the Ovens Valley near Bright, *Pinus radiata* plantations cover
hills on both sides of the valley. The pines, which provide sawlogs for the
local mill, were planted to protect the local pastures from the wort. Mature
pines shade out all undergrowth, even St John's wort. The plantations
provide a wort-free buffer around the valley pastures. But the buffer is not
permanent. Even after 25 years when the pines are harvested, the wort is
the first plant to spring up. It is the last to disappear as the new plantation
grows.

The list of imported weeds is still growing today as the nursery industry
imports botanical specimens from all over the world to satisfy the demands
of suburban gardeners. In Victoria there are approximately 100 species

declared noxious weeds. Each is a plant out of place, uncontained by the natural predators of its homeland. In their new environment these weeds are a costly legacy of past mistakes. However, not all new arrivals were detrimental. Clovers and medics, which migrated from the Mediterranean region, provided a new source of soil nitrogen and were to become the basis for more productive improved pastures.

Homesickness and acclimatisation societies

To some migrants the recreation of home became an obsession. In the 1860s 'acclimatisation societies' sprang up to foster the introduction of *innoxious* plants and animals for productive or ornamental reasons. With evangelical zeal the members set about the task of anglicising the fauna of the land. Members vied for the honour of successfully introducing foxes, hares, deer, blackbirds, thrushes, starlings, finches, sparrows, minahs and pigeons. Particular effort was directed to the skylark, which somehow became a symbol of nostalgia for the old country. Enthusiasts nursed the larks through long voyages across the seas. Newspapers excitedly heralded the imminent arrival of skylarks and travellers recorded joy and excitement on hearing the song of a lark in the bush. Others imported animals for profit, rather than aesthetics. In various parts of Australia donkeys, goats, horses, sheep, cattle, pigs, camels, buffalo, dogs and cats have escaped and run wild. Sometimes they were released when it was no longer possible to use them profitably. Dogs and cats are still being released to the bush today. The most pernicious introduction was the rabbit.

The early whalers who visited the southern Australian coast stowed rabbits for meat. As a form of insurance and a help for fellow sailors, well-meaning whalers released rabbits on scores of islands. A pair released on an island in Corner Inlet near Wilson's Promontory ran wild long before sheep arrived in the colony. The island today is named after the rabbit. Many early settlers imported domestic rabbits for food. John Batman and John Fawkner, two of Melbourne's first residents, fought a legal battle over the accidental killing of a clutch of valued rabbits. The first Victorian graziers, the Hentys, brought rabbits with them. Other squatters bred their own rabbits. Some escaped, but none proliferated. The domestic rabbit could only survive in this alien land in the occasional niche. In the 1840s the Central Victorian squatter Joyce found rabbits surviving in the cavities of rocks on his station. The rabbits never caused him any trouble, providing gun sport for his neighbours.[55]

In 1859, Thomas Austin of the Winchelsea district near Geelong longed for the chance to hunt the game of his homeland. Austin reasoned that the animals would provide an antidote to the homesickness felt by most squatters. He also hoped to entertain the visiting Duke of Edinburgh in a hunting party. Twice Austin imported and released domestic rabbits. They did not survive. In a determined third attempt, Austin commissioned the importation of twenty-four wild Scottish rabbits. Those fulfilling his commission could only catch eighteen wild grey rabbits, so they supplemented their catch with six domesticated brown rabbits. The Australian wild rabbit is browner than its wild Scottish ancestors because of this domesticated addition to the gene pool.

The Austin family earned an unenviable place in Australian history. Austin's efforts achieved a very considerable success. The rabbits bred like rabbits. Rabbits have a breeding cycle adapted to famine and plenty. A doe is stimulated to ovulation by intercourse and is most fertile immediately after giving birth. While there is feed to eat a doe will spend most of her time both lactating and pregnant. Initially, Austin's rabbits provided sport for English visitors. The quality of hunting on Barwon Estate was enthusiastically reported in British magazines. Soon the blue-blood hunters gave way to working class rabbit trappers, paid to control the menacing explosion. Austin reputedly destroyed 20 000 rabbits on his run in the next five years. The rabbit was no respecter of fences, so other landholders shared the fruits of Austin's success.[56] The rabbits spread across the country. The blame should not be placed on Austin alone. Even while Austin and his neighbours were having problems, other aspiring genteel Englishmen in more distant locations were keen to introduce rabbits. They did not quite believe that they would inherit the same problems as the unfortunate Austin.[57] 'It won't happen to me' is a theme to which we will return.

The rabbits ate out pasture, exposing bare soil to erosion. They dug burrows, which exacerbated the uniquely Australian erosion form, tunnel erosion. Tunnel erosion is another result of imbalance in water use resulting from lower plant water use. It occurs in medium rainfall areas where the surface soil is particularly prone to crusting and compaction, but the sub-soil is easily dispersed by water. Water flowing into the holes left by decaying tree roots and rabbit burrows erodes the sub-soil but leaves the crusted surface unbroken. This eventually creates a tunnel. The tunnel is widened with each major rainfall event until it is no longer able to support the roof. The roof collapses, forming pot-holes and, eventually, a large gully. Tunnel erosion is particularly insidious because the initial damage occurs before there is any major sign of erosion on the surface. The only early warning is silt flowing out of soil cracks on lower slopes.

The rabbits began to control the regrowth of the forests, taking over the role of the kangaroo rats. They nibbled the young eucalypt regrowth. Today, rabbit control is one of the biggest costs for tree planters. Young trees must be protected with special rabbit-proof guards. A strange symbiosis developed between weeds and the rabbit. Together they changed the Australian farm landscape. The blackberry helped the invasion of the rabbit. It provided cover for the rabbits and their burrows. In later years the rabbit hindered the establishment of pastures, enabling St John's wort to re-establish after attempts at eradication.

Rabbit control should have become another essential part of the yearly round on the farm. Instead, the rabbits showed up one of the perennial problems of pest control in rural areas: the need for cooperation between landholders. Without total involvement of all landholders, rabbit control was bound to fail. The grazier had to seal his property from his neighbour's

A well developed case of tunnel erosion

rabbits with a wire-netting fence 1 metre high and buried at least a third of a metre into the ground. This could cost five years shearing income.[58] For the large landholder, control was initially exorbitantly expensive. The squatter John Robertson spent £80 000 controlling his problem over a number of years. Hugh Murray claimed to have spent £18 000. Not everyone could afford this expense, yet the whole district could not afford them not to afford it. Joseph Jenkins, an innovative Welsh farmer, spent the later years of his life as an itinerant agricultural labourer in distant Australia to get peace from his wife. His diaries tell of the problems that rabbits gave the early settlers and the difficulties that farmers had in cooperating to control the pest:

> Binding and stooking corn for a local farmer. During the night great damage was done to the sheaves in an eight-acre field by kangaroos, hares and rabbits. The last two vermin are very numerous. The landowners have distributed notices that trespassers in pursuit of game will be prosecuted, so the selectors are unable to get rid of the furry scavengers.[59]

By 1880 Victoria declared rabbits illegal and protected feral cats as recognised enemies of the rabbit. The new law required landowners to act at certain times of the year to control their rabbits. Novel solutions were tried. Farmers organised rabbit drives, in which rabbits were herded into rabbit paddocks and clubbed to death. A canning factory was built to preserve rabbit meat for export, in the hope of turning a pest into profit. The only practical approaches were poisoning, digging and trapping.

By the 1930s, ninety years of grazing had taken a heavy toll on the soil. The stable native pasture, which held the soil together, was in many places severely degraded. Constant harvesting of wool, milk and meat had depleted the soil of minerals, nitrogen and organic matter. Soil structure deteriorated. In many hilly areas, especially on the inland side of the Great Dividing Range, gully erosion was out of control. In Victoria a Committee of Enquiry found that the erosion was serious and requiring immediate attention.[60] Rather than writing a wordy report, the committee used seventy-five photographs to show the extent and severity of the problem. Today these remain some of the most impressive pictures of erosion published in the state. Some are regularly recycled in the current debate on land degradation.

The report had little impact. The government was not interested; weightier issues occupied the public mind as world war was about to break out. Erosion accelerated in the war years. Labour was scarce and so was superphosphate.[61] There was insufficient labour to keep the rabbits in control. The rabbit population underwent another of its cyclical explosions. To make matters worse, the worst drought in living memory deci-

Packing rabbits for export before the introduction of myxomatosis

Gully erosion in a tree grove

mated pasture cover across much of southern Australia. In one season 4.5 million sheep and several thousand cattle were killed in Victoria alone. The damage from overstocking was severe. The soil blew. When the rains came in 1946, the soil flowed. As the soldiers returned home Australian agriculture had reached the depths of degradation.

2

Recreating grasslands

Thirty years after the invasion of the squatters, the grasslands were a shadow of their former glory. Much of the original native grassland was destroyed. To the squatters, with memories of English pastures, the surviving native grasses were unproductive when compared with English grasses. The most obvious idea was to sow English grasses. The idea is as old as European interest in Australia. The explorers Hume and Hovell sowed English clovers as they travelled through the burning Aboriginal grasslands.[1] Twenty years later in the 1840s some squatters were making experimental sowings of English grasses on their leases. Inland of the Great Dividing Range the squatters met with little success.[2] The English grasses were not suited to the Mediterranean climate of much of southern Australia. They needed rain through the summer before seeding in autumn. But the inland summers were dry. The grasses died before they could seed.

Those squatters lucky enough to have country in the wetter coastal and mountain areas had greater success.[3] Some of the English grasses grew well. Most importantly, white clover survived the dry summer. The imported English grasses needed high levels of nitrogen in the soil. Australian soils were mostly very infertile. There were few native pasture legumes that could fix atmospheric nitrogen into the soil. None were palatable to stock. Some, such as Darling pea, were poisonous.[4] White clover could perform this role. It accumulated nitrogen and organic matter in the soil, unlocking the key to the use of many other English grasses. Small farmers in the wetter areas soon set about establishing pastures little different from those on which a dairy cow might feed in England. The farming systems of Europe seemed to work in these areas of the new land. For the graziers in the drier inland districts there was no agricultural system in the world that they could copy. They needed to invent their own system.

Mining the soil

Agriculture is both a form of recycling and mining. Farmers harvest carbon and nitrogen from the air and mine minerals from the soil. Grasses and crops fix the atmospheric carbon, and *Rhizobium* bacteria on the roots of legumes fix the atmospheric nitrogen. Most of the dry matter in plants comes from these sources. A much smaller proportion of plant material comes from minerals in the soil. These minerals are made available to the plant by the gradual breakdown of underlying rocks. Harvesting these minerals without replacing them is akin to mining the soil. This was the nature of the colonial agriculture at the beginning of the twentieth century. Even in the wetter coastal zone the farming system based on English clover pastures was unsustainable. Carbon and, to lesser extent, nitrogen were being recycled, but graziers were not replacing the mineral nutrients. By the 1920s many of the improved white clover pastures in the wetter districts were dominated by weeds such as rib grass, flat carrot weed and dandelion.[5] Clovers were disappearing. Continued harvesting for up to eighty years had depleted the already meagre store of mineral nutrients in the soil. In some dairying areas the mineral phosphorus was all but exhausted and the native grasses were making a minor comeback.[6] The native grasses could survive with less phosphate and their deep roots could tap the unexploited phosphate in the sub-soil, which the shallow rooted clovers could not reach. Graziers were warned they were milking their soil:

> The continual drain that milking cattle and young animals make on the soil is not without its effect upon the pasture. Thousands of acres in Gippsland which once grew excellent English grasses and clovers are slowly reverting to native grasses, flat weeds, dandelion and bracken. Each year the introduced grasses get less and less in numbers. The cause is directly traceable to loss of minerals. The better grasses, and especially the clovers, require soils well stocked in available plant food. These conditions existed for some years after the original timber and scrub were burned, because ashes were returned to the soil. Gradually, however, the essential minerals are being 'milked' out of the soil.[7]

Grass growing on soil with little phosphate is not healthy for sheep or cattle. Cows and ewes produce less milk. This makes life harder for lambs, calves, dairy farmers and graziers. The first squatters grazed their lactating ewes on fresh native-grass pastures growing on recently burnt land. The phosphate levels were highest in this pasture. With the decline of kangaroo-grass pastures and the fencing of the plains, this option was no longer available. By the 1920s there were extensive grazing areas where dry sheep thrived, but where ewes could not rear lambs.[8] In the developed coastal grazing districts stock diseases with names like 'coast sickness', 'cripples' and 'bone chewing' became more common. Each was a symptom of mineral deficiency in the diet.

In 1905 Thomas Cherry spread phosphate fertilisers on pastures (top-dressing) to see if this would improve growth. Nothing is recorded of the results of these experiments. A few years later agricultural scientists noticed phosphate applied to wheat crops improved the pastures that grew after the crop was harvested. In 1914 scientists at the Rutherglen Research Farm began trials to test the effect of superphosphate on pastures. The results were impressive. The legume trefoil, a chance introduction from the Mediterranean region, sprang up from microscopic size to dominate the pasture. Local farmers saw the plots, but were uninterested. Wheat growing held more attraction. Top-dressing pasture was seen as irrelevant.[9] The First World War made a bloody interruption to the experiments. In 1919 the top-dressing experiments moved to the Western District of Victoria. It was thought graziers might show more interest than the grain producers of the north-east. Superphosphate was spread over several trial plots across the district. With one exception, all showed the same impressive result as at Rutherglen.[10] Trefoils dominated the top-dressed pasture. The scientists began promoting top-dressing as a means of improving degraded pastures. The mining analogy was used to make the point clear to farmers:

> The amount of mineral nutrients removed from the land by the annual crops of livestock is considerable. Of these nutrients, phosphate is of special importance on account of the low phosphate content of Australian soils . . . For the past 60 years the average sheep population of Victoria has been approximately 11 000 000 and that of the cattle population 1 400 000. The number of stock slaughtered in Victoria . . . for the past 60 years would probably be 3 000 000 sheep and 280 000 cattle annually. The average amount of phosphoric acid in each sheep is approximately 2½ pounds and in a cattle carcase 15 pounds. Hence each year the total amount of phosphoric acid removed from the farms by animals slaughtered would be 5 223 tons. If we add the amount removed by pigs and rabbits, and also the amount removed as milk, cream, cheese and wool, the total amount would probably exceed 6 000 tons. This amount has been removed from the grasslands during the last 60 years. Thus . . . something like 360 000 tons of phosphoric acid, which is equivalent to 1 800 000 tons of superphosphate, has been annually taken from the soil, and most of this material removed from the pasture lands.[11]

Some of the Western District's dairy farmers took up the idea and began top-dressing. Wool producers viewed the experiments with suspicion.[12] Top-quality wool is a result of breeding and a controlled feeding regime. Excessive feed makes the wool coarser and of lower value. The native pastures were producing top-quality wool. There was uncertainty about the effect of top-dressing on the quality of wool. Producers thought improving the pastures might coarsen the wool. The Hamilton Agricultural Society refused to sponsor field days. Nevertheless, the Victorian Department of Agriculture started a concerted effort to promote top-dressing. A special agricultural train toured the state, taking advisers and demonstrations to inform farmers of a range of improved practices including the use of phosphate. It took ten years for top-dressing to be generally accepted. Farmers were often wary of accepting advice from scientists. Before spending a lot of money on superphosphate, most would try some on a small area of their own farm. Advisers complained that farmers ran these tests without fencing off the test areas. Cattle preferred to graze the top-dressed pasture. This destroyed the evidence of superior pasture growth, leaving nothing to convince the farmer of the value of superphosphate.[12]

Underground clover

While proclaiming the importance of superphosphate to replenish the mined soils of the wetter districts, the agricultural science fraternity had nothing to offer the pastoralists of the inland regions. The only advice seems to have been optimistic praise of the degraded pastures already there:

> Australia's native grasses are justly famous for their grazing and their drought resistant qualities. They are noted for producing the finest and the best wool in the world, and for withstanding great extremes of heat and drought. No other plants have proved the equal of our own for the drier parts of Australia.[13]

The New South Wales Department of Agriculture promoted the 'systematic cultivation' of native kangaroo, wallaby and Mitchell grasses to rejuvenate degraded pastures, even to the extent of recommending the reproduction of kangaroo grass by root division![14] This view through rose coloured glasses is hard to understand. The long-term outlook for the inland pastures was even worse than in the wetter districts. The native grasses were well adapted to the inland climate, but they were not at all adapted to an export oriented agriculture. The harvesting of wool and animals was mining the soil not only of its minerals, but also of its nitrogen and organic matter. In the inland there were no legumes to replace the nitrogen. The long-term outlook was bleak.

In the 1890s an Adelaide Hills farmer named Amos Howard noticed a curiously different clover on his property. It grew through the autumn, winter and spring period and died off in summer. The clover had migrated to Australia from Italy where it was considered little more than a roadside weed. It had adapted to the Italian summer by burying its seeds before the plant died off in summer, hence the name subterranean clover (*Trifolium subterraneum*). The buried seeds survived the desiccating dry summers and germinated again in autumn. Subterranean (commonly called 'sub') clover's growing season coincided with the rainfall season of the inland slopes of the southern Great Dividing Range: moderate winter rainfall and little summer rainfall.

Amos Howard was a man of foresight. The potential importance of this clover for Australian agriculture was clear to him. He invented a seed harvester and began selling subterranean clover seeds. His industriousness was only partially rewarded. Few others saw sub clover as an important pasture plant. To most agriculturalists and graziers it was an agricultural curiosity. In 1910 it was mentioned three times in the *Victorian Journal of Agriculture*, each time merely to identify grasses sent in by readers. The

The subterranean clover plant

commentary was brief and uneffusive: 'An introduction from the Mediterranean regions. Of some use as a pasture plant; it also aids in the suppression of annual weeds.'[15]

The first problem with the adoption of subterranean clover was the lack of phosphorus. While subterranean clover was adapted to the southern Australian climate, like other clovers it was not adapted to the naturally infertile Australian soils. For clovers to successfully replenish nitrogen in the soil the *Rhizobium* bacteria, which colonise their roots, need plentiful supplies of phosphorus in the soil. Top-dressing was still thirty years away when Howard first noticed sub clover. Subterranean clover did not grow well on low-phosphorus soils.

The link between sub clover and phosphorus was made at the new Rutherglen Research Farm in 1912. A small paddock of native pasture was ploughed, sowed to sub clover and broadcast with superphosphate in 1912, the first year of research work on the farm. Apparently the clover grew well, but the result was not promoted with great fervour. The advisers realised that wheat farmers uninterested in top-dressing of native pasture would be unwilling to bear the financial risk of ploughing and sowing pasture before the top-dressing. There was also a concern that ploughing would destroy the native pasture, leading to an increased risk of weed invasion.[16] In 1919 the experiment was briefly mentioned: 'The results show what can be done with artificial grasses to improve the capacity of small paddocks.'[17]

This was hardly a recommendation for wide adoption on farms. Sub clover was still seen as a curiosity in the northern areas of Victoria. In the midst of the top-dressing enthusiasm, most farmers and scientists thought only of native pasture for the drier pastoral districts:

> Over the greater part of the Continent the climatic conditions are too severe for the successful sowing of introduced grasses. Only the native grasses and fodder plants will thrive on the drier areas . . . In the winter rainfall zone of Southern Australia the predominating native grasses are the various species of *Danthonia* (Wallaby grasses), *Agrostis* (the Bent grasses), *Anhistinia* (Kangaroo grasses), *Stipa* (the Spear grasses), *Eragrostis* and *Poa* . . . A great deal of field work awaits the improvement of Australian grasses. The pastoral industry is such a valuable asset that every means should be taken to conserve our native grasses and to aim at improving them in bulk, succulence, seeding capacity and stock carrying capacity.[18]

Many believed top-dressing offered nothing for the pastoralist on the inland slopes of the Great Dividing Range. It was thought that rainfall largely determined the response to superphosphate.[19] Without regular rainfall, superphosphate spread by a grazier would be wasted. It was not the lack of rain that made phosphate unproductive, but the lack of a viable clover that could use the phosphate under the inland rainfall pattern to add nitrogen to the infertile soils.

Another problem with sub clover was that it dried off in summer. To almost everyone, improved pasture meant a perennial pasture that lived throughout the year. Sub clover was regarded as a weed in the improved pastures of the day because it crowded out useful perennial clovers and then died off, leaving only bare ground in summer.[20]

The pieces of the puzzle fell together almost by accident. In the 1919 Western District trials of top-dressing, one site failed to give any response. There were no seeds of trefoil legumes in the soil to exploit the increased availability of phosphate. The research scientist tried to make the most of the situation in the following season by spreading sub clover seed over the experimental plot without cultivating the ground beforehand. The sub clover grew well. The experiment had demonstrated not only the need to apply superphosphate to subterranean clover, but also the possibility of sowing subterranean clover without ploughing the ground. By 1925 the 'curiosity' warranted a major article in the *Journal of Agriculture*.[21] By 1930 the results were 'remarkable'.[22] The clover responded impressively to phosphate. Although sub clover died off in summer, its buried seeds allowed it to quickly reappear after the autumn break. This was the key to improving pastures in a Mediterranean climate.

Instead of identifying the occasional specimen sent in by a farmer, agricultural scientists began actively collecting sub clover seeds from farms. Within a few years, in Victoria, suitable varieties of sub clover had been identified for different areas of the state. The promotion of top-dressing was transformed to a promotion of 'sub and super'. Like the campaign for top-dressing before, success was limited. Widespread use of subterranean clover and superphosphate in pasture development would not occur until after the Second World War.

Golden years of science

The end of the Second World War heralded a transformation of the grasslands of southern Australia. In one of the most serendipitous coincidences of Australian agriculture, solutions appeared for many of the most pressing problems of the grasslands. The combination of subterranean clover and superphosphate had been incubating during the war. It offered the chance to return nitrogen and organic matter to the soil and dramatically increase stocking rates. Without other management changes, this was at the risk of greatly increased vulnerability to drought and long dry summers. But in the 1950s efficient tractors were appearing on many farms. New hay-baling machinery simplified the task of storing fodder for drought. The typical sub clover farm produced more feed in spring than the stock could eat. By cutting, drying and baling the excess the farmer could feed his stock in the lean periods when the sub clover was not productive.[23] In ten years the amount of hay baled in Victoria quadrupled.

Subterranean clover did not grow well in drier areas of the wheat belt, particularly the Mallee. Here, other naturalised immigrant plants had to be used as a basis for improved pastures. Hairy medics were common in Mallee pastures. These small clover-like plants did a little to improve fertility but were often dominated by wild mustard and barley grass.[24] More successful was barrel medic. Like sub clover, barrel medic came to Australia by accident. It was first noticed in Melbourne in 1907, just another garden weed. By 1920 medic had reached the Mallee and established on a number of isolated properties. There it grew well in the good seasons, and produced large prickly seeds, which enabled it to survive the long dry seasons. Medic looked like a clover and fixed nitrogen like a clover. It promised to achieve in the Mallee all that sub clover promised in other areas. By 1938 commercially produced seed was available in small quantities. By the early 1950s researchers had shown that medic-based pasture could dramatically increase soil fertility.[25] It could achieve the same result as sub clover.

The replacement of phosphorus revealed other fertility problems in pastures. In some areas sub and superphosphate did not succeed. The soils lacked one or more of the trace elements needed for the growth of healthy clover plants. Sometimes this deficiency was natural. Sometimes it had been created by many years of exploitative agriculture. These deficiencies allowed agricultural scientists to achieve some of the most impressive demonstrations of the value of scientific agriculture. Where molybdenum was deficient, spreading small amounts of this element on trial pasture plots achieved spectacular improvements in pasture growth. All that was required was 'two ounces to the acre'.

The cost of new pasture establishment in the 1950s

Harvesting oaten hay before mechanisation

Mechanisation of fodder conservation

The sub clover pastures were as much to the liking of rabbits as to sheep and cattle. By the 1950s better methods of rabbit control had appeared. Before the war farmers poisoned with strychnine or arsenic. The poisoning rarely killed more than 70 to 80 per cent of rabbits; so rabbits could quickly reproduce to become a pest again if poisoning was not followed up with gassing and digging out the burrows. Despite government regulations compelling landowners to control their rabbits, there was not enough coordinated landholder action to contain the rabbits. The task was arduous, the results transitory and some landholders and the very poor of society regarded the rabbit as an important source of income.[26] In 1919 the myxomatosis virus had been suggested for controlling the rabbit. In 1936 the virus was imported, but it was not a virulent strain and the results were unimpressive. In 1950 the Commonwealth Scientific and Industrial Research Organisation released a highly virulent standard laboratory strain of the virus in the Murray Valley. The successful new strain was then widely dispersed. Armed with traps, an ampoule of virus and a syringe a farmer could inoculate live rabbits on his property.

The effect of the myxomatosis virus was at first spectacular. Once under the skin, the virus incubated rapidly. Within five days the infection caused swelling of the eyes, anus and genitals. By eight days the swellings had spread over much of the body. Sometimes the genitals burst under the pressure of the swelling. Within ten days a rabbit was dead. Rabbit numbers plummeted. The virus captured public attention. The early 1950s were wet years in southern Australia. Mosquito populations exploded across the riverine plains. The mosquitos spread not only the myxomatosis but also the human disease Murray Valley encephalitis. Outbreaks in the Shepparton district prompted public concern that myxomatosis had crossed the species barrier to humans. To allay public concern three prominent scientists, Frank Fenner, MacFarlane Burnett and Ian Clunies Ross, followed the established historical precedent of immunological research. They publicly injected themselves to demonstrate the safety of the virus.

By the mid 1950s the myxomatosis virus was not as effective as it had been initially. Rabbits were taking longer to die — sometimes up to thirty days. The development of resistance in the rabbit population, and of less virulent strains of the virus, had come about because of natural selection. The small percentage of rabbits that survived the initial infection passed on their resistance to their offspring. In some areas where less virulent strains had been released in the 1930s and 1940s, up to 90 per cent of rabbits were resistant to the laboratory strain. Within eight years of the release of the virulent strain, the average virus in the burrow would kill only 90 per cent of laboratory rabbits. In some areas the virulence was reduced to 50 per cent.[27] Continued releases of virulent strains of the disease helped to keep up the kill rate, but the heady days of the first releases were not to return. By the late 1950s drier weather was reducing mosquito numbers, further limiting the effectiveness of the virus.

Myxomatosis could not be relied on, exclusively, to keep rabbits in check. After the initial high kill rates, poisoning had to continue. Improvement had occurred here as well. The compound sodium fluoroacetate (commonly called 1080) was discovered in 1896. It is an odourless, tasteless poison. Many years after its synthesis it was discovered to be the active compound in a poisonous southern African shrub. In 1946, 1080 was released as a rabbit poison.[28] As a means of poisoning it had a particular advantage. Rabbits are very shy and wary eaters. Previous attempts to poison rabbits met with mixed success. Rabbits would take poisoned baits but, once one or two had died near a trail of baits, the remaining rabbits learnt from the unfortunate experience of the others and avoided the baits. Sodium fluoroacetate is not itself toxic. When absorbed into the body it is converted into fluorocitrate, which is poisonous. The conversion into a poison takes some time. A rabbit that takes sodium fluoroacetate baits does not die immediately. When it takes effect, an hour or two later, the rabbit dies quickly in the burrows rather than near the bait. The first trials with sodium fluoroacetate were carried out in Tasmania in 1953. When used correctly, it could kill up to 98 per cent of a farm's rabbits. The farmer had to lay out two or more trails of unpoisoned carrots for the rabbits and then lay out a trail of poisoned carrots. The unpoisoned carrots overcame the rabbits' natural caution.

The final weapon in the armoury against the rabbit was digging out the burrows; again technology provided an answer. Better tractors, and particularly improved hydraulic systems, meant ripping out the burrows could be done with comparative ease. In a little over ten years rabbit control had become both easier and more effective.

The combination of the new pasture legumes, superphosphate, trace elements, improved fodder conservation and rabbit control promised dramatic improvements in farm production. The sub and super revolution was in the offing. The government decided to hurry the revolution along. Australia faced a balance of payments problem, and increasing agricultural exports was the quickest way to ease it. In 1963 the Commonwealth Government introduced a bounty for superphosphate, subsidising the farm input on which the revolution was based.[29] In 1966 the Commonwealth offered extra money to the state governments to expand their farm advisory and extension services.[30] State governments took up the offer and employed advisers who spread the news of pasture research, rabbit control and fodder conservation. The Korean War wool boom had created high wool prices. Farmers had every incentive to run as many sheep as possible to take advantage of the high wool prices. By the mid 1950s, pasture improvement was the most profitable farm investment a grazier could make. Farms in some districts achieved a fourfold increase in carrying capacity. In twenty years the area in Australia sown to improved species more than doubled.[31] In the thirty years between 1950 and 1980 the total volume of Australian agricultural production doubled.

The agricultural revolution came at just the right time to control the countryside's burgeoning erosion problems. Attempts to control water erosion on grazing properties had been based on mechanical power. Contoured furrows, terraces and grassed waterways controlled the flow of water, but did nothing to attack the root cause of the problem, which was degraded grasslands.

To control sheet erosion and gully erosion the soil needed to be physically protected by plant cover. Bare soil increased both the speed and volume of water runoff, thereby increasing the risk of gully erosion. Bare soil was not as stable as soil held firmly by active roots and was at risk to sheet erosion. The new clover pasture fixed nitrogen into nitrogen-depleted soils. This dramatically increased the fertility of the soils, adding between 50 and 200 kilograms of nitrogen per hectare per year. In some areas the amount of organic matter in soils tripled within a decade. The extra nitrogen allowed pastoralists to grow a thick sward of grass, particularly perennial grasses. The vigorous pasture protected the soil and reduced the peak runoff flows. The other side of maintaining cover was avoiding overstocking; this required the rabbits to be controlled. By eliminating rabbits the risk of all forms of erosion was reduced. The combination of myxomatosis, 1080 and improved tractor hydraulics put effective control of rabbits within the reach of many graziers.

Subterranean clover and superphosphate were also a major boon for weed control. The vigorously growing legume could compete against many of the weeds overpowering the degraded native pastures. For all the effort put into controlling St John's wort with insects, the real solution came with pasture improvement. The wort could not compete against the subterranean clover and was crowded out of paddocks. New herbicides offered some relief from weeds such as blackberry. With good pasture cover, better weed and rabbit control, the task of repairing erosion became relatively simple. With the arrival of the bulldozer the landowner could rectify gully erosion by stabilising the head of an eroded gully and then reshape the gully with earthmoving equipment, before sowing to improved pastures.

The agricultural science community was not the only source of new ideas for agriculture in the heyday of expansion and improvement after the war. P. A. Yeomans proposed his 'keyline' planning both as a solution to the degradation of the grazing lands and as a means of coping with the vagaries of Australian rainfall. The elements of keyline agriculture were water conservation, farm planning and improved soil structure. Yeomans exhorted farmers to build up their soil structure, not only to redress the damage of previous generations, but to improve that which nature provided in the first place. Natural shortages of minerals could be redressed. Rainfall could be controlled more efficiently and new plants could be planted to increase the organic matter in the soil. There was little technically new in this restating of the sub and super revolution. The difference was in the way the new sub clover and superphosphate culture could be integrated into a plan on the farm. Yeomans argued that crucial to the integration of these components in the development of soil was good soil structure:

> Some soil scientists estimate there are 175 tonnes of living organisms . . . in a hectare of fertile soil. Those organisms generally work towards man's health and well being. The importance of fourteen 12.5 tonne truck loads of microbes in a hectare is overshadowed completely by four or five sheep to the hectare.

The sheep or cattle obviously need constant care, but surely that other 'live-stock' warrants some conscious thought when it is so vital. All the elements of growth are made available to us by the various processes of the life cycles of this 'life in the soil'. Soil management can reduce the dynamic force to a low ebb, or tremendously stimulate its activities.[32]

Crucial to keyline farming was the new power of the tractor. Yeomans decried fallow, surface ploughing and fine seedbeds. Instead he advocated a form of deep ploughing, which gradually loosened the compacted deeper layers of soil.

Greater productivity, fewer producers

The tremendous productivity gains and growth of the subterranean clover and superphosphate revolution were part of the luck that earned Australia the 'lucky country' title. Such luck must always run out. Commercial farmers face a perennial marketing problem. There is only so much food that a human being can eat. If farmers produce more food in a small market where everyone has enough to eat, then they will need to ask less for their produce to sell at the expense of their competitors. If every farmer in-creases production, asking less for the product may not produce many more buyers because competitors in a free market will lower their price. The Australian farmer faces this problem on a worldwide scale. Despite starvation and famine in some parts of the world, recurrent 'over produc-tion' of agricultural commodities has plagued agricultural exporters. The market consists only of those with the money to buy. If every exporting country increases its production then prices fall; only those who increase production at a lower unit cost or who are subsidised by a beneficent government will survive.

At the farm level these economic pressures are seen by the farmer as continually rising input costs and declining product prices — the 'cost–price squeeze'. Succeeding generations of Australian farmers have faced the challenge of improving farm production to maintain profitability. Those farm businesses that did not improve farm productivity would often be eventually absorbed by neighbouring farms. Businesses with static pro-ductivity could not remain profitable in the face of continually increasing costs and declining prices for agricultural products.

During the years of the subterranean clover and superphosphate re-volution Australian graziers protected themselves from declining prices of agricultural products by increasing productivity through pasture improve-ment. The lucky streak lasted twenty years. Towards the end of the 1960s the cost–price squeeze caught up with the Australian farmers' 'normal' ability to increase efficiency by increasing output.[33] By the 1970s economic storm clouds that had been growing on the horizon burst upon the Austral-ian farmer. Great Britain, a major market for many of Australia's rural commodities, particularly its butter, retreated into the European Market. The economic condition of the dairy industry in particular slumped dra-

matically. Farmers were forced into the realisation that they needed to
expand their farms to increase their production. The expansion of farms
meant that there were fewer farms and fewer farmers. The number of dairy
farms in Australia decreased by more than 55 per cent in the sixteen years
after 1971; during the same period the total number of farms decreased
from 190 000 to 167 000.[34]

The 1980s have been a microcosm of modern Australian agriculture.
Early in the decade many of Australia's major rural industries suffered a
major drought; in the middle and later years seasons were kindly and
international demand was buoyant, but farmers faced large cost increases
for their inputs. The 1990s began with the prospect of a major rural
recession because of low international demand for the main rural commodities. As real interest rates rose to high levels some farmers who had
borrowed too heavily to fund farm expansion faced financial ruin. Productivity growth in the 1980s was higher than in the 1970s, but considerably
less than in the earlier 'golden age' between 1950 and 1970. Farm numbers
continued to decline.[35] (See Figures 2.1 and 2.2.)

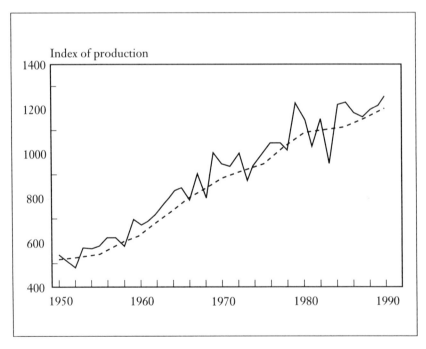

Figure 2.1 Index of the volume of rural production, Australia, annual
and 5 year averages: 1950–90
(*Source*: Australian Bureau of Agricultural and Resource Economics)

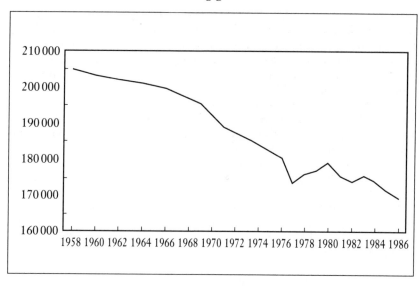

Figure 2.2 Number of agricultural establishments in Australia: 1958–86 (*Source*: Australian Bureau of Agricultural and Resource Economics)

Pests new and old

While the new sub clover pastures overran many of the old pasture weeds, it was inevitable that new weeds would exploit the nitrogen enriched soil. Farmers who invested in pasture improvement were creating a new ecosystem in their paddocks. Capeweed and *Erodium* are two of the weeds that emerged. Capeweed provides good winter feed, but dies off early in summer leaving bare ground vulnerable to erosion. *Erodium* also provides good feed, but the corkscrew seeds it produces contaminate the wool clip. The increased organic matter in the soil provides a suitable environment for new insect pests: cockchafers, crickets, armyworms and cutworms, which live in the soil and emerge at night to eat clover and grasses. The most costly of the new pasture pests was not an insect but a spider, the red-legged earth mite.[36] The mite is a small black bug the size of a pin head. It sucks the sap out of clover plants. Despite its small size, the mite causes serious stunting and can kill young clover plants. In the early days of sub and super, advisers recommended fallowing a paddock before sowing, to eliminate the mite, then surrounding the paddock with a ring of creosote under the boundary fence to prevent the mites migrating back into the new pasture.[37] Cockchafers could only be controlled by spraying arsenic, which was poisonous to the stock. In the 1960s and 1970s organic insecticides replaced arsenic. Their lower toxicity meant that graziers could spray pasture and then graze it, rather than waiting for the arsenic to wash away.

More recent community concern about pesticide residues in food has meant that use of the organic pasture sprays such as the organochlorine lindane has been banned. There is pressure to restrict the use of other pesticides and to rely on biological control. Biological control is not a solution free of unwanted side effects. Biological control of Patterson's curse has been the subject of a prolonged court battle between beekeepers and the Commonwealth Scientific and Industrial Research Organisation. Because Patterson's curse has been a valuable spring nectar source, bee-keepers feared the introduction of biological control would threaten the viability of their businesses. The introduction of virulent strains of the fungus *Phragmidium violaceum* to control the wild blackberry has also been opposed by beekeepers in Tasmania where the wild blackberry is a valuable source of nectar.

More recently the pernicious pest, the rabbit, has again increased in numbers in many parts of Australia. Resistance to myxomatosis has in-creased and in some areas, where there have been favourable seasonal conditions, rabbits are near plague proportions. The release of the rabbit flea in the late 1960s helped spread myxomatosis. But the rabbit flea could not cope with arid conditions and a new strain of the rabbit flea, endemic in Spain, has been considered for release. A new possibility for the war on rabbits is a previously unknown virus that devastated rabbit populations in Asia and Europe in the mid-1980s. Being spread by contact, there is no need for a flea or mosquito to spread the plague.

There is a danger that the promise of an initial spectacular success of a new rabbit disease will convince landholders that the rabbit menace will be controlled forever. The early dramatic impact of myxomatosis convinced some farmers that the rabbit was controlled permanently. The rabbits' return caught some of these landholders unprepared. It is likely that the myxomatosis scenario will be repeated — gradually declining kill rates and a renewed need for poisoning. Controlling rabbits is a task for a commun-ity of landholders, not just concerned individuals. Rabbits breed quickly. A kill rate of less than 95 per cent can be considered a failure as they will quickly repopulate. Little is gained if one landholder kills his rabbits and his neighbour does not. The neighbour's rabbits quickly re-colonise the rabbit-free land. Weed control is also a matter of community cooperation. If one landholder controls the thistles on his property but a neighbour lets them grow, the first landholder's work will be of only passing benefit. The Landcare movement, of which we will see more in a later chapter, is an attempt to increase the social pressure on all landholders to cooperate in this work.

Hidden degradation

The use of sub and super created better soil, made farming more profit-able, and helped to overcome the soil erosion of the 1940s. Despite these dramatic benefits this was not an ideal soil conservation innovation. The crucial difference between the introduced pastures and the original native vegetation it replaced was the deeper roots of the native pastures. The subterranean clover added nitrogen to the soil, which allowed farmers to sow productive grasses. In the 1950s, soil conservationists complained there were too few grasses in many sub clover pastures. Where grasses were planted, not all had deep roots. Farmers often sowed annual grasses such as ryegrass, or perennial grasses with shallow roots. With attention at the time focused on soil erosion, this seemed a minor matter. By the 1980s soil salinity had been recognised as a more serious problem. The improved pastures of the 1950s and 1960s were not improved enough. They did not contain enough deep rooted grasses, which were better able to use rainfall and so reduce the amount of rain seeping down into the watertable. Farmers were encouraged to control salinity by sowing phalaris with their sub clover or to grow lucerne, a legume with deep roots. Unlike the use of subterranean clover and superphosphate, there was little immediate economic incentive to make this change in pasture composition.

There is a more worrying problem than salinity. In some areas of Australia the sub and super advances have stalled. The sub clover is no longer thick and heavy, but is small and stunted. In much of the medium–high rainfall cropping and pasture zones of eastern Australia, where rainfall exceeds 500 millimetres, the problem is caused by soil acidity. Increased acidity weakens the growth of clover. This in turn reduces the nitrogen in the soil, leading to weaker pasture. In the long run the implications are serious for farming and cropping. The pasture revolution that turned around the degradation of the grasslands may be threatened.

For many years European farmers have known that regular applications of lime were beneficial to their soil and crops. Agricultural scientists realised that the regular lime applications were helping to control the acidity of the soil. But, the scientists also realised that harvesting some crops, such as lupins and pasture hay, increased soil acidity. By the turn of the century instructions on the importance of lime balancing appeared regularly in the journals of agriculture. Farmers often had little to show for the application of lime. Pastures did not respond to lime in the way that they responded to superphosphate or trace elements:

> Lime has been used on pastures for many years, but the results achieved have not often been conspicuously successful; and probably more lime has been used here because it was expected to be beneficial, than has been used with the knowledge that it would be beneficial . . . Much of the advocacy of lime for pastures has been based on the free use of lime in England and New Zealand, and not on Victorian experience, and as might have been expected in a state that has some lime responsive and much unresponsive land, a great deal of confusion has arisen.[38]

The lack of obvious response gave little encouragement to farmers to keep liming their paddocks. Now, because of lack of liming, acidity has become a serious problem in the higher rainfall districts of eastern Australia. The success of subterranean clover and superphosphate pastures has exacerbated the problem. Subterranean clover and superphosphate increased the productivity of pastures. Each year farmers harvested more hay, baled more wool and sold more meat. More lime was leaving the farm, leaving the soil more acidic.

Subterranean clover and superphosphate also had the desirable effect of increasing the organic matter in the soil, but this gave rise to an unanticipated risk of increasing soil acidity.[39] To understand how this happens we must consider the cycling of nitrogen between the atmosphere and the soil. Nitrogen is initially fixed from the atmosphere in the form of ammonium salts by various *Rhizobium* bacteria that colonise the roots of legumes. Ammonium compounds are initially taken up by legumes that host the *Rhizobium* bacteria. They are released as the legumes rot and soil bacteria convert them to nitrate compounds (nitrification). Other plants are then able to convert available nitrate compounds to protein; or de-nitrifying bacteria convert the nitrates back into nitrogen. This cycle is pH neutral. At its completion the soil is neither more acid nor more alkaline, because the nitrification process is acidifying and the de-nitrification process is de-acidifying. However, if nitrate compounds are leached from the soil before de-nitrification can occur, the soil is left in a more acid condition.[40] The sub clover farming system, by increasing organic matter in the soil, builds up the ammonium nitrogen levels in the soil through spring. In the dry summer months the pasture dies off while soil bacteria slowly convert the ammonium compounds into nitrate forms. In autumn the first rains leach the accumulated nitrates below the pasture root zone before the annual grasses have time to re-establish and begin de-nitrifying the soil. The net result is an acidification of the root zone. Continued leaching extends the zone of acidity deeper into the subsoil.

As the soil acidifies, the availability of soil nutrients changes. Some, such as phosphorus, molybdenum and calcium, become less available and crops suffer the effects of their shortage. Others become more available — in the case of aluminium and manganese, at toxic levels.[41] Acidity also interferes with the ability of some strains of nitrogen-fixing bacteria to establish themselves on their legume hosts. The sub clover system of farming may not be sustainable without action to control acidity.

The third means by which subterranean clover and superphosphate have acidified the soil is through the acidifying effects of superphosphate itself. The organic agriculture movement places great stress on this aspect of acidification.[42] The first phosphate fertilisers were made of ground phosphate rock. Rock phosphate was often very insoluble and did not prove a very effective way of replacing phosphate in the soil. In the mid-eighteenth century an Englishman created a soluble form of phosphate by mixing ground bones and sulfuric acid, to produce the first superphosphate. When spread on soil, the superphosphate dissolved and seeped through the upper soil layer before being converted to an insoluble form of phosphate much like rock phosphate. The crucial difference was that the insoluble particles from superphosphate were smaller, and therefore more available than the most finely ground rock phosphate. The opponents of superphosphate point to the acidifying nature of the reconversion of superphosphate to insoluble phosphates, and blame this for the increasing acidification of soils. In the past, trials of rock phosphate compared with superphosphate have consistently shown faster responses and higher yields from superphosphate.[43] While there has been some success with rock phosphate in New Zealand, there has been little evidence to support the use of rock phosphate in southern Australia.[44] The argument may well be revealed as ill directed. The direct impact of superphosphate on soil acidity seems to be very small in comparison with the nitrate leaching effect.[45]

The spectre of increasing soil acidity has prompted research scientists to look for new methods of farming that do not accelerate the acidification of the soil. It is unlikely there will be one technique to overcome this problem. Rather, a suite of farm management recommendations will probably emerge. There are likely to be some familiar prescriptions and also a few that turn current common wisdom on its head.[46]

Perhaps the strangest recommendation could be to minimise the accumulation of excess organic matter. For years the wheat industry was plagued by the depletion of both organic matter and available nitrogen. The desperate need for a solution to these problems may have perhaps blinded us to the ecological reality that too much of anything is a bad thing in a living system. We have known for many years that excess nitrogen in pastures leads to rank weed growth and suppression of desirable species, but these excess nitrogen problems arose from the concentration of animal urine and manure in conditions of heavy stocking. Farmers may be forced to look much more closely at achieving a finer balance between too little and too much organic matter.

Deep rooted perennial pastures are likely to reduce the rate of acidification. Deep rooted perennial pastures use more of the soil nitrates as they are produced, and respond to late summer and early autumn rainfall better than annual pastures. This is not new advice for graziers on farms where dryland salting control is a priority. But perennial pastures are not a permanent solution. No pasture will be able to cope with the intense nitrate leaching that occurs under the urine patches associated with livestock; increasing acidity is inevitable where livestock range on pastures. While various stock and crop management strategies may ameliorate this effect, they will only delay the acidification of pasture.

Liming remains the obvious means of correcting an acidified soil. The application of lime costs between $150 and $200 per hectare of land. For most farmers the application of lime is uneconomic; a farmer may spend more money liming the land than the land is worth, with little immediate return on the investment. It takes many years for the lime to leach down into the acidic sub-soil. The response time can be speeded by ploughing the lime into the soil, but this is not always a practical option. The advantage of sub clover lay in its ability to establish without ploughing. This allowed farmers to establish pasture on non-arable land, land on which it is now impractical to cultivate lime into the soil. Yet if liming is not started, sub-soils are likely to become increasingly acidic and recovery will be even more difficult.

For farms with acid soils, the outlook is not very positive. Often the acid soil overlays a naturally acid sub-soil, and liming will have little impact on the problem. As the clovers succumb to the acidity, native grasses will again predominate in the grasslands. At the Rutherglen Research Institute in northern Victoria research scientists are looking to turn a necessity into a virtue. One hundred and fifty years after the squatters arrived, scientists are investigating management systems for paddocks sown with varieties of wallaby grass that are acid tolerant and deep rooted.[47] Native grasses once supported a much smaller pastoral industry, and in parts of Australia they may do so again.

3

Symbolic trees and salinity

The Great Dividing Range sweeps out of the alpine corner of south-eastern Australia in a belt of mountains and hill country, branching westwards across the land once known as *Australia Felix*. The farther west the range extends, the less appropriate the title 'great' becomes, as the slopes become progressively smaller and gentler. The final hills before the range peters out are known as the Pyrenees.[1] The catchments of the adjacent Avoca, Loddon and Campaspe rivers start in the northern watershed of the end of the Great Dividing Range. The Avoca is one of those strange Australian rivers that have no direct link to the sea; it begins in the Pyrenees Range and finishes in the Avoca marshes near the Victorian town of Swan Hill. The Loddon and Campaspe flow into the Murray, but are slowed and diverted along the way by reservoirs. The northern aspect of this range is dominated by old volcanic cones such as Mount Franklin and Smeaton Hill. Many ages ago the cones spewed forth the larva flows that today provide the best agricultural soils of the region.

This large land system, the upper catchments of the Loddon, Campaspe and Avoca rivers, and the hinterland southern basalt plains, provides a study of some of southern Australia's most delicate land systems prone to both soil erosion and salting. The salinity and erosion in this district are not just problems for local landholders. There are implications downstream. Both the Loddon and Campaspe rivers are dammed for irrigation and town water supplies. The Loddon River as it flows to the Murray passes through the area of worst irrigation salinity in Australia. The Avoca is the lifeblood of a series of internationally important wetlands, which are increasingly threatened by salting. We will look at the place of trees in the natural history of this land system, the nature of its land degradation, and the reactions of landholders to salinity and erosion problems.

How green was Australia Felix?

Major Thomas Mitchell came to Australia as Deputy Surveyor-General. Mitchell had a reputation as being a skilful and accurate surveyor and an excellent draughtsman. As the leader of a caravan of twenty-five men, five bullock drays, a boat carriage (Australia's first 'trailer-sailer'), three light carts, eleven horses, fifty-two bullocks and 100 sheep, he set out to survey the Darling River. His expedition took him from Sydney to Portland in south-western Victoria. Mitchell crossed the Murray River at Swan Hill in 1836 and travelled up the Loddon Valley. He turned west at Wedderburn and, crossing the Avoca on his way to Portland, he 'descended into one of the most beautiful spots I ever saw', a land that 'had the appearance of a well kept park'.[2] Mitchell described open grassy vales bounded by undulating hills, with the grasses growing in greater abundance than anywhere else he had seen in Australia. As he moved on he 'discovered a country ready for the immediate reception of civilised man . . . unencumbered by too much wood'.[3] This was the land Mitchell named Australia Felix. As a result of Mitchell's praise of the pastoral country, squatters moved in and took up large tracts of land on the open grassy plains in the 1840s.

As Mitchell had made his way towards Portland he entered south-western Victoria from the north, finding trouble in traversing his carts through the soft, waterlogged soil of the Dundas Tablelands. Sixty kilometres west of the Grampians and about 130 kilometres north of the Henty's Portland settlement he saw 'hills of the finest forms, all clothed with grass to their summits, and many entirely clear of timber'.[4] Mitchell wished to follow the Glenelg River but progress was made difficult by the many tributary creeks, rivers and ravines that had their source on the tableland. North of present-day Casterton, in the Nangella and Wando Vale country of Victoria, Mitchell observed:

> An open grassy country, extending as far as we could see — hills round and smooth as carpet — meadows broad, and either green as emerald, or of a rich golden colour, from the abundance, as we soon afterwards found, of a little ranunculus-like flower. Down into that delightful vale, our vehicles trundled over a gentle slope, the earth being covered with a thick matted turf, apparently superior to anything of the kind, previously seen. That extensive valley was enlivened by a winding stream, the waters of which glittered through trees fringing each bank.[5]

Near present-day Casterton on the land on which the Hentys, Victoria's first permanent white settlers, later established their runs (see *Muntham*, Plate I), Mitchell recounted:

> After fording a stream, we ascended a very steep but grassy mountainside, and on reaching a brow of high land, what a noble aspect appeared! A river winding amongst meadows which were fully a mile broad, and green as an emerald. Above them rose swelling hills of fantastic shapes, but all smooth and thickly covered with rich verdure. Behind these were higher hills, all having grass on

their sides and trees on their summits, and extending east and west, throughout the landscape, as far as I could see.[6]

Granville Chetwynd Stapylton was second in command of the Mitchell expedition. In his journal he commented, 'here we have undulating ground clear of timber except occasional picturesque clumps of trees'.[7] Looking to the east towards present-day Hamilton, Mitchell saw 'beautiful open country', observing 'the whole intervening territory appeared to consist of green hills, partially wooded'.[8]

The northern part of south-western Victoria was not all open savannah woodland or the tussock grassland settled by the Hentys. There were some denser clumps of forest. North of the Dundas Range (near Balmoral) the stringy bark ranges in 1842 'came in upon the river [the Glenelg] so determinedly for many mile' that early settlers thought there was no available country in that direction.[9]

The Hentys had seen some of Portland's pastoral hinterland before Mitchell. In 1835, thirteen months before Mitchell's visit, in a letter from Van Diemen's land, Edward Henty's father, Thomas Henty, wrote:

I have suffered the opinion of those who have taken a bird's eye view of it and who say there is only about 1,500 acres to cultivate. I know better and I believe 200,000 acres of good land may be found not a very long way from the Bay.[10]

The Hentys came to Portland after extensive searching elsewhere in Australia for land suitable for sheep. In December 1834 Edward Henty described the land to the north of Portland Bay:

You can see three-quarters of a mile either way, so this cannot be called thickly timbered. We walked six miles and took a fresh cut back to the river, which we crossed and walked down four miles — land of the finest description, grass as thick as it can grow . . . A more lovely spot I never beheld. Sheep or cattle's sides would soon shake with fat with a taste of the grass here.[11]

The likely locality of Thomas Henty's 200 000 acres (81 000 hectares) was open country inland from Yambuk and Port Fairy.[12] The downs that Mitchell discovered — the tussock grassland and open savannah, which represented another 200 000 acres — were additional to this.[13] On his return from Portland to Sydney, Mitchell traversed Victoria's western plains and the upper sections of both the Avoca and Loddon catchments. His diary provides a description of the original landscape.

To the south of the central catchments of the Great Dividing Range there lay open land — the land most eagerly taken up by the squatters. At Tatyoon, Mitchell described the state of this landscape in September 1836:

We now travelled over a country quite open, slightly undulating, and well covered with grass. To the westward the noble outline of the Grampians terminated a view extending over vast plains, fringed with forests and embellished with lakes. To the northward appeared other more accessible looking hills, some being slightly wooded, some green and quite clear to their summits, long grassy vales and ridges intervening: while to the eastward, open plain extended as far as the eye could reach.[14]

At Challicum, 3 kilometres to the south of Mitchell's line of travel and close to where the above diary entry was made, a painter named Duncan Cooper settled, in 1842, with his squatter partners, as a pastoral tenant of the Crown. Sketches of typical squatters' homesteads are fairly commonplace but there are few comprehensive pictorial records of the early landscape. One is the *Challicum Sketch Book* produced by Cooper, who displayed a high level of technical skill and a fine eye for landscapes.[15] As Mitchell crossed the Challicum plains he travelled 'open downs with wooded hills occasionally to the left', extending to the horizon in the south.[16] We can see how Cooper depicted them more than a decade later in Plate II.

The density of tree cover of the near Challicum hills in the 1840s corresponds to the previously presented, contemporary verbal descriptions of the rolling hill country of central Victoria. Summer fires, referred to in Chapter 1, are in evidence on adjoining runs.

We know that much of the open parkland, which attracted early squatters, did not evolve naturally but was created by the regular fire drives of the Aborigines. The landscape of the southern basalt plains today is somewhat different from that of the early European settlement, but not too different from when Cooper sketched it. The Aboriginal camps and the emus have gone, as have the trees in the lower foothills and foreground, the result of more intensive pastoral activities. The lightly wooded and open plains are sometimes now more open although, in some places, tree-planting schemes have resulted in more tree cover than at the time of European settlement.

The pastoralist settlers made their mark but did not radically change the treescape. Standing in the same spot, today, where Cooper sketched the tree cover on the Mount Cole Range, the cover is at least as dense as in Cooper's sketches. The foothills shown in the left middle ground of Plate II are now more densely forested on the top; the lower foothills and the Challicum Hills (in the left foreground in Plate II) are no longer woodland but given over to pasture, with scattered clumps of trees.[17] The condition of the plains is similar to that in 1850. In other parts of the open plain, tree planting by pastoralists has increased the original cover. For many years of European settlement, for many parts of the plains, the scenes that Cooper painted were typical — an occupied but relatively empty land of open plains and lightly timbered hills.

The colonial artist Eugene von Guérard studied painting at the Dusseldorf Academy in his native Germany. After travelling in Europe he decided to try his luck on the Victorian goldfields, arriving in 1852. However, it was in Australia's richest grazing region, Victoria's Western District, that he received the commissions that provided his income. Later impressionist artists, disgruntled students of von Guérard, sought a freer, more spontaneous interpretation of the Australian landscape than von Guérard's scientifically precise approach to landscape painting. His view from the Bald Hills, near Creswick (Plate III) shows the watershed to the

Avoca catchment as open woodland. The Pyrenees Range appears to have little tree cover, which is confirmed by Mitchell's account.[18]

> We were still however to cross the range at which we had arrived, and which, as I perceived here, not only extended southward, but also broke into bold ravines on the eastern side, being connected with some noble hills, or rather mountains, all grassy to their summits, thinly wooded and consisting wholly of granite. They resembled very much some hills of the lower Pyrenees, in Spain, only they were more grassy and less acclivitous.[19]

Today, much of the southern Pyrenees is denser forest. There are many more trees than when Mitchell looked upon the range.

The region Mitchell was about to enter comprised the upper Avoca and Loddon catchments in central Victoria. He crossed through the narrow waist of the Great Dividing Range. On the northern side of the range he climbed Mount Greenock. Mitchell was pleased to see that in travelling through 'this Eden' there were no barriers to the easy passage of his drays and wagons. His sketch of the scene is Plate IV. This is the view from 'Mitchell's hill', to which we referred in the Introduction. Bared hills abound with scattered clumps of forest. The scene is much the same today (see Plate V). Disbelievers can climb the hill to see for themselves.

The accounts of the first Europeans to walk across central and south-western Victoria indicate large changes in the vegetative cover within relatively short distances. The land was not consistently covered with trees. The land was a mixture of open woodland, bands of thicker forest and of heathlands in the south-west, and large areas of open tussock grasslands; it was not generally densely wooded. The 'forests' were mostly open in nature. The explorers were able to travel through most of the 'forest' either on foot or with horses and drays, without the need to clear tracks. Boggy soil, rather than barriers of timber, hindered progress of the drays. A significant area was comprised of grasslands suitable for pastoral activity. Within ten years of European settlement all the so-called 'forest' in the south-west, apart from some land west of the Glenelg River, had been taken up for grazing under pastoral licences.[20] There is good evidence that after European settlement there was further encroachment of forest and that existing forests increased in density.

Gold fever and selection 'hangover'

The occupation by the squatters was disturbed by the discovery of gold. In Victoria the gold-bearing land stretched over a vast area.[21] The best deposits were concentrated in the central highlands in the triangle between Ballarat, Bendigo and Stawell — much of this in the catchments of the Loddon, Campaspe and Avoca rivers. In 1851 the Mount Alexander gold-field was discovered near Castlemaine. In a few weeks the whole aspect of the country was changed into scenes of intense activity. (Plate VI.) Mount Alexander was more beneficent than the Ballarat goldfields and within a year 50 000 miners were operating more or less successfully in the shallow gullies around the mount.[22]

At the eastern edge of the Loddon catchment lay the Castlemaine and Bendigo goldfields. Travelling from Castlemaine to Bendigo in 1852 William Westgarth, a Member of the Legislative Council for Victoria, described the upper Campaspe catchment as 'the characteristic scenery of Australia under its best aspect . . .':

> The country before and around us was one continued succession of hill and dale, covered plentifully with grass, and more or less open, but never crowded with trees. Even the considerable elevation of Mount Alexander and the line of hills continuous to the north partook of this character, exhibiting grass to the very summits, and dotted over with trees that nowhere concealed the subjacent verdure . . . Beautiful and commanding sites everywhere presented themselves on either side of the roadway; and we speculated on the future of Victoria, the chateaux, the parks, and the picturesque cottages of fortunate gold-diggers rose up before our imagination.[23]

The beautiful scenes were not to last long around the goldfields. The rush of diggers turned the squatters' world upside down. It also turned the soil upside down. Gold diggers were moles, burrowing in their thousands across and under the countryside. Their success depended ultimately on how much soil, clay and rock they could dig up. The landscape held no value to the diggers in their frantic search, unless its features pointed to the presence of gold. The diggers dug and sawed. They excavated creek beds, diverted streams and sunk mine shafts wherever there was the possibility of a strike. First sight of the diggings came as a shock to new arrivals. Dusty, denuded of trees and heavily pock-marked with shafts and mullock heaps, one observer thought Mount Alexander like 'what one might suppose the earth would appear after the day of judgement had emptied all the graves'. As miners prospected the countryside they ransacked 'gullies, creeks, hills, ridges water-courses and ranges'.[24]

The miners needed wood to build huts, cook meals and keep warm in winter. Trees, especially the stringybark, were cleared mercilessly:

> The diggers seem to have two especial propensities, those of firing guns and felling trees . . . No sooner have they done their day's work than they commence felling trees . . . In fact the stringybark is the most useful tree conceivable . . . Its bark peels off from half an inch to an inch thick. These make the sides and roofs of huts. They makes seats and tables . . . and spouts and shutes for water . . . The wood is no less useful. It has the property of easily cleaving, and splits up readily into posts and rails, into slabs for the walls of huts, or into anything else you want. Therefore there is great destruction of this tree.[25]

Initially the miners sought the alluvial gold. Gullies and watercourses were shovelled and discarded for the insatiable cradle. When the alluvial gold ran out, shafts were dug for the deeper alluvial deposits. Timber was needed for the mine shafts. Forests were plundered in Australia's first wave of deforestation. When the gold fever died, in many areas all that remained were despoiled streams, mullock heaps and ground littered with shafts in areas denuded of forest. This was a recipe for erosion. William Howitt lamented the destruction while panning as furiously as the other diggers:

> Little more than a year ago, the whole of this valley on the Bendigo Creek, seven miles long by one and a half wide, was an unbroken wood! It is now perfectly bare of trees, and the whole of it riddled with holes of from ten to eighty feet deep — all one huge chaos of clay, gravel, stones and pipe clay, thrown up out of the bowels of the earth! So much has been done in this one forest in one year; and not only so much, but a dozen of the valleys as large.[26]

The peak of the gold rushes was in 1858. The gold rush soon moved on to more distant parts: New Zealand, Kalgoorlie or the Palmer River. While quiet has now returned to the gullies formerly ringing with the sounds of prospecting and mining, today the mullock heaps, a legacy of the mining, can be seen from a space satellite, appearing as white dots on the satellite photographs.[27] There were other more enduring legacies of the gold rushes: a rapid increase in the population of the colony, protection of forest reserves and erosion on a massive scale.

Once diggings had been worked out they came to resemble old battlefields. Rabbits colonised the eroded creek beds and the broken mining areas. The remaining alluvial miners turned to large hydraulic sluices to work every last ounce of gold out of the river beds. The sluices exacerbated the erosion legacy.

Ironically, the larger gold-mining companies, which survived the diggers, were a force for the preservation of forests. Their deep lead mining needed a continuous supply of timber to line the mines and to generate steam for working the pumps and driving the batteries and winches. Whereas the small digger could only look far enough ahead to plan a move to the next field, the companies realised they needed a continuing forest resource to supply their timber needs. They lobbied to create forest reserves that could provide sustainable yields of timber.[28] The forests needed a protective lobby. The rapidly increasing population was placing pressures on the forests. The increased population needed firewood for heating and cooking. Streets were paved with wood. By the 1880s Melbourne was consuming 400 000 tons of wood a year. There was also the matter of a livelihood for the miners who had not left to try their luck elsewhere. They demanded access to land. The merchant classes of the major towns cried out for land to preserve their vastly increased market. Selection was the answer!

With the discovery of gold the population greatly intensified in the new gold towns built on the Castlemaine, Maldon, Daylesford, Bendigo, Maryborough and Ballarat goldfields. The new population demanded land, and some of the first successful selections were the small holdings released close to these major towns. The people also brought weeds. The district around Maryborough and Maldon must have been the weed capital of south-eastern Australia, with scores of locally declared noxious invaders.

Farther from the towns the squatters used the lax early Selection laws to gain some security. That gave them the freedom to improve the land. In 1869, thirty years after John Hepburn took up the Smeaton Hill run,

Joseph Jenkins had a temporary labouring job for Mr George Hepburn. He recorded the following in his diary:

> I take a dray-load of timber to Hepburn's town house (Smeaton House). That much fire wood is consumed there each day. A regular wood-cutter is kept in the Bush . . . the bailiff and I were cutting wood in the afternoon. This is plentiful. In the paddock of 134 acres, the dead and fallen trees take up one-fifth of the area. Some of this timber has fallen from the effects of time, some from the woodman's axe, some at the hand of the grubber, and others blown down by strong winds. Many of the trees are barked or ringed in order to wither them, so that land can be reclaimed for grazing.[29]

Some selectors followed a similar destructive pattern on their smaller properties. In 1876 the Geological Surveyor, reporting on the Learmonth area north of Ballarat, commented about the southern part of the catchment:

> Before these rich volcanic plains were purchased and put under cultivation they were lightly timbered in clumps, some small remnants of which may still be seen, but the greatest part has been entirely denuded of timber. The Mount Bolton Range still possesses some fine stringybark, with gum, box, black and silver wattle, honeysuckle, and cherry trees, though these are being rapidly destroyed, and, being private property, cannot be prevented.[30]

This district did not see the destruction that was soon to be wrought on the south Gippsland forests. Much of central Victoria was grassy woodland with open, grassy plains to the north and to the south of the Dividing Range. The Geological Surveyor's observation pertains to the more desirable land for cultivation, which was close to centres of settlement. Joseph Jenkins' observation of the Maldon selectors' surrender to the trees indicates the unsustainability of cropping practices associated with early settlement:

> The selectors in general avoid the further clearance of the land, and they exhaust what little they have cleared of scrub. There are in the Shire 260 square miles of land, and two thirds of this is good cultivated land, but it is not being improved, and is again growing timber.[31]

Until the advent of the wheeled tractor the extent of such cultivated land was constrained. Much of central and southern Victoria remained an arcadia well into the 1900s.

Controlling erosion

The story of the Central Highlands of Victoria for the fifty years from the turn of the century mirrors the story of pasture degradation, which has been recounted in previous chapters. Overgrazing, inappropriate cultivation, rabbits and nutrient depletion degraded what was, in places, a fragile land system, causing erosion in its various forms. In the upper catchments of the Loddon, Campaspe and Avoca, the depredation of the miners resulted in huge amounts of silt flowing down the streams and rivers. The Laanecoorie Reservoir built on the Loddon River in the 1890s was reduced to half its volume within fifty years by siltation. A major cause of the problem was the mining companies using dredges to wring the last speck of gold from the river beds. The Sludge Abatement Board, formed to regulate the activities of the dredgers, reported a sorry picture of rampant corporate gold fever destroying the riverine environment:

> . . . long and deep waterholes in the river had been completely filled with sludge and mining debris, that the Loddon flats had been covered four to five feet deep in silt, and that creek beds had been raised, necessitating alterations in the heights of municipal bridges, and the erection, in some cases, of new ones. It was also found necessary to construct a new road between Newstead and Guildford. The general bed of the River, independently of waterholes, had been raised three or four feet. The Shire engineer of Newstead, although then in his 25th year of office, could not remember whether or not there were any fish in the river. He considered the water in the Loddon 'contained too heavy a deposit of sludge, as to render it useless for irrigation'.[32]

The Sludge Abatement Board placed greater controls on the dredgers, fining and prosecuting those who failed to take precautions to protect the stream. The Board felt uneasy that similar stream damage by farmers was continuing regardless:

> It seems anomalous that one set of people engaged in making their living by the recovery of metals from the earth should, and rightly so, be prosecuted and fined for injuring streams and land, while another set, merely as a result of mismanagement and neglect, are allowed with impunity to cause damage of the very same kind.[33]

Despite the grumbling of the Board, there was a marked reluctance by the government to prosecute farmers. Just as today, corporations with greater size and anonymity make less sensitive targets. The same reluctance to apply penalties for poor soil management occurred half a century later when the government was considering a new dam on the Campaspe River.

In 1959 the Victorian Government decided to build a water supply dam on the middle reaches of the Campaspe River to create what today is called Lake Eppalock. Farmers in the lower reaches of the catchment had been calling for the dam at regular intervals from the turn of the century — regularly after every drought. By 1959 the government was happy to oblige them, but there was a major problem in the upper catchment above the dam. The dam was not viable while the then high levels of erosion continued in the upper Campaspe catchment. The fledgling Victorian Soil Conservation Authority was directed to control the erosion and safeguard the state's investment in the new dam.

The Soil Conservation Authority had a mammoth task to control a vast erosion problem in a very few years. The existing physical damage had to be recovered. Erosion gullies had to be stabilised and battered down. The

underlying cause of the problem, degraded soils and pastures, had to be
solved. The answer was to establish improved sub clover and phalaris
pastures on the hillsides. Luckily for the government, two local farmers
had solved the puzzle of how this could be done easily by sowing with a
chisel plough. The Ross brothers of Mia Mia had struggled with physical
control of severe gully, sheet and tunnel erosion on their farm through the
early 1950s, but found that mechanical means of control offered only a
short-term solution. They realised the only long-term control would come
from revegetating the hillsides. They modified a chisel plough by adding a
seedbox, and sowed a mixture of lime, superphosphate, phalaris, ryegrass
and subterranean clover using a modified 'keyline' method. By the early
1970s they had treated all their property, trebling the carrying capacity and
virtually eliminating erosion.[34] The Soil Conservation Authority saw the

Improved pasture and gully battering of an Eppalock gully — before and after

Ross brothers' new technique in 1959 and quickly adopted it as the best solution for the catchment.

Landholder cooperation was essential if the Eppalock catchment project was to succeed. The Soil Conservation Authority called public meetings in the catchment, where they explained the problem, their role, their plans and what they expected from the landholder. The Authority was offering a deal to the farmer, a deal with something for everyone. The government placed no blame on current landholders for the problems of the catchment. The state would bear the full cost of reclaiming existing erosion damage and part of the cost of establishing perennial pasture. In return, the farmer had to agree to manage the land to minimise future erosion damage. The plan recognised that many farmers had little to gain from erosion reclamation works, though the state and downstream farmers had much to gain. The only practical method of getting speedy action was for the state to pay the full cost of reclamation. Sowing improved pasture had a double benefit. It increased the productivity of the farm and reduced runoff. Farmers were expected to bear some of the cost of this work.

The first and tangible part of the farmer's commitment was to buy the seed and fertiliser that were to create the new pasture. The Soil Conservation Authority would then pay contractors to sow the new pastures. Improved pastures tripled the stocking rate on some farms. This allowed farmers to consider radical changes to the way they managed their properties. Instead of growing wool they could produce fat lambs or cattle. They needed new dams to ensure adequate water supplies for the greater numbers of stock. Because a given paddock could produce three times as much grass, farmers also needed to consider making their paddocks smaller. The new management options implied a new farm layout.

This opportunity to replan the farm was where the Authority expected farmers to fufil their side of the bargain. The Authority wanted farms to be planned with paddocks fenced according to land capability, appropriately sited tree plantations to minimise soil damage caused by cattle camping at night on fragile soils, and fragile areas with steep slopes or salted land fenced off. It was expected that, eventually, the project would be managed by the participating farmers after the Soil Conservation Officers had moved on to tackle other problems. This meant continuing to fertilise improved pastures, maintaining the fences around fragile soils and managing them sympathetically, controlling the rabbits and maintaining the erosion control structures built by the Authority.[35]

By 1970 the Eppalock project was considered to be a success. Three hundred and seventy-three landholders had been involved. Three hundred and twenty square miles had been planned. Over one thousand chutes, silt traps and drop structures were built and over one thousand gully heads had been battered. Chisel seeders had sown improved pasture over 45 000 acres. One hundred and twelve thousand trees had been sown in 50 miles of gully.[36] Improved pastures and erosion-control structures had eliminated the worst of the erosion and silt flow in the Campaspe River.

The initial success masked a deeper difficulty — the social sustainability of the pact between the Soil Conservation Authority and the landholders.

The Authority had provided the impetus and commitment to the project. Would that commitment transfer to the landholders when the Authority's officers moved on to other projects? The reports of the period indicate that officers took every opportunity to involve landholders in planning for the project. The officers were working under pressure to produce quick results; and waiting for landholders to be involved was a slow business. Soil conservation officers often drew up plans for the farm while designing the engineering works. Farmers were consulted in the process, but the farmers saw the plans more as government documents, rather than plans towards which they felt any commitment. This problem was more obvious with the erosion control structures. As the project ended the effort to transfer responsibility to the landholders was stressed in the final report:

> Care is taken to impress on landholders that while many of the works done by the Authority are at no cost to them, they are in fact works which belong to them and that their purpose and maintenance are understood. To impress this fact, as parcels of work become due for handing over future responsibility to the landholder, inspections are arranged to discuss each item of work in the field and to discuss a works maintenance statement.[37]

As time went by it became increasingly clear that the local landholders did not accept that they owned the erosion control structures. Maintenance was not carried out. Farmers waited for the Authority work gangs to arrive to do the maintenance. By the late 1970s the same problem was emerging in many of the subsequent projects undertaken by the Soil Conservation Authority. Landholders drew a clear distinction between what were seen as non-productive erosion control works and productive activities for erosion control.[38] At the time, such perceptions by landholders were not willingly acknowledged by the senior officers of the Soil Conservation Authority.

Subterranean pressures

Silt was not the only downstream cost of upper-catchment land degradation in the Central Highlands of Victoria. Salinity can also be destructive for water supplies. The Campaspe River supplies comparatively salty water for the Campaspe West Irrigation System. When water from the Campaspe River is used to augment the Bendigo water supply it has to be mixed with other water. In high-flow conditions the Loddon and Avoca rivers pour salty water into the Torrumbarry Irrigation System between Kerang and Swan Hill. Saline water in the Avoca River threatens valuable wetlands in the Avoca marshes near Kerang.

During the Eppalock Project salinity caused concern because salty patches of land were prone to erosion. These salt-affected areas were fenced, to control grazing, and planted with salt-tolerant grasses. Controlling the saline discharge reduced erosion but had less impact on the flow of saline water into streams.[39] Salt flow in streams was not the pervasive concern that it is today. Downstream irrigators have realised that their saline water supply can be traced in part to the salt flowing from salty seeps in the Central Highlands.

Governments are now more interested in reducing soil salinity than at the time of the Eppalock project. During the 1970s and 1980s, the emphasis for salinity control shifted from being a necessary component of the control of soil erosion to an attack on the cause of salting — the recharge of the underground aquifers. This seemed the only approach that could offer long-term solutions to both the problem of loss of land and salinisation of streams.

Dryland salinity is caused by saline seepage and salt scalding. Salt scalding occurs where the sub-soil is naturally salty and is exposed by the erosion of top-soil. Seepage salting, caused by rising watertables, is the greater threat to Australia's non-irrigated land. Seepage salting in dryland occurs because land clearing and modern agricultural processes of cropping and grazing have resulted in less water being used by the current ecosystem than was used by the pre-existing natural ecosystems. We should not assume that all salt in this environment is a reflection solely of European farming. Salt lakes were not uncommon at the time of European settlement in the less well drained volcanic plains. They were common in many parts of Australia.[40]

The basis of dryland salinity is a major change in the use of water. New plants are using less water than the original vegetation. Unused water becomes the source of either increased runoff or higher watertables. Part of any long-term solution must be to grow plants that use a greater amount of water where water is seeping down to the watertable. In the Central Highlands of Victoria, recharge and discharge areas can occur in close proximity, with localised salting caused by water percolation from nearby hills. These areas are known as local groundwater recharge zones. However, in other situations, involving regional groundwater systems, discharge areas can be far removed from their corresponding recharge areas, with the underground flow of water through aquifers and deep leads.[41] The direction of these aquifers may be independent of surface topography, making it difficult to identify the recharge areas for the respective discharge areas.

Where there is selective recharge, the obvious solution has been to recommend planting trees on the recharge areas. This will most often be on hilltops where the rock is fractured, the soil shallow and conditions are not very suitable for deep rooted grasses. The deeper roots of the trees will intercept more water than grasses, leaving less water to escape to the watertable. This view is the basis of a national tree planting scheme — further dryland salinity will be prevented by planting trees on water intake areas. The exact location of these areas is often difficult to determine.[42] Each localised catchment will require an extensive geological study to produce a still uncertain answer about exactly where trees should be planted. A full understanding of the recharge and discharge characteristics of a particular area of localised salting requires a significant investment in exploratory drilling and investigation. The results of this research are not

always the expected answers. For example, the recharge may not be coming from the nearest hills, but from more distant hills. It is not economically feasible to undertake drilling on all salt-affected farms. Advisers are
forced to give 'best bet' advice, relying on the evidence that drilling can
confirm expectations more often than not. Some scientists say that we only
know 10 per cent of what we need to know to solve such problems.[43]

The difference between conceptual, geological, or agronomic understanding of the causes of salinity and the detailed empirical knowledge to
explain the cause of a given local incidence of salting is revealed in the draft
Salinity Control Strategy for the region that includes the upper catchments
of the Avoca and Loddon rivers. In this region, 127 salinity discharge sites
have been identified. Each of these sites is categorised on whether the
salinity is caused by local, regional, or semi-regional groundwater systems.
The discussion of the causes of salinity at each of these sites, in nearly every
case, is couched in terms such as 'the cause of salting at site . . . appears to
be', ' . . . seems to be', ' . . . is thought to be', ' . . . is probably . . .', or ' .
. . is possibly'.[44] A great deal more evidence from groundwater monitoring
is required to accurately understand the size and location of recharge areas
so that rational and economically targeted action can be undertaken.

When a preferential recharge area can be identified, the management
issues are relatively simple. But how do you go about controlling recharge
when all of a farm is a recharge area? The simple solution of planting trees
is not acceptable. If you cover the farm with trees, the farm disappears. At
present the recommended solution is to sow deep rooted perennial pastures that grow through most of the year and use more water in a year than
normal pastures. In effect, they are productive exotic replacements for the
displaced native perennials such as wallaby grass. If planted over a large
area they will reduce or even eliminate recharge. Lucerne and phalaris are
deep rooted perennials, both of which have high water use potential. Soil
conservation officers recommend them as another arm of their 'best bet'
strategies for salt control.

It is wise not to become too carried away with 'best bet' advice. It has
only been in recent years that lucerne's ability to lower watertables has
been demonstrated in the paddock. We are still waiting for such a demonstration for phalaris.[45] It is not yet known whether these measures will
result in sufficient watertable control to eliminate salt damage. Even the
old method of discharge control is being recommended again. In Western
Australia tree planting around saline seeps is being used to control
watertables. In the Wimmera area of Victoria lucerne is being planted
around discharge areas and early evidence is surprisingly favourable. The
lucerne seems to have pushed the salt seep into retreat. Continued planting
on progressively recovered ground may eliminate the seep entirely. We
should not become overly excited about single successes like this in one
area. We do not know how sustainable the single gain is, or how applicable
it is to other areas. For the moment, all we can say is that any farmer
hoping to control salt will need to wait a long time to recover land from
salt damage. It follows that salinity control will require at least a small act
of faith by each participating landholder.

Faith is a matter of belief and perception. The key to salinity control will be farmers' perceptions of the solutions, so we shall spend a little time considering landholders' perceptions of the problem. The Loddon and Avoca catchments are a good place to do the exploration. The area contains local, semi-regional and regional groundwater systems that contribute to salting. The mechanisms of salinisation in this area are better understood than elsewhere in south-eastern Australia.[46] Here there is more evidence for farmers to digest than elsewhere.

In the upper, southern end of the Loddon and Avoca catchments one-quarter of 1 per cent of the land has been identified as being salt affected.[47] When driving along the Sunraysia or Pyrenees highways in Victoria you could be forgiven if you did not see any salt. Salinity appears to be caused by local groundwater systems.[48] Its incidence is localised. You have to drive away from the main road to see the signs. Each outbreak has a different story to tell.

Earlier we recounted Joseph Jenkins' recollections of Smeaton Hill, in the south of the catchment. Today, at Smeaton there are about 40 hectares of salt-affected land on a band of sedimentary country sandwiched between ancient volcanic cones. The salting is caused by a local groundwater system. The watertables may well be recharging through the treeless basalt volcanic cones. Although the salt areas have good vegetative cover many trees have died as a result of rising ground water.[49] Perhaps the omens were there 150 years ago when John Hepburn advised Governor La Trobe: 'On my land there is much good land to appearance, but the crop is below the average of my neighbours, which I can only attribute to the want of proper drainage, the land being very wet in winter'.[50] There is an active tree planting group in this district. It is ironic that some of the landholders with salt affected ground would like to see the volcanic hilltops around them covered with trees although they were naturally treeless. Here, salinity control does not coincide with recreation of a natural habitat.

South of the Great Dividing Range is the Haddon District. The salt here is localised, occurring in seeps on the southern side of hills or in the occasional patch on the valley floor. A little over 1 per cent of the area is salinised, with the average affected area being 4 hectares. Haddon is a rural residential dormitory for the Victorian city of Ballarat. It is populated by hobby farmers and the occasional full-time farmer. The new residents bought their land not to make a better living, but to improve the quality of their life. Salting for them is a matter of aesthetics rather than production. Salt presents different problems for the shire, which must site new subdivisions so septic-tank systems are not built where there are shallow watertables. It appears some of the recently subdivided land was salty. Perhaps the previous owners reasoned that unproductive land was better put under unproductive housing.[51]

In the Moonambel area of the Pyrenees exceptionally good wines are made. The style is usually big and full, and the red wines often have a eucalyptus flavour typical in central Victoria; the shiraz is peppery and the cabernet is minty. Adjacent to the vineyards there is extensive salting in the Moonambel valley. This first became apparent in the early 1960s.[52] The salt has spread rapidly up and down the bottom of the valley and now covers 160 hectares, including 50 hectares of farm land. The vines are on the hillsides so we are unlikely to get a pepper and salt shiraz. The local road needs reconstructing every eight years because of the lack of structure of the saline soil.[53] The seepage of groundwater and increased surface runoff has also led to severe water erosion in areas where salt has destroyed the ground cover and soil structure. The Moonambel valley salinity is caused by a local groundwater system with fractured rock intake areas on hilltops. The recommended control strategy combines tree planting on the hilltops with pasture on the lower slopes. There is a LandCare group in this district, but the strategy has not been embraced with excessive enthusiasm by local land holders.[54]

Travelling further north in the catchment one encounters semi-regional and regional watertable problems. Within 15 kilometres of Moonambel is the upland plain of Natte Yallock. The plain was formed by a lava flow that cut off the Avoca River, forming a lake. Erosion in the upper catchment eventually filled the lake with sediment, forming the plain. Obviously erosion is not a new phenomenon in the district. Salinity at Natte Yallock is thought to be the result of a regional watertable based on the Loddon deep lead — an earlier river bed of the Loddon river that has been buried well below the surface by layers of volcanic lava.[55] Farmers in this district cannot hope to control salt on their properties by planting trees on areas of selective recharge. If the LandCare group in this area is to control its salt, it will only be through extensive planting of deep rooted crops over most of the plain and most of the farms. Trees are not a practical solution. The most promising option is lucerne. It may also be possible to lower the watertable by pumping water from the deep lead. This will increase drainage from the watertable to the deep lead. This pumping will allow some farmers to irrigate their lucerne crops.

To the north beyond the edge of the Great Dividing Range the deep leads from the upper catchments spread out underneath the flat riverine plain. The riverine plains are also experiencing rising watertables, and the implications are more severe than in the upper catchments. This is a regional groundwater problem. The flatness of the country means that large areas of farmland will be salinised. It was once thought that the deep leads were the cause of the problem and that the upper catchments were the cause of the rising pressure in the deep leads. It is now clear that the whole of the riverine plain, including the major irrigation areas, is a recharge area. The problem and solutions are similar to the situation at Natte Yallock, only the scale is vastly greater.

At Burkes Flat, 80 kilometres west of Bendigo in the mid-Loddon valley in Victoria, cropping and grazing replaced goldmining late last century. Here salinity is caused by local catchment groundwater and 5 per cent of the valley floor has been salt affected.[56] Burkes Flat is at the upland front, the junction of the uplands and the riverine plain of the Murray Valley, which is particularly vulnerable to salting as a small rise in groundwater levels affects large areas of land. Burkes Flat is the only location in Victoria where there is quantitative evidence that recharge works appear to be lowering groundwater levels. Burkes Flat has become a mini Eppalock project, but this time with the emphasis on salinity control.

Landholders in the Burkes Flat catchment first approached the former Soil Conservation Authority to assist with erosion and salinity problems in the early 1960s.[57] Revegetation of the saline discharge areas and erosion control works were undertaken in the early 1970s but saline areas continued to spread. During the late 1970s and early 1980s investigations showed that the groundwater causing salinity came from the immediate catchment through fractures in the underlying bedrock. The management within the catchment was then directed towards control of the sources of the groundwater problem rather than the effects. In 1983 a catchment of 800 hectares, incorporating land owned by four landholders and including about 120 hectares of valley floor salting, was established as a demonstration site. Trees were planted on the high recharge areas — the shallow stony soils and the rock outcrop areas on the ridges — and deep rooted pastures were planted on the moderate or low recharge mid and lower slopes. Salt tolerant grasses were then planted on the saline discharge areas to stabilise the land.[58] The trees have yet to achieve maximum water uptake but the deep rooted pastures have reached maturity. To date, monitoring of the groundwater levels has shown that under the pasture on the mid slopes there has been a 2–5 metre drop in groundwater over a period of five years. Other areas of the catchment were still showing typical seasonal fluctuations, but lower net fluctuations. Groundwater levels within the salt-affected lower areas have not yet dropped, most likely because, as yet, the trees have had little impact and because a large slow-moving wall of underground water remains higher up the slope.[59] Despite the early signs of some interim success, there is unlikely to be an immediate rush of landholders to take up the Burkes Flat approach in other catchments.

A reticence for reafforestation

In Australia a small but significant number of farmers have become enthusiastic planters of trees and have undertaken other activities aimed at reducing soil degradation when it exists on their land. Most farmers, however, express a concern about such matters that is not matched by action. Most farmers who plant trees to control salinity plant token numbers of trees. There are several reasons for farmers' lack of activity in relation to tree planting for salinity control.[60]

Less than 1 per cent of the land of the upper Loddon and Avoca catchment areas in Victoria has been identified as being salt affected; therefore it is not surprising that local landholders don't consider salting to be one of their most pressing problems.[61] Farmers across the Central Highlands believe there is a significant salinity problem in the region, but far fewer think there is a problem on their farm. In 1989, although nearly half the farmers in the district saw some salt on their properties, it was a long way down the list of farm concerns. Low farm income, high farm costs and inadequate local services were the dominant issues. Ironically, the highest levels of concern about salt were expressed by landholders in the Haddon district, where two-thirds of landholders believed there was a very great need for salinity control on their properties. Nearly all these concerned landholders were hobby farmers or people with rural retreats who had less direct dependence on the land for their income. The few full-time farmers of Haddon who depended on their land for income found salinity a much less worrisome problem. Having lived in the area all their lives, they believed the existing salt patches had always been a feature of the district.[62] Perhaps the recent migrants to the district had brought with them a concern for salt derived from the media, rather than from observation of the land on which they lived. They had a much less detailed understanding of the mechanics of salinity; a quarter of them believed excessive use of chemicals and fertilisers was the major cause of the salinity problem.

If farmers are going to plant trees to control salt, they must see or anticipate salt damage. Seeing salt is a complex process. The evidence is there for some to see, but it can be interpreted in many ways. Salinity rarely appears overnight; it creeps into a locality slowly and insidiously — a little like old age. The worst end of soil salting is bare land covered with a salt crust. However, there is a considerable lead-up to this state. Only the very perceptive notice the early first signs; and the early signs seem too insignificant to be concerned about. In 1989, 40 per cent of Loddon–Avoca farmers were unaware of how to recognise the early signs of encroaching soil salinity.[63] Because a problem such as salinity takes a long while to have a serious effect on a given farm, landholders tend to see it as a problem for the future rather than a problem of the here and now: it tends to be a problem for someone else, elsewhere than 'on my farm'.[64] The other side of the time equation is a growing familiarity with salt. Does familiarity lead to concern or to apathy? When obvious salinity exists in an area a frequent landholder reaction is that 'its always been there' and 'it doesn't seem to be getting much worse'. Given the difficulties of controlling salt, it is easy for landholders to conclude that they must live with salinity.

As we have seen, there is the lack of specific localised scientific knowledge about tree planting, water tables, and salinity. The relationship between planting deep rooted species and groundwater might be acknowledged as correct in principle, but there is little substantive information specific to given catchments. There is virtually no information about the time that will elapse between the investment in deep rooted species, such as trees, and the control of groundwater discharge. In the nearby Goulburn River catchment the State Government had considerable difficulty in getting scientists to agree on practical solutions to salinity. It has had even greater difficulty convincing some government farm advisers that it is reasonable to ask farmers to control salinity considering the uncertainty about the proposed solutions. When people are confronted with uncertainty they tend to delay any action. It is traditional for many farmers to wait and see if a new idea works on a neighbour's farm before trying it themselves. There will be a long wait before the answers to the salinity questions are known. Because of slender current knowledge farmers are loath to expend money and effort.

Planting trees is a simple task; but the effort is considerable. Planting trees that will survive on the tops of hills is harder work. The area to be planted must be fenced, the site must be ripped or ploughed to create suitable planting sites. Weeds need to be controlled before planting. In Victoria, trees must be planted in winter with tree guards to protect against rabbit attack. Rabbits, if sufficiently hungry, will eat the bark off young trees. In winter they are often hungry. Before summer the trees must be mulched to preserve moisture and prevent competition from weeds. Weeds must be controlled through spring. The trees may need to be watered during summer and early autumn. The following winter the failures will be replaced and the whole process repeated. The work of planting can now be eased by direct drilling of tree seeds using recently developed seed drills.

Tree planting is also an expensive business. The cost of the trees is a small proportion of the total cost. Fencing is the major direct monetary cost, together with the production foregone from land planted to trees. At one site on the Black Range in the upper Loddon catchment, $44 000 has been spent to plant trees on a selective recharge site a little over 40 hectares in size. Even that sum underestimates the full cost. Much of the labour was voluntary. Caring for the trees over the following years has not been included in the cost, nor has the loss of future production from the site. The establishment cost of the Burkes Flat catchment work was of the order of $55 000 (1983 dollars), excluding farm labour and operating costs such as cultivation and fertiliser.[65] This represents about $70 per catchment hectare. The Eppalock project promised immediate and sure returns for a farmer's participation. There is no such prospect in Burkes Flat. Victorian landholders are unlikely to see the unsure long-term returns from the Burkes Flat project as justifying such an outlay unless subsidised by the government. It is difficult to ask farmers to bear these costs when there is, yet, no direct evidence of successful reclamation of dryland salting in south-eastern Australia achieved by tree planting. In north-eastern Victoria, in the upper Goulburn catchment, a community group developing a plan to control dryland salting did not think farmers were willing to incur such costs. They suggested that an 85 per cent subsidy was required to encourage tree planting in recharge areas.[66] Considering the costs we have recounted, this policy would entail large sums of taxpayers' money.

The problems of uncertainty and cost aside, most farmers tend to believe that trees should be planted where the salinity occurs (the discharge area) rather than where it is caused (the recharge area).[67] In the upper catchment hill country the farmers who experience soil salting — those who are in discharge areas — rarely have the recharge area causing the salinity on their properties. Strategies that might reduce watertable levels, or prevent further increases in the watertable level, require farmers in the recharge areas to modify their farming practices. These are not farmers experiencing the soil salting so there is little incentive for them to adopt salting-control practices.

These reasons, and the long delay between investment costs and any future return, mean that most farmers do not undertake activities to reduce salinity problems unless the activity produces some other productive benefit besides the salinity control. For tree planting these other advantages are, most commonly, shade, shelter and beauty. Tree planting for shade and shelter is done along fence lines and in gullies. Tree planting for beauty is done around the farmhouse. This is where most of the farm tree planting has been undertaken. Very few farmers in the Loddon and Avoca catchments have planted trees on hilltops.[68] Planting trees away from hilltops has been re-interpreted as a gesture towards the control of salinity. Farmers professing concern about salinity are most likely to have planted trees along fence lines, scattered across the farm, around the house, along gullies or creeks, or around dams.[69] This may be due partly to a failure to grasp the idea of trees as pumps. In the upper Loddon–Avoca area, although nearly 90 per cent of the farmers agreed that planting trees on hilltops helped salinity control, over 40 per cent did not believe trees could be used to reduce the amount of rainwater seeping to the watertable. Scientists' and farmers' understanding of the recharge and discharge phenomena is superficial and the use of trees by farmers has been symbolic.[70]

Where there is less need for shade and shelter, trees are viewed with less enthusiasm. The cropping farmers on the lower Avoca and Loddon plains have been less positive in their beliefs about trees than livestock farmers in the upper catchment. Planted trees are seen to encroach on valuable cropping land, compete for soil moisture and get in the way of cultivation machinery. Trees drop branches, which foul harvesting equipment and make management of the farm more difficult. Cropping farmers are much less inclined to plant trees than livestock farmers.[71]

The relationships between the sometimes conflicting beliefs of farmers about trees and salinity can be seen in a 'belief' map. A belief map depicts an *average* view of how people see issues. (See box: *Belief Maps*.) The belief map (Figure 3.1) shows upper Loddon–Avoca landholders' beliefs about reducing salinity. In this map of landholders' beliefs three practices are depicted that can help to control salinity: *planting trees*, *growing lucerne* and *improving pasture*. The practices are represented by the lightly hatched circles. In the map there are five attributes that farmers may associate with tree planting, pasture improvement or lucerne growing. These attributes are *reducing soil salting, increasing the farm's capital value, having a good farm, profitability in the long run* and *profitability in the next few years*. They are represented by unhatched circles. In the middle of the map, represented by the darkly hatched circle, are *Loddon–Avoca landholders*.[72]

Belief maps

A belief map is a method of illustrating, on paper, the way people view complex issues. Relationships between the parts of a complex issue have many dimensions, and a belief map displays as much of this complexity as is possible to display in two dimensions.

We could ask a person to estimate the distances between each of the six state capital cities of Australia. For the six cities we would obtain 15 inter-city distances. Standard mathematical methods could then be used to convert these distances into a map that showed the perceived location of each capital city.

When a group of people is asked to estimate these distances, the estimates can be averaged and the same mathematical techniques can be used to produce an average map to show how the group perceives the location of the capital cities.

The same method is used to build belief maps. In a belief map, a belief is defined as the perceived distance between the objects that comprise the belief. We determine which objects are central to people's beliefs about a particular complex issue. Then we ask them to estimate the distances between these various objects. A belief map can be produced from the average perceived distances obtained.

When people are asked to estimate how far they see themselves from various belief objects the position of the average person can be mapped in relation to the belief objects. We then have a map of the average person's attitudes towards the belief objects.

Belief maps show how people understand complex relationships. For example, belief maps can be used to help us understand the often complex decision about which car to purchase. If people believe car X to be luxurious and expensive, in a belief map 'car X', 'luxurious' and 'expensive' will all be close together. If people believe also that car X is fuel inefficient, 'car X' will be some distance away from 'fuel efficient'. The closer people place themselves to 'car X', the more favourable their attitude to 'car X'. Belief maps also help us to show people's understanding of even more complex and abstract issues such as land management.

We can use the belief map to explain the token, or symbolic, tree-planting behaviour of pastoral landholders. Planting trees is believed to be related to controlling salinity, but the relationship is not close. Tree planting is not seen as a panacea. More importantly, tree planting is not closely associated with important farm goals. Landholders closely associate themselves with 'profitability in the long run', 'increasing the farm's capital value', and 'having a good farm'. Planting trees is not believed to be strongly associated with these goals.

Salinity control will be most successfully encouraged among land-holders in the upper catchments by promoting improved pastures incorporating the deep rooted pasture species. Improved pastures are associated with 'having a good farm', 'profitability in the long run', 'increasing the farm's capital value' and the 'Loddon–Avoca landholder'. The attraction of pasture improvement (using deep rooted species) is that it is closely associated with the important farm goals of 'having a good farm' and 'long-term profitability'. It will be easier to promote pasture improvement to farmers; and salinity control can be achieved without all farmers

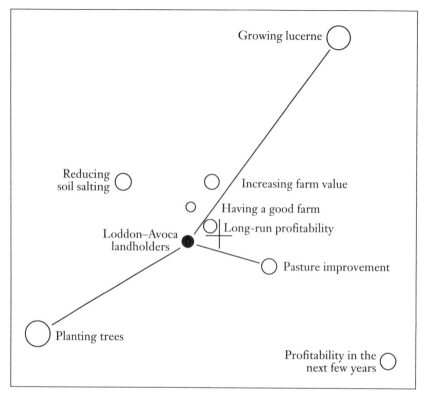

Growing lucerne

Reducing
soil salting

Increasing farm value

Having a good farm

Long-run profitability

Loddon–Avoca
landholders

Pasture improvement

Planting trees

Profitability in the
next few years

Figure 3.1 A belief map of Loddon–Avoca landholders' perceptions of salinity control

necessarily believing it to be associated with pasture improvement. As with trees, there are some doubts about the effectiveness of improved pastures to control salinity in all situations. Unlike trees, there are fewer doubts about the likely success of any promotional campaign based on improved pastures.

Historically, soil conservation advisers had only been able to recommend closing off the hills to stock and leaving them to regenerate as bush. Where this worked erosion was reduced, but the hills harboured rabbits, which fed on the pasture of the mid-slopes and valleys. Often the hills did not regenerate because there was no bank of tree seeds in the soil or because sheet erosion and crusting prevented germination. A small group of farmers solved two problems by combining rabbit control and pasture sowing methods. They welded a pasture seed box and air seeder onto a ripper dragged by a bulldozer. The bulldozer could travel where the hills were too steep for a tractor. The ripper destroyed rabbit burrows and prepared soil for a rough sowing. The seeder dropped seed and super into the rip line.

Pasture establishment is an activity well suited to Landcare group efforts. The first step is rabbit control, which is only achievable by group cooperation. The second step is sowing pastures. Where a group of landholders use a bulldozer the cost of its transportation between properties is minimised. Pasture can be sown with superphosphate, but the pasture must be regularly top-dressed to ensure good establishment and maintenance. This is most effectively done with a crop dusting plane. It is even more effective to fertilise all the paddocks along a range rather than trying to spread the superphosphate only on selected paddocks. The group approach may be less useful in encouraging continuing good management of the improved pasture. Deep rooted perennial pasture species require careful management. The management difficulties associated with species such as phalaris have convinced a significant minority of farmers that it is not worth the effort to grow it. There is little sense in trying to push these people into establishing improved pasture because of salinity control. Without good management of improved pasture there is little likelihood of gaining any of the possible watertable benefits.

A hard headed view of the salinity problems of the Loddon, Avoca and Campaspe catchments suggests that the really significant future losses to salinity will occur farther north on the riverine plain and the upland plains — problems we consider in later chapters. The localised salting at the top of these catchments has caught the public's attention; relatively, it is insignificant. There is little likelihood of salt affecting large areas of land here. The hilly nature of the country will prevent this because most of the land will always remain well above the watertable. This cannot be said of the plains country. On the plains the best solution seems to be a combination of growing lucerne and pumping groundwater. The impact of lucerne on watertables is the success story of the Burkes Flat experiment. Lucerne was a common crop in the Natte Yallock district of Victoria until it was devastated by the lucerne flea in the 1970s and disappeared from local farm culture. Despite the release of new flea-resistant lucerne, a significant number of farmers show little interest in returning to the crop.

The position of 'growing lucerne' in the belief map (Figure 3.1), isolated from 'reducing soil salting', confirms that there will be great problems in encouraging farmers to plant lucerne for salinity control. Growing lucerne is more distant from the Loddon–Avoca landholders than is planting trees. Lucerne is not seen as profitable in the short or long run, or as relevant to good farming or salinity control. Part of the problem is that many soils in the district are unsuitable for lucerne. This cannot explain why farmers who believe they have soils suitable for lucerne still place lucerne in the same relative position. These farmers still see other disadvantages. A lucerne paddock cannot be managed like a normal pasture paddock. The grazing of lucerne needs to be strictly controlled and this means managing the farm in a different way. Farms may need to be refenced. It seems the old ways of management are preferred.

Beyond symbolic tree planting

Campaigns to improve land management are not a new phenomenon. Nor is their high risk of failure. For most people there is a discrepancy between their attitudes to the environment and their behaviour related to the environment. This should be kept in mind when seeking to change attitudes to bring about better land-management practices. Expressing a positive attitude to controlling salinity, for example, usually doesn't take account of the costs that might have to be borne to bring about its control. As we will see in Chapter 4, community opinion and social pressure are often harnessed to bring about more appropriate environmental behaviour that might not otherwise follow from a belief that it is the right thing to do. Such an approach can set up a cycle of reinforcement for environmental beliefs, because humans are good at finding reasons for what they do and not very good at doing what they find reasons for. Local community support and local peer pressure may influence farmers in recharge areas to engage in tree planting for salinity control but this is only likely to occur in situations where the problem is localised. Most property owners do not believe extensive tree planting to be in their own narrow economic interest. A much more concerted effort will be required to encourage tree planting on a widespread basis. Such an approach would require the whole community to contribute a very large subsidy to landholders who participate.

Tree planting is not a panacea for salinity control. We do not know how many trees are required to redress the current groundwater problems, exactly where the trees should be planted, or how long it will take for the trees to have a beneficial effect. The movement of salt, via underground water systems, is slow — sometimes very slow. Where removal of trees has contributed to dryland salinity the effects seen today are caused by removal of tree cover twenty to fifty years ago. Although we are unclear about the size of the time lag from cause to effect, we are even more unsure about the length of the time lag from when trees are replanted to when watertables are sufficiently lowered to control salinity. Tree planting needs to be strategically located. We know at a general and conceptual level the geologically defined areas of preferential groundwater intake. For given catchments, we do not know with sufficient precision the specific localities where trees should be planted to reduce groundwater intake. We do not know exactly which pieces of land should be resumed into public ownership for reafforestation. This ignorance would not be a problem if there were no cost involved. There are significant costs in gaining information, planting trees and in production lost from the land on which trees are planted.

There are similar technical doubts about the use of improved pasture to reduce recharge of watertables in some situations. Despite these doubts there are obvious conservation and production benefits from improving pasture. Improved pasture reduces runoff, which reduces soil erosion. It encourages rabbit control.

Improved pasture and the associated livestock production will provide economic benefits in the short and long term. Improving pasture is more acceptable to farmers than tree planting. It has already been done in Eppalock in Victoria. From these perspectives the reward for the promotion of pasture improvement is likely to be far greater than the reward from tree planting. Trees are not seen to belong in the middle of the paddock.

When we look at regional groundwater problems it is clear this is where the greatest future losses from salinity will occur. Trees, by themselves, offer no real solution to this danger. Solutions will be based largely on deep rooted crops like lucerne. Lucerne will be as difficult to promote as trees but, as a means of salinity control, it is more likely to be a viable solution than trees. We have seen that many of the plains over which Thomas Mitchell and early European settlers travelled had only sparse tree cover. The potential gains from lucerne on the plains are likely to be far greater than from tree planting in the hilly upper catchments.

The question has to be asked whether government investment in extensive tree planting is in the wider community's interest. In some cases it is a misguided attempt to recreate an arcadia of trees that did not formerly exist. We need to disentangle the symbolism of trees from the practicality of tree planting. Simplistic solutions with broad-scale revegetation programs are unlikely to be economically feasible. With the current fuzzy knowledge on the locality requirements for tree planting to control salinity, the resumption of land would need to be on such a scale that, in the medium term, the cost would have to be borne in a lower community standard of living. In 1921 the rural correspondent in the newspaper *The Australasian* observed that the value of tree plantations 'can hardly be described in pounds, shillings, and pence'.[73] This is still the nub of the problem; our currency is now decimalised, but the value of trees for reducing land degradation is still difficult to determine.

4

Visions of timbered farmscapes

The number plates of motor vehicles registered in the State of Victoria proudly proclaim Victoria as the 'garden state'. The genesis of the theme had less to do with the rural countryside than with the parklands of the capital city, within which metropolis reside two-thirds of that state's population. New South Wales may well have been able to claim itself the parkland state if it had taken the advice of W. A. de Beuzeville. As forest ecologist for the Department of Forests he planned a grid of timbered highways for New South Wales, with the showpiece 1.5 kilometres wide and stretching 1000 kilometres from Tocumwal to Goondiwindi.[1] When the first Europeans arrived the description 'garden state' could well have applied to much of the countryside. In Chapter 1 we recounted the enthusiastic observations of the early squatters, who marvelled at the grassland and likened their new homeland to a park. Later, in 1854, William Westgarth described the Australian countryside:

> The general resemblance is still more striking in the Australian landscape. The open forest, free from underwood, with its grassy carpet beneath and its park-like aspect, is common to all latitudes of the country . . . to the pastoral facilities which this kind of country offers, ready-made as it were at the hand of Nature, is to be traced the rapid progress of the Australian colonies.[2]

Other observers in the 1840s and 1850s had similar descriptions of the landscape. Nearly all these observers use the phrase 'park-like' to describe the landscape that was the legacy of the Aboriginal burning regime.

The squatters occupied the lightly timbered woodland and open grassland. Occasionally they cleared some forest, but the essence of their business was using as little hired labour as possible. Clearing was too labour intensive. They left the trees where they were, except for harvesting timber to build homesteads, fuel fires and construct the early rudimentary fencing of home paddocks. While their runs were unfenced they could burn their pastures and unintentionally maintain the open forests in the same manner as the Aborigines. Abundant grass was available for minimum effort. Government leases positively discouraged effort by providing no security of tenure and no compensation for improvements to the leased land.

Fall and rise of the eucalypt

The first Selection Acts gave Western District squatters in Victoria the opportunity to secure their land tenure. Security of tenure had contradictory impacts on the woodlands. Some squatters engaged in ringbarking native trees to extend the grazing area.[3] James Bonwick was an early advocate of 'girdling', or ringbarking, trees to increase stock production:

> . . . the quantity of stock maintained in timber is far below that fed on the plains. The growth of good nutritive grass is prevented; first, by the shadow of the trees preventing the sun from developing the saccharine juices of the grass; secondly, by absorption of food for the trees themselves; thirdly by the increase of scrubby and rough vegetation fostered in the shade; and fourthly, by the accumulation of the wreck of fallen leaves and bark, which are, also, injurious in affording sustenance to, and being the existing cause of, our bushfires. All sheep and cattle farmers admit the force of the argument. The only practical remedy proposed is that of girdling the trees. By this simple process we cut off communication between the root and the branches . . . The sap ceases to run, the leaves fall but once more, and the grass is open to sun and air.[4]

Today, many farmers would use these same arguments as reasons for not planting more trees.

The main impact of the squatters on the forests was less from direct than from indirect destruction. In 1861 the squatter Neil Black observed that the eucalypts around Terang were dying off in great numbers. The same observations could be made in various localities across Victoria, from the Western District to East Gippsland. Many fully grown trees were slowly dying. Security of tenure had allowed the squatters to build fences. Fences meant the end of intentional grassfires. This sounded a death knell for many of the older trees. The agency was a form of eucalyptus 'dieback'. The regular fire regime under the Aborigines had acted as a break on the soil insect population. When the fires stopped, the insect population increased and mature trees, weakened by water stress due to soil compaction, lacked the strength to withstand the increased numbers of leaf eaters. This was particularly true of red gums in heavily trodden areas near stock watering points. Squatters like Black blamed an explosion in the possum population. He wondered if when the 'aborigines die away the opossums on which they live chiefly increase and destroy the Gum trees'.[5] The experienced bushman Alfred Howitt blamed a combination of soil compaction and insect population explosions.[6]

The regular burning had not stopped in the grasslands only. Diseases such as smallpox had decimated the Aboriginal people living in the uninvaded lands in the hilly country. They were no longer able to farm the open forests with fire. Forest litter began building up to proportions capable of sustaining much less regular, but much hotter, bushfires. The hot fires were capable of killing mature trees.

Though the end of the burning meant death for older trees, it also meant new life for re-emerging eucalypt saplings across the Colony. The eucalypts' regrowth had been controlled by a combination of a regular burning and grazing by the small marsupials that lived on the native grasslands. The decimation of the clumpy native grasses destroyed the habitat of these animals and their numbers declined. Investment in fencing drastically reduced the frequency of burning. The man-made ecosystem, which kept the eucalypts in check, was shattered. In Chapter 1 we re-counted J. C. Hamilton's experience with regrowth in the Victorian Wimmera where he saw parts of his lease revert to forest, destroying the understorey grasses.

The story was similar in Australia's south-eastern corner. Alfred Howitt was a keen anthropologist and natural scientist. Over thirty years he observed the forests and plains of Gippsland as a bushman, gold commis-sioner and official protector of the Aborigines. In 1890 he wrote a lengthy article for the Royal Society of Victoria in which he concluded that the forests in Gippsland were more widely spread and denser than at the time of white settlement, even taking into account the most extreme difficulties faced by the selectors in South Gippsland. His observations are pertinent to the debate over logging practices and land clearing:

[Before the arrival of the first squatters], the marsupials had been so few in comparative number, that they could not materially affect the annual crop of grass which covered the country, and which was more or less burnt off by the aborigines, either accidentally or purposefully when travelling, for the purpose of hunting game.

These annual fires tended to keep the forest open, and prevent the country from being overgrown, for they not only consumed much of the standing or fallen timber, but in a great measure destroyed the seedlings which had sprung up since the last conflagrations . . . The incidence of bushfires acted . . . as a check on insect life, destroying among others, those insects which preyed upon eucalypts . . . The results [of stopping fire] were twofold. Young seedlings now had a chance of life, and a severe check was removed from insect pests . . .

The immediate valley [of the Snowy River] was a series of grassy alluvial flats, through which the river meandered. After some years of occupation, the whole country became covered with forests of saplings . . . and at the present time has so much increased . . . that it is difficult to ride over parts which one can see by the few scattered old giants was at one time open grassy country . . . Similar observations may be made . . . in the Omeo district, namely that young forests of eucalypts of various kinds are growing where a quarter of a century ago the hills were open and park like. In the mountains from Mount Wellington to Castle Hill, in which the sources of the Avon River take their rise, the increase in the eucalypt forests has been very marked. Since the settlement of the country, ranges which were then only covered by open forest are now grown up with saplings of E. sieberiana, and others, as well as dense growths of Acacia . . . These mountains were, as a whole, according to accounts given to me by surviving aborigines, much more open than they are now . . . Again at the Caledonia River, as at the Moroka, the ranges are in many parts covered with forests not more than 20 years old. On Mount Wellington . . . I observed one range upon which stood scattered gigantic trees of E. sieberiana, now all dead, while a forest of young trees of the same species, all of the same approximate age, which may probably be 12 years, growing so densely one could not force a way through on horseback. The same observations may be made in Western and South Gippsland . . .

I venture indeed, to say with a feeling of certainty, produced by long observation, that, taking Gippsland as a whole, from the Great Dividing Range to the sea, and from the boundary of Western Port to that of N.S.W., that, in spite of the clearings which have been made by selectors and others, and in spite of the destruction of eucalypts [by insect pests] the forests are now more widely extended and more dense than they were when Angus McMillan first descended from the Omeo plateau into the low country.[6]

The regrowth seemed to be particularly heavy after heavy rains that ended a prolonged dry season — the early 1860s and 1880s being most noticeable. We will never know the full extent of this regrowth. It did not happen over the whole of the open forests, but it was not an isolated problem for the early agriculturalists. In some districts whole new forests appeared. In northern New South Wales the creation of the Pilliga forest has been described by Eric Rolls as a direct result of European settlement.[7]

Symbolism of the 'scrub'

The early Selection Acts of the 1860s failed to bring closer settlement and more intensive use of former pastoral land. Under the later Acts the government made more direct attempts to settle the countryside with small farmers rather than leave the land in the hands of a small number of powerful squatters.[8] In an attempt to prevent squatters monopolising the land by duchessing the selectors with dummy selections, the later Selection Acts required selectors to improve their runs. Improvement meant fencing and clearing the selection, building a hut and growing crops. Many selectors did not need much incentive to clear their land. Each selector had far less land than the squatter and survival was a struggle against overwhelming odds. He needed every piece of available land for production. He needed timber for buildings, fences and fuel. There was no place for trees on many of the small blocks of the selectors.

In some districts the legal requirements made little difference. In the Victorian Western District cultivation, more often than not, was a waste of time because the best soils were already being farmed — a fact not immediately obvious to the Lands Department. On the western tablelands, selection occurred on an area that was partly heath-covered and partly moderately timbered: ' . . . an area that, for its size, grows more ferns and supports more kangaroos, than any other ten areas in the Colony'.[9] Settlement in such an area was slow to cause major modification to the existing vegetation. In other areas the enclosures that came about with selection would have taken a greater toll of the tree cover. In many places economic pressures defeated the dream of selection. Joseph Jenkins complained that selectors would not expend effort to clear their land around the gold-fields.[10]

Ironically, in the thick forest of South Gippsland no struggle was too great for the persistent selectors. Here was the forest to end all forests. It was dense, perhaps as dense as any other forest on the earth. Why the forest was not open like the rest of Gippsland is not known. Perhaps the land was too wet to burn. Perhaps some catastrophe had befallen the Aborigines who had lived in the area. The forest almost destroyed the explorer Count Strzelecki. He was lulled into misjudgment by the open forests he had traversed elsewhere in Gippsland. Entering the South Gippsland forest he emerged many days later at the other end, less his horses, supplies and most of his clothing. The next explorer to pass through was Brodribb, in 1840. He tried to skirt round the forest to the north, and still had a difficult time:

> The thick, high scrub was entwined with strong vines. At last we had to take it in turns to cut away the vines and scrub, to allow our horses to proceed, and one of the party always had a compass in hand, so as to direct our course . . . Sometimes we came upon an immense tree that had fallen down across our course, perhaps five or six feet in diameter and one hundred yards long . . . After journeying through this scrub for nine days, with much hardship and suffering, our clothes torn from our backs, and our shoes nearly worn out, as well as being short of provisions, we got through the scrub.[11]

Despite his struggles, Brodribb had expressed confidence that the land would eventually be settled. He described the land's two advantages. The first was rainfall; it rained on Brodribb's party nearly every day as they struggled through the scrub. He also observed that the country 'although thickly timbered, is very rich'.[12]

The squatters made no impact on the scrub. A succession of leasehold-ers quickly realised the futility of settling the country. In 1851 the scrub experienced a major bushfire. Many old trees died, but the regrowth sprung up, thicker than ever.

By the late 1870s selectors had realised the value of regular rain. Selectors in the north of the state were abandoning their selections because of a severe drought. Wet South Gippsland was very attractive. English grasses unsuited to the north of the state would grow here. The second advantage was the soil. It was deep and fertile. Settlers who took up forested land did not necessarily choose the land that was easiest to clear; they assumed that the forest cover indicated something about the fertility of the soil and the amount and reliability of the rainfall.

The selectors faced a huge task. Mr Williams of Moyarra walked away from the forest twice before finally gaining the courage to select on his third attempt.[13] What he had seen was daunting. In some places there were 1000 to 1200 trees to the hectare. Between the trees were vines, sword grass and bushes. No one dared venture off the track without a compass.

Surveyors needed to clear a line of sight to make a survey of a block. There was no room to swing an axe without clearing the low scrub, so surveying marks were often dispensed with. Farmers often had to clear their land before they gained an idea of the topography of their selection.

Felling was only the first step in clearing the land. The timber was green and axes needed to be honed razor sharp. When cut, the trees often did not fall but remained upright, held by vines. Whole masses of cut trees would remain standing until the combined weight would break the supporting trunks. The whole mass would then come crashing down. The work was dangerous and injuries were common. The selectors learnt new methods of felling:

> Being novices at the work we had much to learn, but soon found that the best, quickest and safest way was to fall scrub in batches; that is, nick everything on both sides, leaving sufficient uncut wood to keep the trees standing, the nick on the back being a little higher than the one on the front. This nicking would be continued to the top of the hill, when a tree would be selected by each man, and these cut so as to fall into others, which in turn fell into those in front of them, and the whole four or five chains would come to the ground with a tremendous crash. We would then go down into the gully and repeat the process.[14]

This method of felling left all the smaller timber felled and stacked in the same direction, pointing downhill, in the best possible formation for the burning stage of the clearing process. The larger trees were either ringbarked or left standing. The selector waited for a hot dry day at the end of a dry spell to light his logs. A good burn was necessary to eliminate the hard, dirty work of clearing up after the burn. In some seasons these conditions did not occur. Then the selector was forced to wait another year before sowing his pastures. The economic necessity of achieving a good burn forced the selectors to learn how to light a good fire:

> The game was, in lighting scrub, to get it lit round as quickly as possible. The more men you have for this purpose the better. Some people would light on the windy side only; this was a mistake, for in the first place the wind might change just after you lighted it, and in the next, unless the wind be very strong, the heat of the fire will still it, or, rather, cause an inrush of air from all sides. This draws the fire upwards, but prevents it from travelling, and should there be a ridge across the line of march, the fire will not travel well down the opposite side. But if the scrub be lit all around, each fire draws the other to it by reason of the upward rush of rarefied air and the consequent inrush from all sides.[15]

After the burning came the picking up. The remaining half-burnt logs were piled into heaps and relit. This was hard work. Charcoal rubbed the skin off hands, and the smoke and soot irritated the eyes. Of the whole process of clearing, this was the most disliked task. After a spring and summer of felling, burning in summer and picking up in the autumn and early winter, the paddock was finally ready for sowing. Even then the forest was not vanquished. Within two years the regrowth would appear and the paddocks would need to be recut.

The forest the Gippsland pioneers destroyed was exceptional. Their labour of destruction was also exceptional. In the development philosophy of the early twentieth century they became a legend, an icon of perseverance and hard work. These pioneers and their work have become a new legend in current land-degradation literature. Now their labours symbolise ignorance. The striking photographs of their tree felling are used to demonstrate the work of the selectors.[16] They are new icons for the destructiveness of our pioneers. The South Gippsland selectors were not typical of all the selectors. Selectors on the northern plains often had so little usable timber on some selections that timber was purchased to build huts. Those selectors unfortunate enough to be farming on the often poor land around the goldfields did not need to worry about clearing. Much of their clearing work had been done by the miners. Selectors on the river flats of Gippsland would have been forced to clear many trees, much of it new regrowth forests. On the western plains there were large areas where there were no trees. Selectors there also had to buy wood to build their huts.

The most telling push for tree clearing was the era of pasture improvement that gathered momentum in the 1950s and accelerated in the 1960s. Until this time on many properties, other than those principally devoted to cropping, there was remnant vegetation providing a reasonable water balance, an acceptable aesthetic vista and a residual ecological habitat; although the understorey species that provide diversity and balance to interdependent plants and animals were lost. The 1950s ushered in a 'golden age' of agriculture. There were long periods of high prices for many rural commodities, general national growth and intensification of agricultural production. The expansion of sown pasture meant that grazing land was cultivated; trees were often removed and hilly country was more intensively farmed because of a combination of high product prices, high returns to the investment of pasture improvement and government taxation incentives. In the 1950s departments of agriculture were publishing advice on clearing, ringbarking and poisoning trees.[17]

Most of the agricultural land in New South Wales and Victoria was already settled; development merely brought about more intensive use. Large areas of land were being cleared in Western Australia and in the brigalow country in Queensland. Much of this land was cleared by heavy machinery. Tracked diesel tractors equipped with tree-pushers and bulldozer blades cleared land much more quickly than the old hand methods. Large areas were cleared by two tracked tractors operating as a team dragging a chain or wire between them and pulling over the lighter trees and scrub in their path. In Victoria, as recently as the 1960s the government rural settlement authority cleared large areas of the Heytesbury forest in the Otway area using these methods.[18]

In Western Australia in the late 1950s and the 1960s large tracts of Mallee land in the south-west were cleared for cropping. Previously, because of trace element deficiencies and excessive rabbit infestations, this land was seen as having little agricultural use. High wool prices in the 1950s provided a catalyst for land development. Under conditional

purchase agreements new landholders were required by the Western Australia Lands Board to clear at least 100 hectares each year for the first four
years after taking possession of a block. Western Australian Farmer Bob
Twigg recalls taking up his settlement block:

> For someone like me coming off a 200 acre dairy farm, quite frankly I wasn't
> game to walk too far into this 4000 acre piece of bush because I'd get lost, apart
> from anything else . . . After the bush was knocked down and burnt, the
> contractors came. The next operation was to begin the initial cultivation . . .
> there was an initial cultivation and subsequent operations — reploughing,
> raking, seeding. I tackled my 4000 acres in six pieces, over six years, and it was
> all done the same way: it was rolled or chained in the winter time, burnt and
> fallowed and then cropped the following year. After two crops the land was then
> sown to pasture which was the beginning of a normal clover ley rotation.[19]

Western Australia has more than its share of soil salinity and land
degradation problems, many of which can be traced to this relatively
recent excessive clearing of lands that in many ways were marginal for
agricultural development.

The manner in which community attitudes, and government legislation, can dramatically change direction can be seen in taxation concessions
for land clearing. In 1989 the Australian Government announced a scheme
to plant one billion trees within ten years; and since 1985 certain capital
expenditures to prevent land degradation have been allowable as an outright deduction against taxable income.[20] In contrast, between 1947 and
1983 farmers could write off against their taxable income the expenditure
incurred in the destruction and removal of timber, scrub or undergrowth
indigenous to the land. Such a taxation allowance encouraged high levels
of expenditure on land development. The government's rationale for such
policies was postwar agricultural expansion to encourage economic growth,
full employment and equilibrium in the balance of payments.[21] In 1952,
John McEwen, the Minister for Commerce and Agriculture, proposed that
the program of agricultural expansion was:

> . . . not only to meet direct defence requirements, but also to produce food for
> the growing population, to maintain our capacity to import, and to make our
> proper contribution to relieving the dollar problem . . . the Commonwealth
> Government has decided that activities directly concerned with the production
> of essential items of food and agricultural products shall be classified in import-
> ance with defence and coal production.[22]

In recent times ministers of the Crown have equated tree planting with
sustainable agriculture.

Pasture improvement, subterranean clover and superphosphate did not
sound a death knell for all trees on agricultural land. An unexpected
development of the sub and super period was the expansion of the tree
population in large areas of outback New South Wales. The short period
of myxomatosis's total domination of the rabbit coincided with a series of
very wet years in south-eastern Australia. Tree and shrub seedlings sprouted

in their millions, extending the cover of the regrowth forests, particularly the native pine forests of western New South Wales.

Ambivalent attitudes to the bush

The rural attitude to trees has been ambivalent. There has not been a total hatred of the tree as some commentators would have us believe.[23] For those whose job it is to promote farm tree planting today, it is encouraging to realise there is an historical tradition of native and exotic tree appreciation on Australian farms. Appeals to the aesthetic sense of the landholder have an historic precedent on which to build. Even the herculean destructive efforts of the South Gippsland pioneers were tinged with a bittersweet sense of loss at the passing of the forest. The selectors Hansen and Gillan both thought many would come to regret excess clearing of their land and would come to long for stands of forest on their land, for shelter, for timber and for beauty. Other South Gippsland settlers saw beauty in the forest and cleared it with regret.[24] Some of the clearers did leave forest patches of timber for beauty, but the patches were doomed by the clearing of the rest of the forest. The parts were not as strong as the whole. Magnificent tree ferns left unprotected in otherwise clear valleys had little chance of long-term survival. The open land and dead ringbarked giants

An early South Gippsland farm among the dead forest giants

were now ripe for major fire. Those tree plots and gullies protected from the clearer's axe were destroyed by the great fires of 1898. Some settlers regretted the loss of the remaining trees in addition to the loss of stock and buildings. For others the job of clearing had been magnificently completed. William Johnstone helped his father clear a block in the Poowong district of South Gippsland. He expressed the equivocal feelings of the selector in a poem, *Retrospect*.[25]

Retrospect

When first I came to Gippsland, no seer could foretell
That the light tapping axe rang the forests deathknell;
It spread like an ocean and rolled like a tide
Whenever King Storm on the treetops did ride.

From the ridge to the gully no break could be found,
And the keenest observer could not see the ground;
But the axes and fire great havoc have played
With grim forest giant and lovely fern glade.

Ever gone are the hunters that covered the hills,
Ever gone are the fern glades that sheltered the rills,
And gone are the dells where I oft loved to roam
And bring in wild flowers to garland my home.

Never more shall I see the green forest again
Wave free in the sunshine, droop sullen in rain;
No more shall I sway to each altering whim
The laughing, the tearful, the wanton, the prim.

Never more shall I list to the lyrebird's song
That boldly he trolled forth, so clear, and so strong.
Or listen, mazed, as he mocked every bird,
And mimicked to life every sound that he heard.

Never more shall I wander, awe-struck and subdued,
While the shades of deep night on the forest did brood,
And feel, when along those great aisles I have trod,
I worshipped alone in a temple of God.

But away with these fancies, 'Tis better today
Where the forest encumbered, the children now play
In meadows bespangled with flowers whose hue
Is brighter than those that the pioneers knew.

> *Where the forest delighted, perchance two or three,*
> *The present rich meadows fill hundreds with glee.*
> *Our wives and our children, our homes and our farms*
> *Are dearer and better than natures wild charms.*
>
>
> W. W. Johnstone, 1917, in *Land of the Lyrebird*

Elsewhere, where there was naturally little tree cover, or where the cover had been removed, some landholders and government scientists felt the need for more trees. At the turn of the century government advisers exhorted farmers to plant trees on the treeless plains of South Australia. Planted along fencelines, the trees would beautify the landscape, shelter stock, host insectivorous birds, provide fuel and eventually replace fenceposts.[26] Some in South Australia also believed the appearance of trees would bring rain.[27]

On the basalt plains in south-western Victoria many graziers planted trees where formerly there were none. Besides planting oaks, elms and pines around homesteads, pines and similar trees were planted as windbreaks in paddocks. On the treeless plains at Lismore between 1873 and 1886 eucalypts were broadcast on 48 hectares of ploughed ground; and in the following three years a further 207 hectares were treated in this way.[28] On three estates covering 28 000 hectares there were 690 hectares of tree plantations along fence lines. Initially the trees were planted as seedlings; later seeds were broadcast directly into the ground.[29] Farther north at Baangal Estate, at Skipton, also on the treeless Western District plains, the 40-hectare paddocks were bordered by a wide belt of tall sugar gums and southern mahogany mixed with bushy pines, wattles and pepper trees extending for 67 kilometres. These plantations provided firewood, fencing timber and shelter for stock.[30] Similarly at nearby Carranballac a 34-hectare plantation of pines and eucalypts established in 1914 was a long established district landmark. Many of these trees are now dying, having either reached the end of their useful life or, perhaps, because they were not indigenously suited as narrow plantations in an area in which trees have not naturally survived in the past. Their role, in an environment modified by man, was to provide stock shelter and aesthetic attractiveness in a windswept plain.

In Western Australia foresters regularly exhorted farmers to plant trees on their farms for the usual reasons of shade, shelter, stock and wood products, but also to control salinity.[31] In New South Wales Dr Tepper advocated trees to control erosion on hilly country:

> Now the hills are more or less completely denuded of the despised scrub, the ground firm and hard, the surface bare . . . The rain or thunderstorm descends; down it rushes unobstructed from the hillsides, carrying with it part of the still remaining fertile soil . . . we have ruined our country by denuding hills and

plains of all trees and shrubs under the idea of improving pastures and fields, and have destroyed their fertility by overlooking the fact that trees and shrubs are not only in the world for firewood, timber, for shade, or ornament, at our convenience, but to create and maintain the conditions permitting man to exist.[32]

Perhaps because of its relatively small land area and its large urban population Victoria has been a leader in promoting the establishment of trees on farm land. Before the First World War government advisers were exhorting farmers to plant trees on their farms:

> One of the most striking features of the present-day scenery of many of the farming districts of our State is the general absence of well-cared-for plantations of shelter and timber trees. We are more familiar with the gaunt dead forest giants that stand like mystic sentinels of the past. Farmers! Cannot you, who are bound by every tie of gratitude to the trees for their friendly shade during the hot noonday, and their warm shelter from the cold winter blasts, do something to encourage the growing of them?[33]

The early interest in farm tree planting was for predominantly utilitarian purposes, with less emphasis on aesthetic value and virtually no concern for ecological purposes. Today the emphasis has changed but utilitarian purposes are still paramount, particularly for the majority of landholders.

In 1883 Dr L. L. Smith lamented the fate of many farm trees and in a paper to the National Agricultural Society of Victoria he proposed a farm tree-planting competition.[34] In 1911 the Victorian Cabinet approved the awarding of prizes of £100 in each of six classes of a farm tree-planting competition.[35] The competition was not universally successful; all the Mallee country competitors withdrew because of the unfavourable 1912–13 season. The emphasis of the competition was the growing of plantations for shelter, shade, windbreaks, timber supply and general purposes of ornament and utility.[36] Information was provided in agricultural journals on appropriate species and management techniques. For decades the same theme was regularly repeated as foresters sang the praises of trees.[37] Trees around farmhouses, sheds and stockyards played an important part in the judging of the properties of entrants in the farm competitions conducted by agricultural societies in the 1920s.

The sub and super revolution dampened official ardour for farm tree planting. It placed the departmental advisers in a difficult position trying to reconcile increased production and tree conservation. In the 1950s writers in the various journals of agriculture gave advice on clearing, but tempered this with articles advocating caution in taking clearing too far. Foresters continued to exhort the protection and planting of trees in the same journals.[38] In Western Australia the Commissioner for Soil Conservation controlled clearing by issuing permits where he was satisfied clearing would not increase the risk of soil erosion.[39]

In Victoria an independent voluntary organisation was stepping into the breach. Just as the shocking wind erosion of the Mallee was to galvanise a small group of committed individuals to lobby for soil erosion prevention, another disaster was to galvanise a band of dedicated tree planters. The 1939 bushfires had devastated an area of forests equivalent to one-seventh of today's forest cover.[40] From these ashes a 'Save the Forests Campaign' emerged, with the aim of creating greater public interest in forestry and involvement in tree growing. Government grants were supplied and a nursery was established. After eight years 65 000 trees were being distributed annually. This campaign evolved into the Natural Resources Conservation League and twenty-five years later 800 000 trees were provided annually for sale and for free distribution. By the early 1980s the league was distributing more than a million trees annually; more than a third of these were being planted by farmers.[41]

Land management planning

Farmers have always been exhorted by various experts to plan their farm operations. Farmers do plan their farm activity to a lesser or greater extent. When farmers are exhorted to plan, what is at issue is the degree of formality of planning, its quantification and precision, the length of the planning horizon, and the degree to which expert advice or knowledge is incorporated in the plan. Planning a farm can focus on future financial outcomes, a stock-breeding plan, a program of pasture improvement or, as is currently popular, planning farm layout to reduce or prevent land degradation. This last planning activity in recent times has often been called 'farm planning'; it is more appropriately called land management planning.

The currently popular farm planning had a precursor in the 1950s. Then the emphasis of farm planning was on soil conservation. It was epitomised in the Westgate Farm Planning Project near Ararat in western Victoria.[42] Westgate had been farmed since 1865; excessive cropping, overstocking, the depredations of rabbits and declining fertility had caused sheet and gully erosion, and some localised soil salinity. The farm plan involved subdivisional fencing of different classes of land established by using aerial photographs and other maps, the provision of contour banks, diversion banks to carry water away from gullies and the creation of grassed waterways and new dams. The aim was to improve the productivity of the soil by farming within the limits of ecological safety for each soil type. In this plan, trees were included only as shelter around the farm homestead and as a woodlot. This form of farm planning obviously created a more sustainable farm, both ecologically and financially. Today the property is farmed by a descendant of its first owner; to maintain economic viability additional land has been purchased, and a vineyard has been established in addition to the traditional enterprises.

The more recent vogue of tree planting was not evident in the Westgate plan. The one planned woodlot was on a lower flat. Salinity was treated by planting salt-tolerant grasses on the small saline area of the farm. The emphasis was on sustainability through grassland rehabilitation. This vision of trees was not limited to conventional agriculture. At the same time that the Westgate experiment was in progress, Australia's first indigenous alternative agriculture advocate, P. A. Yeomans, had just published a book on an alternative form of farm planning, 'keyline' agriculture.[43] The key to healthy pasture was a planned system of farm dams, keyline channels to increase the effective catchment area and distribution channels for farm watering. Trees also had a place on a 'keyline' farm:

> There is probably no other land development work which has been so completely unplanned and haphazard as that of timber killing and clearing, and no factor of fertility so completely ignored. To grow crops and satisfactory pasture on forested country clearing of timber is necessary. . . However, like cultivation, clearing has been overdone, with the result that soil fertility eventually suffered and crop and pasture yields were affected.[44]

Yeomans' prescription for tree preservation or planting was not conventional. He advocated keeping trees off the hilltops. Trees on hilltops made rabbit control more difficult and increased the erosion risk:

> The practice of leaving all the steep country in timber to protect it from erosion has not been successful . . . Steep country left, fully timbered, is often the greatest bushfire hazard and the worst area for pests. A fire in a timbered area, followed by heavy rain, is one of the causes of widespread land erosion.[45]

Instead, Yeomans advocated placing trees on the flatter and lower parts of the farm, along fences and tracks laid across the farm contour, and along property boundaries. Apart from bringing deep minerals to the surface for the pasture, the trees protected the pasture from stock in wet weather because the cattle camped on the firmer ground in the undisturbed soil of the timber belt.[46] In 'keyline' agriculture the tree was justified for the good it could do for the pasture.

A more recent variant of land-management planning is the Potter Farmland Plan, an idea that grew from Victoria's garden-state ideal. The Potter Farmland Plan began in 1984 when the Ian Potter Foundation, a philanthropic trust, sought to contribute to the reduction of land degradation by funding a project that treated the underlying causes of salinity, erosion and dying trees by combining ecology and agriculture. A Potter whole-farm plan involves examining the natural boundaries for farm layout and the subsequent modification of inappropriate subdivisions to take account of management and land degradation problems, which allows better land use. Stock are excluded from eroded and saline areas, which, together with recharge areas, are revegetated. A revegetation plan incorporates shelter belts, clumps and individual trees, farm wood supply and wildlife habitat.[47] The Potter Farmland Plan has been demonstrated on fifteen farms in three localities near Hamilton in south-western Victoria. The landholders agreed to work with the Potter Farmland Plan for four years, sharing the costs. The Potter Foundation contributed $900 000.

The Potter version of farm planning has similar aims to that of the earlier whole-farm plans, but places more emphasis on the place of trees on the farm. One of the concerns of the proponents of the Potter Farmland Plan was to reduce rural tree decline, which is happening in many pastoral areas in Australia. When the European settlers arrived open woodland dominated much of southern Australia. Today this woodland occurs only in small patches on pastoral land and many of these remnants are dying. Sometimes the remnants are dying because of old age but often they are dying prematurely. The problem does not have a single cause.

The severity of tree decline increased greatly in the 1970s. This was most likely connected with changes in agricultural practices in the 1950s and 1960s such as aerial application of fertiliser and the improvement of pasture by sowing with clover and introduced grasses together with intensified stocking rates.[48] Rural tree decline has been most obvious on the New England Tablelands in northern New South Wales. The original ecosystems for many tree species have been changed with settlement. In grazing areas some of the large eucalypts remained but the associated understorey species, which provided a habitat for a complex system of predator insects and birds, did not survive. Nor did younger seedlings and saplings, which would normally rejuvenate the older stands. Imbalances in insect populations and the lack of natural predators, because of habitat removal, are thought to be one of the causes of tree dieback. The improved fertility of the soil under subterranean clover pastures has created a friendly environment for some ground-dwelling grubs. One, the christmas beetle, now emerges in plague numbers on warm summer nights in search of a meal of eucalypt. The senile old woodland giants often do not cope with the stress of such defoliation. Today, farmers are advised to replant balanced habitats of appropriate mixes of tree and plant species planted in shelter belts and linked in corridors.

The Milne's farm is one of the Potter Farmland Plan demonstration farms. The Milne family run stud cattle and sheep at Melville Forest, near Coleraine in western Victoria. At the turn of the century Melville Forest was a fairly dense woodland of red gum trees. We have no record of tree cover at the time of white settlement. With the cessation of Aboriginal burning the density of treecover is likely to have increased markedly between 1840 and the late 1880s, when timber mills were established to provide sleepers for the newly establishing railways. In the 1850s William Moodie was able to lose his way, probably in the new red gum regrowth.[49] Bruce Milne's father, who came to the property in 1946, recalls that in the 1920s men were paid 10 pence a tree for ringbarking. On his best day a man could ringbark seventy trees. The earlier mills had left plenty of timber — the trees that were too big to take away or those with fine spreading habits, which were regarded as seconds. The Milnes' farm has a sparse remnant of old red gum vegetation; today ringbarked trees remain as stark grey sentinels of the landscape.

Bruce Milne believes that the best land managers care for their land because they appreciate that European-style farming practices have degraded the land and because they understand that there are practical and economic ways to treat land degradation. The Milne family prepared a farmland plan for the farm. The plan was the focus for thinking about the way the land on the farm was organised. Each different management area on the farm is managed according to its capacity for sustained production. Bruce Milne believes optimum production will be achieved with the aid of healthy soils, productive pastures, trees, shrubs, birds and insects. Bruce Milne is not a typical farmer. He believes that one of the best measures to raise awareness would be the establishment of hundreds of key demonstration farms in different regional areas across the country.[50]

Ross Kitchin is also a Potter farmer. In 1981, when Ross and Annabel Kitchin moved to their farm at Wando Vale, 30 kilometres from the Milnes, the farm, (a former settlement farm) was run down, over-grazed and under-fertilised. Some of the improved pastures had reverted to native grass. Originally, this area did not carry many trees; it was an open savannah, with manna gums clumped together in patches. It was described in 1836 by Major Mitchell (see Chapter 3) as open grassy country. James Bonwick, in 1857 described Wando Vale as being covered with luxuriant grass and a few trees.[51] Ross Kitchin's farm is a little to the east of where Mitchell travelled but is near the original Wando Dale Estate.[52] In 1876 William Tibbits depicted Wando Dale as park-like with a good covering of young-looking manna gums. (See Plate VII.) Today, on the hill vista there are fewer trees; some are very old and some are dead. It is a tired and

Wando Vale today — the Kitchins' farm at Wando Vale not far from the Wando Dale homestead

somewhat senile savannah. On the closer slopes by the creek there are more trees than are depicted in Tibbits' watercolour.

The Potter Farmland Plan approach is aimed at changing an ethic regarding land management. What is proposed, on the basis of medium-horizon economic analysis, is uneconomic without a subsidy.[53] Any economic analysis is complicated by questions involving future earnings and future land values. If the activity was clearly *economic* for farmers there would be no need for the scheme. Rather, the aim has been to develop a social responsibility among landholders for developing and protecting a more satisfactory land ecosystem unconstrained solely by immediate economic gains. The Potter demonstration was started by selecting farmers who held these values.

The Potter Farmland Plan demonstration farmers were not typical farmers; they were younger and better educated than the average farmer in the Hamilton district and were more environmentally aware. In a survey,

Stark grey sentinels — ringbarked red gums and regrowth red gums at Melville Forest

undertaken soon after the Potter demonstration farms were initiated, the demonstration farmers responded positively to participating in the whole-farm planning process, indicating that they had become more aware of land degradation and the ways to combat it, including the planting of trees.[54] Early in the program the demonstration farmers' attitudes towards the program were much more positive than those of the pastoral farmers in the surrounding locality. Using belief maps (see Chapter 3) we can compare the beliefs of Potter demonstration farmers with the beliefs of surrounding farmers concerning the Potter land-management planning approach and the control of land degradation.

The Potter farmers' beliefs (Figure 4.1) were more tightly held than the beliefs of surrounding locality farmers (Figure 4.2). The Potter farmers believed the Potter Farmland Plan was closely associated with planting trees and reversing land degradation. The Potter farmers saw themselves as being closely associated with these things and with the Potter Farmland Plan. The locality farmers' beliefs about reversing land degradation, plant-ing trees and improving productivity were less strongly held. They saw themselves as much more distant from the Potter Plan and from tree planting. Early in the program the locality farmers only associated the Potter Farmland Plan with planting trees.[55]

The Potter Farmland Plan was a demonstration that is not easy to replicate on individual farms because of the large external subsidy it incor-porated. Only one-third to one-half of the on-farm costs such as fencing, ground preparation and tree planting or seeding was met by the participat-ing landowners. The remaining costs were funded by the Potter Founda-tion. Additional planting labour was provided by conservation volunteers. Many local landholders may be sceptical of the Potter approach, particu-larly the financial support given to the demonstration farmers. The scep-ticism is not without foundation because of the uncertainty surrounding the medium-term economic payoff to the investment involved. There has been a mixed local reaction to the Potter approach and to some of the

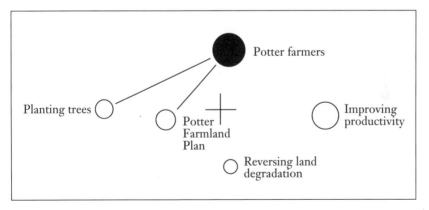

Figure 4.1 The Potter demonstration farmers' beliefs about the Potter Farmland Plan

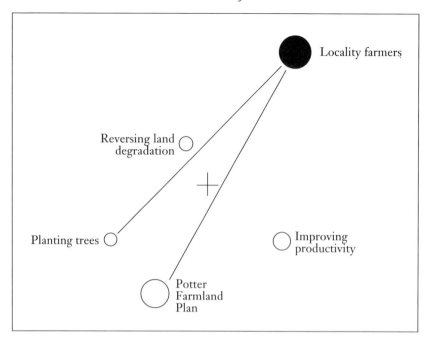

Figure 4.2 The surrounding locality farmers' beliefs about the Potter Farmland Plan

recipients of apparent largesse. There does seem to have been a changing local perception, not always publicly acknowledged but often expressed in changed behaviour, of the need to re-establish trees in the landscape as a solution to soil erosion and salinity problems as well as for utilitarian and aesthetic reasons. Formerly, a farmer who planted several thousand trees a year would have been considered silly by the neighbours. Ross Kitchin credits the Potter Farmland Plan with giving direction to a group of farmers who could see a problem but needed direction and social support to solve it.

The personal scepticism seems to diminish as one travels farther from the demonstration farms. The farms have been highly successful as demonstration venues. Many hundreds of farmers have inspected the properties. But as one leaves the Western District, other problems emerge. The farmers in the Hamilton area are predominantly pastoralists who graze sheep and cattle with an occasional cropping sideline. Tree planting and land-management planning are more problematic for farmers whose enterprises are predominantly cropping because trees are seen to compete directly with crops for soil nutrients and to encroach on land that might otherwise be cultivated. The beliefs of cropping farmers about the linkages between whole-farm planning, planting trees and reversing land degradation or improving farm productivity are less strong than those of pastoralists.[56] They see tree planting as decreasing farm productivity and unrelated to reversing land degradation.

A more widely diffused and less expensive program for landholder participation in the control of land degradation is 'Landcare'. The national Landcare program is modelled on the Victorian LandCare groups established in the late 1980s.[57] Landcare groups have been a growth industry in environmental extension; the rate of formation of groups outstripped the growth in technical and support resources available to service them. In a Landcare group, landholders in a locality work together to tackle local land protection problems such as salinity, soil erosion, rabbits or tree decline. The groups identify local land degradation problems, creating awareness about them, and undertake works, such as tree planting, to overcome them. The role of government departments has been primarily as technical consultants. The group approach means that locality problems extending beyond one farm, or extending onto public land, can be tackled in the common interest.

The Australian Government announced that it planned for the 200 Landcare groups existing at the start of the 'Decade of Landcare' to be increased by a further 1000 groups by the year 2000. The government proposed a community program for planting 400 million trees in its one billion tree planting program.[58] Enthusiasm has often run ahead of understanding what is involved. (See box: *A Billion Trees*.) At a planting density of fifty trees to the hectare, a billion trees would cover 20 million hectares — almost equal to the area of Victoria. Such hyperbole suggests that the government reaction to land degradation is more concerned with public relations than with a considered understanding of the problem.

A billion trees?

'A billion? That's just a start,' said the national manager of Greening Australia, Mr Gerry Morvell. 'Mind you, a billion won't solve the problem. It's just symbolic. But we'll go on to a billion billion in 30 years. A billion? It wouldn't even cover a large farm in Queensland. Wouldn't even cover the Australian Capital Territory. Look. At three metres spacing that's a thousand stems per hectare. That means . . . let me see . . . yes, that means you only need a million hectares.' [The ACT is a mere 140 000 hectares.] 'Well you could sit and play with figures forever. Planting might have to be denser in some parts.'
'A billion just a beginning', *The Age*, 3 Jan. 1990, p. 11

In the seventy-two forests declared in New South Wales by 1879 the tree count of those assessed varied from two-and-a-half mature trees to the hectare inland to eighty on the tablelands and coast.
Eric Rolls, *A Million Wild Acres*, Nelson, Melbourne, 1981, p. 400

Tree plantings following the straight lines along paddock or farm boundaries

Direct drilling of tree seeds and fencing off areas for regrowth will produce many more trees. In recent times many more trees have appeared along farm boundary fences, in paddock corners, and occasionally on hilltops. We have seen that it makes little economic sense for farmers to plant trees anywhere else. The 'parkscape' which so attracted the first European visitors is being recreated, though now the trees do not follow the sinuous lines of the drainage courses, but follow straight lines along paddock or farm boundaries. Rather than parks, a new farmscape is being created, a farmscape more akin to a golf course with wider fairways and fewer sand traps.[59] In later chapters we shall see that there are some land degradation problems that will not be solved by returning the land to an earlier park-like state.

Agroforestry: enthusiasm and doubt

Agroforestry takes tree planting on farms another major step beyond the work of the Potter Farmland Plan. It is an attempt to make it profitable for farmers to plant trees beyond the fence line into the middle of the paddock. Some people think of commercial forestry as a form of agriculture in which the trees are merely a very long-rotation crop. In agroforestry trees are grown in conjunction with conventional forms of agriculture. Agroforestry embraces trees grown in discrete plantations, in belts of timber separated by pasture, so-called 'ribbons of green', or widely spaced trees surrounded by pasture. In the latter form agroforestry is a more formal return to the open woodlands where many squatters first grazed their sheep.

Agroforestry is promoted for a mix of utilitarian, economic, ecological and aesthetic motives. Proponents claim a wide range of advantages: timber production, stock shelter, pasture shelter, watertable or erosion control, grassfire control, increased honey production, provision of a wildlife habitat, increased farm value and beautification. The competitive relationship between these benefits is less commonly discussed. A plantation designed to capture one benefit will be poorly designed for other benefits. Pruning for timber production reduces shelter and grassfire protection. Plantations with high habitat value will mostly have a low timber value. In many regions trees planted for watertable control will be poor timber producers.[60] The most crucial relationship is between pasture and timber production. This relationship may be competitive (more of one means less of the other) or may be complementary (more of one means more of the other). The relationship will not be the same for different situations and localities and is certain to change from being complementary to being competitive as the proportion of forestry to agriculture increases. James Bonwick's earlier cited 1857 observation about the competitive nature of timber and grass holds when there is more forest than grass. The ideal is to carefully design plantations to achieve the most efficient compromise between competing objectives. Forest economists have been able to show, using a reasonable range of interest rates to discount for earnings to be received in the future, that in medium and high rainfall districts it is possible to gain moderate though not spectacular returns from some forms of farm forestry. Narrow pine timber belts and specialty timbers promise the best returns, with widely spaced plantations being less profitable.[61]

In Australia it is rare for farmers to have established forest plantations for commercial timber production. Landholders have planted trees predominantly for shade and shelter, aesthetics, ecological advantages, or the control of land degradation. When a tree is planted for log production it will not be harvested for somewhere between twenty and forty years. There is much uncertainty for an individual in waiting for thirty years to receive an income. Over this long period prices are likely to change, technology may change, local mills may close and bushfires or pests may damage the trees and reduce their value. More importantly, although the future discounted cash flows may suggest a handsome profit, few individuals have a planning horizon where they are willing to make an investment and then defer income for thirty years. For larger companies with a portfolio of investments, or a range of forests at different stages of maturity, circumstances are different. However, the companies planting trees are generally only those ensuring a supply of timber for their own milling operations. For an individual, making predictions about markets far into the future is a risky business. Two examples, one recent, one historic, make these risks clearer.

The first recorded example of agroforestry occurred sometime during the 1860s and 1870s. Western District pastoralist William Moodie developed a commercial enterprise of producing wattle for its bark:

> . . . backing my opinion I had planted about half a million young trees on which I quite expected to get a net return of not less than 1/- per tree or about £25,000. But like a good many things that promised well we were doomed to disappointment. A small green caterpillar was the first enemy which denuded the trees of their leaves and retarded their growth. Following them a boring grub took

possession and simply destroyed the trees for stripping. So what seemed a small fortune dwindled to about £500.[62]

In Moodie's day little was known of wattle production so a Royal Commission was appointed to inquire into the subject. Over the next fifty years there were similar outbursts of enthusiasm for the wattle industry across Australia.[63]

Farmers today face a similar problem to Moodie. Not a great deal is known about growing trees in conjunction with conventional forms of agriculture. Much of the data on which economic evaluations of agroforestry have been based is necessarily drawn from overseas sources.[64] The minimal research in Australia has established that shelter provided by windbreaks increases the survival rates and productivity of lambs on the windswept plains of western Victoria. There appears to be no Australian research information establishing increases in cereal crop yields when grown in association with forest plantation.[65] The accepted wisdom of crop farmers is that yields will decline because of competition for water and nutrients.

There also remains Moodie's problem with insect predators. In districts where eucalypt dieback is a problem, tree plantations need to be large and to include an understorey of shrubs to provide habitat for predators of the predators of the trees. Without this protection, agroforests run the same risk of dieback. In these areas there is no future for agroforestry with native trees when the only stable formations, large plantations, are the least profitable.[66] The future of agroforestry in these districts must be with exotic trees, probably pines.

In 1963, a century after Moodie's venture, a group of private investors funded the planting of a softwood plantation in the Cathedral Valley of Victoria. One thousand people, mainly farmers, invested $750 000 in the 560-hectare plantation. Three years after planting the investors were promised that their first dividend from the thinning would materialise in eight years time in 1974, and that thinning would continue regularly from that date, producing regular dividends until the final harvest. By the time of the first thinning the local pulp mill had closed. Thinning was therefore unprofitable, so the plantation was left to grow unpruned. Twenty-seven years after the first planting the company has finally paid its first thinning dividend. A pulp mill in Benalla took the thinnings. After meeting transport costs, thinnings valued at $900 produced a mere $50 in royalties.[67]

In the meantime, agroforestry is more likely to be successfully promoted to the average farmer if its proponents concentrate on the aesthetic and ecological advantages of trees and create community acceptance of the strategic planting of trees to control land degradation. Today, besides utility and beauty, trees are seen to have a much more pervasive role in maintaining stable land systems. Even here the details are yet to be fully worked out. It is widely accepted that strategic tree planting will control rising watertables and wind erosion in sandy-soiled cropping districts. These benefits will only be captured with well planned plantations. Trees planted in areas of low recharge are a very expensive solution, and the benefits may not be noticed for generations. Badly designed planting strategies can increase wind turbulence, enhancing the erosive power of the wind. Where trees compete with pasture, they can increase the risk of water erosion by reducing ground cover and increasing the size of water droplets collecting and then dripping from leaves.[68]

In the face of uncertainty, earlier advocates of farm trees promoted plantations for areas of the farm of no otherwise productive value for agriculture, or for shelter, ornamentation and farm wood supplies.[69] Despite contemporary uncertainty, some proponents have ardently promoted the commercial benefits of agroforestry. Pamphlets glibly promise a harvest of 'prime beef one day, prime logs the next'. Such enthusiasm is driven more by the desire of foresters to see trees planted on farms than the desire to see farmers exploit a proven commercial opportunity.

Taking the case of Victoria, today forests cover one-third of the state. This forest cover is of varying density and type, including Mallee scrub. Only 4 per cent of Victoria's forest is privately owned. In the fifteen years to 1987 reforestation of public forest land was in balance with forest clearing. In the same period there was a net loss of 15 per cent in the area of private forest land.[70] There has been an understandable concern about the further loss of tree cover on private land. At the same time, the historic motivation of farm-tree advocates was being challenged by a new conservation ethic: valuing trees for their own sake, rather than for the protection they could give productive pasture or the aesthetic beauty they could add to a farmscape. The conservationists valued trees where they could preserve natural habitat and ecosystems. From this point of view, a piece of remnant vegetation on a farm has more value than an artificial plantation. South Australia, and more recently Victoria, has acted to control the clearing of remnant vegetation. The wheel of government policy has turned full circle.

5

Pastoralism in the rangelands

There are two sorts of land in Australia. In South Australia they are divided by a line drawn across the map by an early Surveyor-General, G. W. Goyder. On one side of the line the rainfall will support non-pastoral agriculture, and on the other side of the line it is too dry for anything but grazing. In the Flinders Ranges in South Australia ruined stone walls at Simmonston mark the site of a town that never was. Together with other planned towns, only laid out on paper, it marked an optimistic hope for a wheat frontier on the wrong side of Goyder's Line. South Australians talk about 'inside' country and 'outside' country.[1]

Today the outside country is more commonly called rangeland. In stories and film it is often called the outback. The rangelands are not a single entity. The vast arid and semi-arid rangelands cover 70 per cent of Australia's land area, two-thirds of which is grazed by sheep, cattle and kangaroos. There are at least five different ecological and agricultural systems within the rangelands. They comprise the most finely balanced agro-ecological systems in Australian agriculture. In contrast with the *improved* agriculture in the rest of Australia, rangelands pastoralism is characterised by low use of inputs, the use of natural pastures and low intended modification of the land. There have been few scientific advances to extend the agricultural capacities of the rangelands. Stocking rates are low, often only one sheep every 10 hectares. Pastoralists must operate within the constraints of drought and without the capacity, which comes with development, to repair physical damage to the pasture environment. The systems of rangelands agriculture are closely linked with ecology.

Managing drought

The first European explorers dismissed the rangelands as without value; but subsequent entrepreneurs found that grazing the rangelands with domestic sheep and cattle could be extremely profitable.[2] European occupation of the area beyond Goyder's Line in South Australia and of the central Queensland plains and the Gulf country began in the late 1850s and early 1860s. Later, the plains and ranges west of the Darling River became one of the fastest growing pastoral regions of New South Wales. Finally, in the 1890s, the Northern Territory and the Kimberley in Western Australia were occupied. Much of the occupation was by leasehold. Since the occupation, there have been relatively few opportunities for scientific advances to extend the capacities of the rangelands. The rangeland grazier has had to learn to live with the limitations of the land and the climate. Managing drought has been the key to survival in the rangelands.

In Australia's rangelands droughts come with a fairly regular frequency, and often extend for considerable periods. The opinion of veteran residents of western New South Wales is that drought comes in seven-year cycles — not regularly at seven-year intervals but at that average rate. Meteorological records suggest a somewhat less frequent average interval. Because of the availability of drought subsidies in some states, graziers in shires such as Paroo in Western Queensland officially experienced nearly perpetual drought between 1964 and 1982.[3]

Two of the biggest early landholders, Sidney Kidman (1857–1935) and James Tyson (1819–1898), developed their pastoral empires by establishing butchers' businesses in mining towns and beat the droughts by owning properties over a wide area of country in several states. Kidman had a string of stations across Queensland, New South Wales and South Australia.[4] They were able to move their livestock over vast distances in response to local seasonal conditions as well as take advantage of, and manipulate, the local markets for livestock. Earlier large landholders were not so astute, or so fortunate. Hugh Glass, a successful merchant, stock agent and large landholder in Victoria, leased nearly 1.25 million hectares in the dry country between the Lachlan and the Darling rivers in the 1850s and 1860s. In a good season the feed from Mitchell grass and saltbush was prodigious, but in a drought the surface water soon vanished. In the droughts of the 1860s Glass lost 14 000 sheep in one week on the road near Balranald. A subsequent fall in wool prices and his large borrowings hastened his business demise.[5] Large pastoral finance companies, which predominated in the Western Division of New South Wales after 1880, were also hit hard by drought but, with their many holdings, had an advantage over individual owners by being able to move sheep and by their scale of operations.[6]

Without the advantage of widespread properties, the small pastoralists must respond with much acuity and few degrees of freedom to the vagaries of season and livestock markets. Small pastoralists generally limit their stock to the number that the native vegetation will be able to support, with the amount of feed in the dry period of the year determining the stocking rate. Nearly every year there are months in which the animals are forced on to the standing haystack of dried-up perennial grasses. During a drought these reserves become exhausted; the pastoralists may then cut down mulga and other edible shrubs to provide feed for sheep. Natural pasture such as the Mitchell grass is eaten down, saltbushes are reduced in size, and gradually the pastoralists are forced to the harder choices. They may sell or send away some or all of the stock; they may occasionally buy in feed; or they may slaughter the less valuable stock and endeavour to hold the rest, or let them die gradually as the drought continues. Producers have to decide early whether to sell stock or transport them elsewhere for agistment. A few producers may borrow money to feed stock, which often die anyway. During an extended drought the price of livestock often becomes so low that the costs of transport and selling are greater than the selling value.

At Bourke in western New South Wales in August 1965 the rainfall for the preceding nine months was 42 millimetres — the second lowest since 1871, the normal average being 279 millimetres. White Cliffs, nearby, had only 150 millimetres in the preceding two years. This was one of the two most serious droughts since records had been kept. The traditional rules for managing in this sort of drought are simple, if brutal. An old pastoralist gave salutary advice to a novice grazier at the beginning of this drought:

> I went down the road and asked my neighbour for advice. He was an old fellow who had spent his life on the place so I reckoned he should know what was best. He basically said 'let them run or die but don't go into debt to feed them'. In this day and age I reckoned his advice was rubbish and more or less told him so. I fed the sheep for over a year and lost most of them one way or another. I finished with a few sheep and owing a lot of money. I only wish I had taken the old chap's advice. I should have known better because he's a pretty wealthy man and must have weathered this sort of thing in the past.[7]

Graziers have few options for business growth; there are few opportunities to develop land to improve its productivity, other than indirectly through better stock control and water facilities. The major opportunity to develop a sufficient income base to survive a drought is via property expansion. The geographer J. Macdonald Holmes earlier observed sanguinely that 'the small man must overstock to make a living and the big man overstocks to make enough to get out to better lands near the coast'.[8] After a drought there is an increased activity in land transactions and lease transfers. In keeping with the brutality of good management decisions, the successful grazier who has made the right choices will have the opportunity to buy the property of his neighbour who may have made the wrong decisions. Following the drought in 1965, which lasted into the early seventies, 44 per cent of the surviving graziers around Charleville in south-western Queensland either purchased a larger property, extra land or extra leasehold. Their principal goal was to ensure a satisfactory income in the future.[9] Most purchases were at the expense of another grazier who failed to survive the drought.

Buying land must be approached cautiously. As well as enough land, graziers also need to have sufficient capital to survive the seasonal and economic fluctuations. Successful early pastoralists knew the importance of a conservative balance between debt and equity. William Brodribb, a squatter we introduced earlier, was not averse to taking risks. He was the first settler at Port Albert in Victoria and had stations in northern Victoria, the Riverina, and the Western Division of New South Wales. Having survived, somewhat scathed, the economic depression of the early 1840s, Brodribb was wary of excessive debt. In 1858 he sold one-fifth of his *Wanganella* Riverina run (210 square kilometres) rather than borrow the capital needed to develop it. At the time of writing the current owner of this station, Rupert Murdoch, was trying to follow Brodribb's advice in his business empire by selling assets to reduce debt. Brodribb did not find his own precepts easy to follow and a drought on his station at Booligal on the Lachlan River brought him into difficulties in 1864–66 from which he took 'quite twelve years' to recover.[10]

Undoing the outback

Traditionally, pastoralists have been skilled at stock management and stock husbandry but have had less understanding and skill in managing the pasture, vegetation and soil resources. The earliest settlers focused on stock condition as the indicator of the quality of a pastoral run. Brodribb's advice was:

> . . . the way I judge of the country, if stocked, is, to notice the appearance and condition of the horses, working oxen, cattle, and sheep; if they appear healthy and in good condition, one may be sure he cannot do wrong in buying, provided the price is not too high.[11]

Stock management, for immediate economic reasons, has been based on the condition of the animals and rarely on the state of the pasture. Aboriginals have observed that 'the European people are very good at managing cattle and sheep but not so good at managing land'.[12] Early pastoralists experimented to obtain suitable animals for the environment, suitable systems of managing the animals to exploit the vegetation, and suitable techniques of wool preparation. It was clear that little could be done to manipulate the vegetation other than indirectly, and very often not obviously, by pressure of grazing. On many northern cattle properties the annual round-up of cattle had more in common with hunting than with farming. And so stock management rather than pasture management became the culture.

Under rangeland conditions it is not feasible to save and store fodder. Feed supplies must be conserved on the ground; hence the level of stocking has to be in balance with major seasonal changes — with some of the same risks that attend playing the stockmarket. Droughts have major ecological

Severe gully and sheet erosion in the Kimberley rangeland in northern Australia

as well as economic ramifications. The pressure of too many stock on the dwindling grazing fodder has ecological impacts in a delicately balanced natural system, which may not return to its former state when the seasons again became favourable. The ground will be left uncovered. The soil will blow or erosion gullies will develop.

Overstocking has been a feature of rangelands pastoralism from the start. In evidence to a 1900 Royal Commission an Inspector of Stations observed:

> I knew the West Darling country first in 1878 . . . before it was fenced in and occupied by sheep. I knew it in its aboriginal state. There is no doubt in my mind that the carrying capacity of the country was greatly overestimated by the early settlers. In its virgin state, with saltbush and other edible bushes in their prime, there was hardly a limit, except as regards water, in the opinion of settlers then as to what the country would carry, and in many cases it was overstocked accordingly, but they forgot in doing this they were eating the haystack, and there was no crop growing to build another.[13]

Paradoxically, in the west of New South Wales and elsewhere in the rangelands there is a problem of not too few trees, but too many. Woody weed or shrub infestation of the rangelands has been a problem since soon after the first pastoralists introduced their flocks onto this country. In 1901 the Western Lands Commission remarked on the problem. At the same time, the manager of a government research station in the New South Wales Western Division observed the entire absence of grasses or herbage among the dense growth of injurious scrub on land that fifteen years previously was well grassed, open forest.[14] Half a century earlier, when the first explorers were traversing the areas now more densely covered with scrub, much of the country was grassland or open woodland. The cover of foliage provided by increased woody scrub shades out the grasses; the woody scrub reduces the presence of the grasses by competing with them for water. The pastoral value of the land for grazing is reduced. More importantly, the soil is exposed to severe erosion because the increased shrubs and trees result in bared soil surface, which becomes vulnerable to sheet erosion after rainfall. The worst erosion happens around watering points, particularly when there are only a few watering points on a property.

In arid regions the population of woody plants waxes and wanes, but in semi-arid eastern Australia there is an increasing dominance by shrubs to the exclusion of the native grasses. The ecological balance between the woody shrubs and the grass species fluctuates delicately with changing natural conditions. In the mulga lands there seem to be three states of equilibrium: an open grassland with scattered shrubs, open grassland with patches of dense shrubs, and widespread dominance by shrubs. Over a long period a mulga grassland may move from one state of equilibrium to another. Explanations of the changing balance are not always simple. Overgrazing of the grass species is as much an effect of the shrub invasion as it is a cause.[15] Shrub seedlings germinate and develop after heavy rainfall years in conditions of *excess* growth when there is usually ample grass and little opportunity for grazing control of the shrub seedlings by stock. As we

Using a bulldozer to clear young woody scrub in the rangelands

saw with the increase of the eucalypts in southern grasslands after the cessation of burning by the Aborigines and the squatters, lack of fire is a key ingredient in tipping the balance of woody species and grass. In the natural sequence of events the grass build-up would eventually provide dry fuel for bush fires, which would eliminate the developing scrub seedlings. Under pastoral conditions grazing may increase the frequency of shrub establishment by reducing soil moisture utilised by the grass and by reducing the potential of fires, thus bringing on a more shrub-dominated vegetation. Fire exclusion has created a plant community dominated by long lived, eventually fire-resistant shrubs. In the semi-arid lands such woodlands often take hundreds of years to develop.[16]

The settlers unwittingly encouraged shrub growth by putting an end to the system of regular burning that had been used to manage the country by the region's former Aboriginal occupants. The pastoralists also increased the grazing pressure on weed-competing native pastures, not only by introducing sheep but also by sponsoring a kangaroo build-up through the provision of watering points in previously waterless areas and by exterminating the dingo. During the past thirty years the woody-weed problem has burgeoned. Rabbits have contributed to this. Rabbits bared much of the rangeland in the 1890s and early 1900s. In the 1940s it was thought that they would eradicate the last stands of mulga. When myxomatosis diminished the rabbit population from 1951 the mulga came back and, as a result of high rainfall in the early to mid 1950s, established at much greater density than in the past. Previously the problem seemed too big, or its spread was too insidious, for landholders to devote much attention to it.

Drought protection or land protection?

Australia's rangelands are vast plains with shimmering distortion of objects in the distance, ever-present mirages, and flat horizons in any direction to which you cast your eye. Henry Lawson dryly observed that there were no mountains out west, only 'ridges on the floors of hell'. The rangelands are geographically flat; but when the government becomes involved in dealing with drought, using the contemporary metaphor of political economy, the rangelands are not 'level playing fields'. The ravages of serious drought have always produced hardship and a level of financial crisis for some, if not many, pastoralists. Selected reports of the worst of these situations — farmers killing livestock, or farmers unable to meet financial commitments — provide emotional stories in the mass media. Under pressure to alleviate the hardship of drought, governments have not always been able to adopt the style of brutal management that graziers have found the rangelands demand. Governments have generally readily agreed to provide public funds for drought assistance, usually as subsidies on the transport of livestock, the purchase and transport of fodder, or interest rate subsidies. Such approaches are based on the rationale of lessening the impact of the drought on those affected. The aid is seen as a short-term measure to attempt to restore the status quo as it was before the disruption by drought. Relief has not been seen as either positive in the sense of improving land use, or negative in the sense of reducing intensity of use, but with the neutral aim of trying to maintain business as usual.

This short-term view has provided incentives, in the longer term, for graziers to lessen their efforts to cope with future droughts from their own resources. When the transport of fodder and interest rates are subsidised, conservative stocking policies and the building-up of financial reserves are discouraged because these strategies receive no subsidy.[17] The rangeland 'playing field' is not level because the drought subsidies are not paid to all graziers in a drought area. They are paid to the less prudent who have been slow to reduce stock numbers to the level of food supply and those who are forced to borrow rather than draw on previously built-up financial reserves. Such theoretical conjectures are borne out in practice: a government stock inspector at Cunnamulla in Queensland's south-west observed that from past experience he could nominate the first fifteen landholders to make application for relief after a drought is declared. The long-term result is an acceleration of the rate of rangeland degradation.

Attempted scientific revolutions

In the rangelands there has been no sub and super pasture improvement revolution — the major technological development of twentieth century Australian agriculture affecting stocking rates. In the 1950s and 1960s, there were big leaps in productivity in southern Australia's agriculture; the same has not been true in the rangelands. (See Figure 5.1.) Between 1900 and 1970 the long-term livestock carrying capacity in Australian rangelands has been remarkably stable. Annual sheep numbers have fluctuated between 12 and 30 million; but only in ten of 63 years to 1963 did they exceed the 27 million sheep that were carried in 1893.[18] The rangeland sheep population was about a quarter of the Australian flock between 1910 and the mid 1930s; today it is about 15 per cent and continuing to decline. Just over 4 million cattle were run in 1895; between 1910 and 1970 annual cattle numbers rose as high as 4.4 million and did not fall below 3 million. From the mid-1970s to the early 1980s cattle numbers were much higher than the long-term levels. This was a result of very high cattle prices in the early 1970s followed by a major slump. Good seasonal conditions in the mid-1970s meant that cattle were retained until prices recovered.

That the development successes of temperate Australia would not apply in the rangelands was a lesson that was not easily or happily accepted. One of the more extraordinary schemes to develop the rangeland for cropping took place in Western Australia. It was the Ord River Scheme, 1600 kilometres away from the main settlement areas of Western Australia. The final stage of the Ord River Dam, opened in 1972, stored a volume of water abut nine times the volume of Sydney Harbour. It had the potential to irrigate more than 75 000 hectares of land. It was planned to grow irrigated

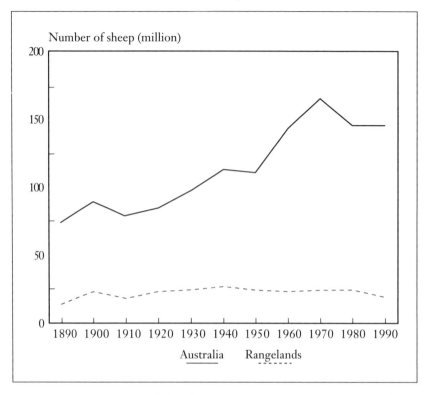

Figure 5.1 Expansion of the sheep-carrying capacity of the rangelands compared with that of Australia as a whole for 10 year periods: 1890–1990 (*Sources*: Barnard *The Simple Fleece*, 1962, ABS, and ABARE Pastoral zone surveys)

cotton, maize, barley, wheat, safflower, soybeans and pasture. Only cotton was grown, unsuccessfully, for about ten years. The area sown to cotton did not exceed 5000 hectares. In 1988 a little more than 5000 hectares was being irrigated to grow a range of other crops.[19] The failure of the Ord River Scheme was one of Australia's most costly development lessons.[20]

Twenty years ago there were high hopes that the South American legumes of the *Stylosanthes* genus would be able to do for at least part of the tropical rangelands what subterranean clover did for the temperate farmlands of Australia. The grass could take advantage of the irregular wet and dry seasons of the tropical rangelands, fixing nitrogen into the soil. It could be oversown into existing native pasture, producing at times an impressive increase in grass cover. Scientists' enthusiasm for improved tropical pasture species has not been matched by the successful use of these species by pastoralists.[21]

Another innovation, the introduction of zebu cattle, met with greater acceptance by graziers. In the 1870s ticks had invaded the northern rangeland pastures. They presented a serious problem for graziers. European breeds of cattle were devastated. The zebu cattle from Africa had a greater tolerance of ticks. They were also hardy, better adapted to the harsh conditions on the rangelands. Their hardiness has been enhanced by feed supplementation to overcome the mineral-deficient diet available from the native grasslands. The shortage of minerals in the grasslands diet contributes to a high death rate in both calves and cows. It could be argued that the zebu has been the greatest contribution of science to rangeland agriculture. However, the zebu has been partially superseded by Brahman cattle, which are more productive. The Brahman is a mixed blessing. Those concerned about the deterioration of the rangelands complain that the Brahman does not have the decency to die when the grass does. The Brahman survives droughts longer than both zebu and European cattle, and so can do more damage to the native grasses. The feeding of supplements to cattle has also led to more cattle being run on some properties.

Public land and private management

Today the managers of the rangelands are faced with a choice between keeping on the current path, conservation of the native grasslands or development of the rangelands. The essence of the conservation path is conservative stocking management. This could be achieved with either the stick or the carrot. The stick approach has an initial attractiveness. The rangelands of Australia are publicly owned Crown land. The states with rangeland are, in combination, the largest landlords outside the former Soviet Russia. The grazing rights to specific areas of land are issued by land administrators on behalf of the state governments. Grazing land is held under pastoral leases, which are either term leases or perpetual leases. Most term leases grant a lessee the right to use an area of land for between thirty and fifty years, while all perpetual leases grant a lessee and his successors the right to use the leased land forever but with no right of conversion to freehold. Perpetual leases are the most common form of lease in New South Wales. While grazing is the predominant use, in different states varying amounts of land are given over for aboriginal use or for conservation (See Table 5.1: Land use in the rangelands.) The tenure system allows governments to control the use of land by attaching covenants and penalties to leases. The state governments have two main functions: the allocation of land and its control. The emotional question of ownership of Aboriginal lands has presented more difficult decisions in recent times, but the control of land use has always presented difficult problems.

In the past, closer settlement policies in Queensland and New South Wales restricted the area of land that could be leased by a person to an area sufficient to support one family. Big holdings were cut up into areas that were calculated to carry 7000 sheep, with little account for droughts and with a stipulation that the property be fully stocked within three years. In South Australia leases have contained covenants that require lessees to 'not

Table 5.1   Land use in the rangelands: 1990 ('000 sq. km)

Land use	State					
	NSW	*NT*	*Qld*	*SA*	*WA*	*Aust.*
Grazing	309	761	1254	419	958	3701
Vacant Crown land	0	214	14	198	828	1254
Conservation	6	4	26	31	109	176
Aboriginal land	0	357	26	117	218	718
Other uses	0	0	0	2	65	67
Total	315	1336	1320	767	2178	5916

overstock the land . . . which in the opinion of the Minister or the Pastoral Board would have the effect of depreciating the ordinary capacity of the land for depasturing stock'.[22] South Australia is proposing to introduce new controls on stocking in areas within a lease as well as the creation of a pastoral land-management fund into which a proportion of leasehold rentals will be directed. Such action reflects new community concerns that control of rangeland management needs to be better than it has been in the past.

Jurisdiction, regulation, and enforcement powers are all obvious components of the mechanism for government control to ensure appropriate rangeland management. In the past penalties have not been easy to administer. Throughout Australia the only penalty for breaching the lease regulations has been forfeiture of the lease. Although breaches are not uncommon very few leases have been forfeited, because the forfeiture penalty is an exceedingly severe and administratively difficult procedure to implement.[23] There are parallels with impotence in the administration and control of television licences by the Australian Broadcasting Tribunal. In the rangelands lease forfeiture usually requires the eviction of a manager and his family from their home, the seizure of the lessee's property and the assumption of the immediate responsibility for the ongoing management of the property. Graded penalties, such as shortening the length of a lease, would allow more administrative flexibility but not necessarily result in the encouragement of management action that only has long-term benefits. South Australia's proposed legislation provides for sanctions of compulsory musters to determine stocking rates and compulsory property plans to ensure an agreed program to overcome problems of land deterioration.

The more insurmountable difficulty is that standards of land management are not easily or cheaply monitored. There are three problems: the cost of gathering the evidence, the seasonal variability of the rangeland ecology within and between years, and the difficulty of establishing whether the damage was caused by the current leaseholder or a predecessor who may have brought about the damage at some considerable time previously. From an evidence point of view, there will always be a range of opinion as to the relative contribution of management practices and seasonal effects on the status of the rangeland ecology at a particular time. In the short term, only the extreme cases of overgrazing may be beyond doubt. In the longer term, when the evidence is clearer, it is often too late. Public interest in the management of the rangelands is today sufficiently developed that basic threshold standards of appropriate land management should be expected. In 1966 Campbell observed:

> . . . it might be necessary to terminate the leases of managers who are patently incompetent or who repeatedly overstock their leases, provided that they are not forced to such practices through the external pressure of inadequately sized holdings.[24]

The rangelands are not a place for pastoralists without adequate financial reserves; and rangeland administrators have the power to establish the minimum size of leaseholds. The evidence of pastoral inspectors, supported by photographs, should be sufficient to establish the cases to which Campbell referred. Western Australia has begun vegetation monitoring of its rangelands to record changes in the plant populations and densities and changes in the soil condition. Monitoring sites have been established at an average density of one site per 7500 hectares. A comprehensive vegetative monitoring program for Western Australia would involve about 12 600 sites. It is likely that the cost of universal, objective monitoring of vegetation change, as well as changes in the density of domestic, native and feral animals, and change in climatic conditions will be disproportional to the contribution of the rangelands pastoral industry to the national economy.[25] This will be so at least until there is a clearer knowledge of the economic cost of land degradation in the rangelands.

Given the variety of rainfall and the diversity of vegetation on different properties it is difficult to see how, for any but the more extreme cases of mismanagement, it is possible to have a feasible arrangement that doesn't put the responsibility on the manager. The past emphasis on government controls through leaseholds and covenants has had little success.[26] Monitoring will not solve the problem; it will simply measure that the problem has happened. The emphasis of any monitoring should be to put the onus on the manager to be more sensitive to the ecological implications of property management.

There are problems of economic balances in establishing appropriate property size — a judgment where governments have not necessarily known what is best. Closer settlement between the 1920s and 1940s broke up the large station runs carrying 50 000 to 100 000 or more sheep. These sheep were run in mobs of 5000 to 7000, placing enormous grazing pressures at shearing and crutching times on the paddocks surrounding woolsheds and watering points. But leaseholders on larger properties tend to receive higher incomes and, more importantly, proportionally less variable incomes than those on smaller properties. The larger properties tend to be stocked less intensively and, being less vulnerable to economic fluctuations, can be better managed for longer term objectives. In New South Wales and Queensland closer settlement produced properties of approximately similar size, which has limited the ability of property holders to readily expand by buying out those properties that have become uneconomic. Over time the size of property needed to make a living has increased. In areas where all properties are the same size the only expansion option may be to double property size. The rangeland property market differs greatly from a city property market where land is easily bought and sold.[27] Within a given locality there are very few properties and buyers may have to wait for years for a block they want and then compete for it with others.

Gentle encouragement, using the carrot approach, would ease the administrative burden of the 'big stick' approach. This could mean the use of government advisers in the traditional agricultural extension role. But there are problems here as well. Being an agricultural adviser on the rangelands is not a simple job. Pastoralists in the rangelands place a high value on their way of life, particularly being their own boss, being independent and free from supervision.[28]

Pastoralists must rely on their own understanding of local problems and local grazing conditions. There is little scope to call on outside expertise and there may be little outside expertise available. Pastoralists tend to be conservative towards innovation and new ideas — a conservatism that has generally stood them in good stead.[29] As we saw earlier from the old pastoralist's advice the characteristics of successful pastoralists are different from those associated with successful farmers in southern Australia. Most pastoralists place high value on the advice and experience of long-time local residents and don't associate themselves closely with the advice of government advisers. Figure 5.2 is a belief map of a typical south-western Queensland pastoralist's perception of management advice.[30] The map shows that the pastoralist sees the experienced pastoralist as a good source of advice. The outsider, whether a pastoralist new to the area, a stock agent or an extension officer, is not associated with good advice.

When people seek technical advice it is generally to help solve current or imminent problems. There is little technical advice, beyond animal husbandry and management practices, that can be extended to pastoralists because most rangelands research is associated with long-term rather than short-term problems of rangeland use. Pastoral management is craft oriented and is partially learnt by sharing experiences and asking advice of others when the opportunity arises. This works satisfactorily in good times but breaks down when there is a rapid decline in economic or climatic circumstances.[31] Then pastoralists may be faced with extra work, less time and less money for travel and reduced personal contacts. When economic conditions deteriorate smaller pastoralists often take casual employment away from their properties. When conditions are bad people are less inclined to socialise. Feelings of uncertainty and personal threat intensify the natural isolation just when management decisions most critical to the long-term health of the rangelands need to be made. At this stage external

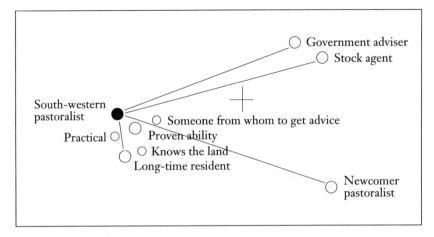

South-western pastoralist

Government adviser

Stock agent

Someone from whom to get advice

Practical

Proven ability

Knows the land

Long-time resident

Newcomer pastoralist

Figure 5.2 Belief map showing a typical south-western Queensland pastoralist's perception of sources of management advice

advice is unlikely to influence landholder decisions unless direct subsidies are involved. Where advice is to be directed to balancing stocking levels with long-term carrying capacity of the land it needs to be given when economic and seasonal conditions are good — the very time many pastoralists are least interested because they think they have no problems to worry about. This leads us to consider what controls exist for the management of the rangelands; and to ask whether the carrot or the stick is the better way to attempt to tackle such problems.

Landcare groups are being formed in the rangelands. It is the hope of the sponsors of the Landcare movement that graziers as a group will be able to provide support and even social pressure to encourage better land management. Is the Landcare group appropriate for the rangelands? If you have an aversion to company the rangelands is the place to be. Pastoralists living on the 4000 properties spread across the 70 per cent of Australia called the rangelands are not jostling with each other for personal space. The pervasive characteristic of rangelands pastoralism is its isolation. Even with modern forms of communication most pastoralists and their families live in virtual isolation on their properties. They rarely meet or see other people except on weekly or less frequent trips to town or other properties. How will Landcare deal with this situation? In the Northern Territory the Landcare model may be even less appropriate. Nearly half the stations are owned by absentee owners. It is hard to see the Holmes a'Court family, the Vesteys or the Murdochs having time to spend at Landcare meetings. The properties are run by managers, and the managers may not stay long. The Landcare incentives are unlikely to be appropriate for these properties. The maximum Landcare grant for a group is $20 000. How far will this much money go on a group of properties measured in hundreds of square kilometres? Not all the land is in the hand of corporate owners. One-third of the Northern Territory land is controlled by the indigenous people, the Aborigines. Here we have a community, but at the time of writing there has been concern by the Australian Conservation Foundation that Landcare funds have not been accessible to the Aboriginal communities.[32] We must wait to see how effective Landcare will be in the rangelands, but the omens are less than propitious.

If the carrot approach seems too difficult, the least we can do is make sure there are no incentives encouraging overstocking. Drought assistance is an inappropriate carrot. In a drought, cases of genuine hardship are better treated as welfare support rather than by the government providing assistance by input subsidies, which, in the longer run, discourages sensible land management and encourages the retention of stock when the grass and other species are at a high risk of longer term damage. The government should provide advice and infrastructure, such as transport systems and market reporting services. Most of the longer term land degradation problems of the rangelands — soil erosion and loss of edible plants — have their genesis in the big droughts. The real cause of the problem is that managers are often inadequately prepared because we cannot predict these droughts.[33] Drought problems can be reduced by timely action by graziers. Government sponsored research into the prediction of drought may in the long run be the best form of assistance for protecting the rangeland and the livelihoods of the pastoralists.

An ecological balancing act

Conservative stocking rates may be a luxury available only to those who can afford them.[34] Recently in northern Queensland rangeland graziers with high debts have been forced to either overgraze their leases or consider some form of development in order to meet property loan repayments. What does property development have to offer? For much of the rangelands, very little. Stocking rates of only one cattle beast per square kilometre mean many could not even justify another watering point. In the tropical rangelands problems of loss of herbaceous cover and soil erosion may be overcome by pasture improvement but, often, only if stocking rates are not again increased to recoup the investment. In Queensland the use of *Stylosanthes* species for pasture development is still being investigated. Since 1950 nearly fifty legumes and fifty grasses have been introduced to the rangelands. Phosphorus supplementation of stock feed has been shown to improve the health and growth of cattle. Tree and woody-weed clearing technology has developed. Fire remains the most satisfactory control method for the large areas of woody weed.[35] However, it has not been used widely as it requires ideal seasonal conditions to create a body of dry fuel, and also because drought-wise landholders are loath to set fire to valuable feed. Less feed means more grazing pressure in the short term. Some people are placing their hope in using one-pass blade ploughing with lightweight machines, or using newly formulated cheaper herbicides. New chemical pellets can be dropped by aeroplane to kill trees and clear land with little risk of regrowth. When is a woodland a woody weed? There has been debate in agricultural science and conservation circles over the appropriate use of these chemicals, and it is difficult to arrive at hard and fast conclusions.

Agricultural scientists are debating whether there is any need to clear the woodlands, and the evidence is mixed.[36] CSIRO scientists have observed that in the open woodlands of dry tropical northern Queensland killing trees and sowing pasture species can be an effective method of overcoming grassland degradation provided certain conditions are met:

> Firstly, landscapes likely to suffer from salinisation should not be cleared, although such landscapes may not be easy to recognise. Secondly, land with steep slopes, unstable soils or economically uncontrollable regrowth also should not be deforested. Thirdly, the graziers' philosophy of overgrazing must also change.[37]

One must conclude that the answer to the question of clearing woody scrub and treecover depends on where the work is done.

The conservation movement is unsettled at the thought of destruction of trees in northern Australia while governments are encouraging the planting of trees in southern Australia. This highlights a fundamental ideological difference. The Queensland Department of Primary Industries has released guidelines for woodland development. It has advocated that trees be retained on all steep land and watercourses, that 20 per cent of flat land be left covered in strips for shade and shelter, and interconnected corridors of trees between watercourses across neighbouring properties be developed.[38] The result would be similar to the outcome the Potter Farmland Plan has tried to achieve in Victoria, but achieved by removing trees rather than planting them. The Queensland Department of Primary Industries sees this approach as a step towards sustainable agriculture. To others it is seen as destruction of natural ecosystems; retaining the trees being synonymous with maintaining the ecosystem — an ecosystem that has been changed considerably as a result of influences we have already discussed. If the trees remain and pasture is developed, the ecosystem also will be changed markedly.

Southern observers will have to take account of rapidly changing ecosystems and regrowth history before engaging in automatic condemnation of clearing activity by pastoralists in woodland areas of the rangelands that have become excessively dominated by woody shrubs. Pastoralists in woodland areas can learn from the mistakes of some of the excessive clearing in southern Australia that has contributed to dieback, rising watertables and soil erosion. Some pastoralists in the more productive northern rangelands now have the opportunity, with cheaper arboricides and heavy machinery, to clear land excessively. In a system finely balanced ecologically and economically it is all too easy to move to a situation of too many or too few trees.

Most interest in property development has come from companies or from farmers under pressure to service property debt. Well established graziers are often less interested in development. If there is enough income to maintain the traditional muster on undeveloped rangelands, most pastoralists will find little attraction in development.[39] Pastoralism in the rangelands has survived economically because purchased inputs have been kept to a minimum. Higher cost development is generally only successful on the 'inside' land along Australia's green margin. Higher stocking rates inevitably mean greater risk of loss when drought arrives. An alternative view of property development has been put forward by CSIRO, which has concluded that there is a greater profit for graziers in running lower stocking rates to maintain the standing feed and to provide flexibility to respond to the fluctuations in the beef market.[40]

Harmony with nature in the rangelands is not easily achieved. It is very difficult for pastoralists to match livestock numbers to seasonal fluctuations in feed supply and severe overstocking has been an endemic corollary to drought, particularly where property sizes are too small to be economically viable or where managers are ignorant of the long-term consequences. However, the effects of drought are now less serious than in the 1890s when stock were left to die or were walked out along barren stock routes. Indirectly, technology has had some influence: now stock are trucked out, unless the value of stock is less than the cost of transport. Rabbits no longer cause the environmental degradation they wrought in the southern rangelands in earlier times.

The vast space of the rangeland is both its strength and its weakness. A rangelands drought that lasts several years and covers a good part of several states is likely to attract less attention in the capital cities than a train strike that lasts a few days. However, the environmental consequences of events such as the droughts are increasingly catching people's attention. All the signs point to the forces for change in rangeland management coming from those outside the rangelands.

Some groups have suggested a retreat from at least part of the rangelands as the best solution to deteriorating condition.[41] The practicality of this is questionable. It would not be possible to return the rangelands to their 1850s condition. The land would not return, naturally, to its former condition. There are large populations of feral and other introduced animals, which it would be economically impossible to eradicate. Use of the former Aboriginal regimen of 'firestick' management would be untenable for such a large area. We are slowly experimenting and learning how to apply the firestick in Kakadu and the Tanami. Maintaining a management system for this vast area would incur a cost that no government would willingly bear. The economic returns from the rangelands under today's system of management are finely balanced. There are no massive economic surpluses and this, in turn, means that governments are unwilling to provide the monitoring to police regulations that ensure a sensible and stable system of land use. More novel solutions, based on creating markets for kangaroo meat and creating rights for leaseholders to harvest free-range populations of kangaroos, free to move with seasonal food supplies, are unlikely to be feasible in the near future.

The pastoral use of the rangelands provides a contrast to man's involvement with the land elsewhere in Australia where, for the most part, agricultural productivity was enhanced and the land was developed. The constant stock numbers in the rangelands point to the nature of the ecological balance between pastoralism and the rangelands environment. The rangeland productivity gains have been limited to genetic improvements in stock and to fewer people running more stock per person, as a result of improved techniques of stock handling. Without the advances of science and the capacity for development, the rest of Australia's agriculture would operate on a similar level. The rangelands are exhibiting the degradation we created in the southern grasslands a century ago and there are no simple technological solutions, only politically unpalatable choices.

Part Two

Fallow and beyond

6

Sustaining the wheat crop

In 1788 the first governor of Australia, Captain Arthur Phillip, established a wheat farm on the narrow coastal plain of Port Jackson. Self-sufficiency in food production was one of the first priorities for the governor of this colony isolated from the rest of the world by a sea voyage of many months. To those in London who had sent Phillip on his mission, there seemed good reason to expect the successful establishment of a farm to feed the colony. The country was only lightly timbered and the convicts were a ready supply of farm labour. To Arthur Phillip the prospects seemed less bright. The land was strange, the seasons and soil unknown. On the First Fleet there were ploughs, but no animals to pull them. Unwilling convicts would have to do the work of the absent horses. Neither were there experienced farmers among the first settlers. The inexperienced would be directing the unwilling to farm the unknown.

To feed the fledgling colony Phillip had no choice but to organise rudimentary state farms manned by convicts. The hoe, the plough and the scythe were the limits of technology used on these farms. Cultivation was an integral part of the farming cycle. It was used to control weeds and create a seed bed for the wheat among the tree roots and stones of the poor sedimentary soil of Port Jackson. The cultivation was not overdone. Today Australian farmers are experimenting with minimum-tillage agriculture to protect the soil. The first Australian farmers also experimented with minimum tillage, but for entirely egocentric reasons — determining how little effort one could put into hoeing without suffering at the hand of the overseer. The same experiment extended to all other tasks on the convict farm: sowing, reaping, stooking and harvesting. Phillip's forebodings proved accurate. The first farms failed. Despite the dependence of convicts and overseers on the wheat produced by these farms, they did not understand how to farm the land and there was no incentive to encourage those who worked on the farms. The English wheats matured too late to avoid summer drought and fungal infection. The settlement faced slow starvation.

Phillip rejected state farms and tried instead to create a peasant agriculture. He granted implements and small parcels of land to well behaved convicts. The convicts had no overseer. They were their own task masters, but they did not survive long as farmers. They did not know enough about farming. Their land was often poor and infertile. Even in the good years when they produced wheat sufficient for the colony's need, the markets were glutted and the farmers were exploited by the powerful merchant class, mostly members of the corrupt New South Wales Corps. There was still no incentive to farm wheat.

In his third attempt at establishing wheat farming Phillip abandoned the peasant path. He granted officers in the New South Wales Corps larger tracts of land to farm using convict labour. These farms did not fail, but they did not grow wheat. The officers realised the economic futility of wheat, which cost more to grow and transport from the Bathurst Plains to Sydney than to import from overseas. They abandoned wheat and produced sheep for the meat market; later, they followed the example of the Macarthurs, producing wool for export. As the farmlands of the colony advanced across the Blue Mountains, wool production became the dominant farm activity of the colony. While the area of wheat grown in New South Wales increased steadily from 1797 to 1821, production was insufficient to meet the expanding local demand. The colony increasingly relied on wheat grown elsewhere.[1]

While the officers of the New South Wales Corps were rearing sheep, in Tasmania there was a greater opportunity to grow wheat profitably. The cool wet Tasmanian climate was better suited to the English varieties of wheat. Wheat was sown close to the northern river estuaries, which provided the key to cheap shipping transport to Sydney. Tasmania could compete with overseas producers to supply Sydney with wheat. The wheat industry in Tasmania was still predominantly human powered and initially relied on cheap exploited convict labour. It also exploited the soil. In England farmers had developed a system of rotations to preserve soil fertility. The introduction of superphosphate had ushered in the era of high English agriculture and English farmers were producing wheat on the most efficient wheat farms in the world. Australian wheat farming was seen as a failure by comparison. Farming techniques were backward and the farms had low productivity. Australian farms did not look like prosperous English farms.

From hand hoe to stripper

Various schemes were proposed to create Australian farming in the image of England. These proposals took little heed of the limitations imposed by the Australian climate, soils and markets. The farther removed the schemers were from Australia, the wilder their proposals became. The most distant of schemers was Wakefield. He argued that Australia did not have a proper balance between capital, land and labour and was therefore unable to achieve balanced (English) development. Because there was too much land, he argued there was no incentive to invest in land improvement. The solution was to set a minimum land price to attract to land ownership only those with sufficient capital for land development. While firmly resident in an English jail because of his elopement with a minor, Wakefield wrote letters purportedly from Australia, arguing his case. Without the limitations of proximity to Australian reality to expose the schemes, the English governing classes found his proposals attractive. Wakefield's theory became the model for colonial development.

The colony of South Australia was founded as an experiment in Wakefieldism. The site on the shores of St Vincent Gulf was chosen because of its obvious advantages for the wheat grower. There were fertile, treeless plains adjacent to a good anchorage. This solved the problems of transportation and clearing. However, the local interpretation of Wakefield's theory did not allow the first South Australian farmers to take advantage of the abundant cleared land around the settlement. The farmers were expected to farm in English style on small blocks of land, employing farm workers to the same extent as in England. But the price of labour in Australia was higher than in England. There were not enough workers. South Australian farmers could not afford to employ large numbers of Australian workers. Wakefield's vision floundered.

While the civil administration gradually put aside Wakefield's principles, the farming community set about solving their labour problems by mechanical means. The hoe had been replaced by the ox and plough in the 1820s. The number of furrows on ploughs gradually increased. Settler farmers followed the example of their convict predecessors and kept the

Ploughing fallow

amount of ploughing to a minimum. Early wheat farmers boasted of achieving two crops for one ploughing, a feat that would be admired today, but was judged harshly at the time by self-appointed English experts such as William Howitt:

> The stubble was standing up about a yard high, as if the reapers had been too lazy to stoop, and had only cut off the heads of the corn; while in some places it was pulled up by bullocks that were in the field. A portion of it, ploughed up again, lay in huge lumps, which farmers in England call 'horse's heads' . . . altogether one of the most wretched attempts at tillage I ever saw.[2]

The major labour problem was the harvest. The need for harvesting labour can be better understood from a contemporary description of the travails of harvest time using only the scythe and flail:

> . . . we began to think of our coming harvest, which we made up our minds to tackle with our own labour, but being quite unused to that kind of work, and the crop a very heavy one, we jibbed when about half way and hired an experienced hand to finish it, whom we paid at the rate of ten shillings an acre. We in our reaping had followed the English style of stooping to a low stubble, which severely strained our backs, but the colonial expert cut a little less than half way up the straw and crammed the short stuff into a long band. The sheaves, a mass of heads, were so heavy they could scarcely be lifted on to the dray when loaded, but that was the way all bush reaping was done. Straw was of no value and the long stubble facilitated the subsequent burning off to get a clean surface for the next ploughing; but in this instance, considering there was a good sprinkling of grain shed on the ground, we folded the sheep by night over the whole, the following autumn, which gave it such a good manure as yielded a crop of nearly 30 bushels the following season . . . This, with the succeeding self sown crop, made up over 60 bushels per acre for one ploughing and sowing, not a bad result.
> . . . for the threshing the tarpaulin was called into requisition and laid on the ground by the side of the stack and a man hired to knock it out with a flail, for which we paid him sixpence per bushel. Separating the grain from the chaff was effected by the man standing on the stool and pouring it from a dish as it was handed to him, trusting to the wind carrying the chaff clear of the tarpaulin and the clear grain dropping in a heap at his feet. This [was] securely bagged.[3]

South Australian farmers felt the need for labour acutely during the short harvest time. The colony could not support a permanent population large enough to harvest the wheat crop, there being little other work available for the rest of the year. Farmers could not be sure they would get their wheat harvested. The inventors among them got to work. John Bull suffered this problem as much as any wheat farmer, but in 1842 a solution came to him in a moment of inspired frustration at the drunken unreliability of his workers. With a crop ready for reaping he faced the usual difficulty of procuring harvest labour:

> On December 24, 1842, I was able to induce five men to accompany me and I conveyed them to my farm at Mt. Barker. I did not allow them to work on

Christmas day, but they had Christmas fare. I engaged to give 15/- and one bottle of rum per acre, with rations, for hand reaping. The crop was dead ripe, the heads drooping with the full weight of the plump grain. On the 25th a fiery hot wind was blowing and continued on the following day when I expected the reapers to start work, but they were missing. I found them at the nearest grog shop. After some trouble I got them away to start work on the following morning. Before a sickle was put into the crop the loss of shed wheat was one bushel to the acre and a further loss necessarily followed in harvesting.

Immediately on my return I took one of the men, the most sober of the lot, to see the over-ripeness of the crop . . . and pointed out to him how careful they would have to be in performing their work in handling the standing crop and in binding. Calling his attention to the shed grain on the ground and to show him how tender the heads were with the full grain staring us in the face out of the gaping chaff, I passed my left hand with my fingers spread under and just below the ears, allowing the straws to pass between my fingers, the ears being close to the palm of my hand. I then struck the heads with a sweep of my right hand and held out my open hand for the man to see the clean threshed wheat in the hollow of it, most of the chaff having been carried away. Before I moved from my position in the standing corn I stood in a sort of amazement and looked along and across the fine even crop of wheat. The ideas I had sought in vain now suddenly occurred to me and I felt an almost overwhelming thankfulness. I did not move, but sent the man for a reaping hook and caused him to cut me a small sheaf of wheat which I took into the barn. There, holding a bunch of it in in a perpendicular position, I struck the ears with a circular sweeping blow upwards using a flat and narrowed piece of wood and found the threshed grain to fly upwards and across the floor; and thus I satisfied myself that the grain would bodily fly at a tangent up an inclined plane when struck by beaters and that a drum as in a threshing machine would not be required to complete the threshing and so felt I had gained the correct idea for a field thresher and that a segment below the beaters would be apt to cause the wheat to be carried around and so be lost.[4]

In 1843 an Adelaide committee offered a prize for the best mechanical harvesting machine. Bull exhibited his machine, but did not win. No prize was given. The committee could see no merit in Bull's invention. His machine was never reproduced, but the ideas behind it were quickly adapted by John Ridley who created the first stripper. At first the wheat farmers of South Australia were suspicious of the new machine. They feared it would waste grain by spilling it on the ground. No farmer would allow Ridley to test his machine on their wheat. Ridley demonstrated it by buying wheat still standing in the paddock and then harvesting it himself using the stripper. That convinced the wheat growers. The stripper saved time and money and enabled farmers to harvest crops that were too short for the sickle. Workshops sprang up across the South Australian wheat plain to fabricate the stripper. Neither Bull or Ridley took out patents on their inventions. The stripper was a gift to the farmers of Australia; it helped South Australia became an exporter of wheat.

While South Australia was exporting wheat, the colony of Victoria was struggling with the pressure for land reform. The Victorian wheat industry produced only a small portion of that colony's domestic requirements. Wakefield-inspired land settlement policies that had tried to contain settlement within the easy reach of the law in Melbourne had failed. Squatters had illegally occupied the majority of the agricultural land. The squatters were not permitted to grow crops for sale. This restriction was unnecessary. Even if the squatters had wanted to grow wheat, the cost of transport and labour were exorbitant; there was no point in growing wheat without cheap transport. The commercial wheat industry was confined to limited areas with ready access to markets. Some commercial wheat was grown around Melbourne and a small wheat export industry was established on the rich volcanic plains adjacent to Port Fairy. Across the rest of the colony wheat was confined to small homestead fields sufficient to feed each squatter and his shepherds.[5]

The gold rush of the 1850s changed the wheat industry. Those with the right to crop the land near the diggings made spectacular profits. The digger and chronicler William Howitt observed, 'at present . . . land is the truest gold mine'.[6] Farmers and squatters supplying the goldfields sold wheat in a sellers' market.[7] They were protected from imports by the cost of transport from Melbourne to the diggings. Howitt was not the only digger to recognise the prosperity of these farmers, though fewer recognised this as a passing prosperity. The observations of the diggers helped to fuel the demand for land reform that found expression in the Victorian Selection Acts.

Limits of fertility and rainfall

Those who framed the various Selection Acts believed that by redistributing the squatters' land to the working class they would create a prosperous society of farmers to replace the grazing squattocracy. The yeoman farmer of England, who grew a mixture of farm produce, would be transplanted to the Australian countryside. In Victoria the first two Selection Acts allowed the selection of 640-acre (259-hectare) farms in isolated areas and smaller selections around the goldfields and major towns. The only areas where these Acts achieved any success was in the goldfield districts. There, selectors were able to combine farming with digging, labouring or trade work in the new towns. Unfortunately, the politicians misinterpreted the success of the smaller properties and the failure of the larger selections. It was seen as justification for lowering the selection size for all properties to 320 acres. Hopes of successful selection farming on farms of this size were woefully ill founded. The original 640-acre limit was based on the surveying convenience of one square mile and estimates of the productive capacity of the most fertile areas of the colony; but many of these fertile areas were already lost to the squatters, who dummied and peacocked their way around the first failed Selection Acts. Most selectors had to farm instead on the drier and less fertile northern plains.

The selectors faced marketing and transport problems. Wheat was not profitable. The only marketable product was wool, but the selectors' farms were too small and the selectors lacked the capital to build financially

viable grazing properties. The selection laws required selectors to make
£320 worth of improvements to each selection and crop 32 acres, as well as
paying back the selection price over the first seven years of occupation. To
survive, most selectors needed to take work away from their selections.
The only paying work for farmers in the new selection areas was with
squatters who had managed to evade the intentions of the Selection Acts,
a pattern of farm labour relationships that survives to this day.[8] Even with
selection sizes of 640 acres, only those with no debt could expect to be
better off on their selection than working for wages for the squatters.[9]

To save the selection scheme, the Victorian Government built an inland
railway network. The first railway through north-central Victoria to Echuca
was aimed at taking trade from the South Australian riverboats. Later,
other rail lines provided access to western Victoria, the Wimmera and
eventually parts of the Mallee. The railways of other colonies also snaked
inland. In the same decades, machinery innovation allowed wheat producers to crop more land. Ploughs increased in size to five or six furrows. The
elusive challenge of a machine that combined the tasks of reaping, threshing and winnowing was finally solved with McKay's 1885 harvester. The
key to profitable wheat production became owning a farm large enough to
take advantage of these inventions and lying within 50 kilometres of a
railway station. The only way for most farmers to achieve the necessary
farm size was to borrow from financiers willing to break laws prohibiting
lending to selectors.[10] Under the pressure of the debt, selectors sowed most
of their selection to wheat to maximise their cash flow. A wheat industry
based on continuous cropping was created. The selectors were soon to
realise the limitations of their soils under this exploitation.

Early Sunshine Harvester

Most Australian soils were old long before the first man, black or white, arrived on the continent. There had been little geologically recent glaciation, volcanic activity or geological uplift to enrich the soil with new minerals. Many soils had been continuously leached and weathered, and were low in nitrogen, phosphorus and some trace elements.[11] Most nutrients were held in the plant matter growing above the soil or in a shallow top-soil. To farm these soils for more than a few years it was important to maintain these reserves of nutrients. Growing wheat mined the soil of nutrients. Each wheat sack transported from a selection took with it some of the soil's nutrients. In Europe farmers had learned to replenish the soil by growing legume crops in crop rotations, resting the soil under pasture or regularly spreading animal manure. The ancient Greeks had mastered these practices, interspersing crops of corn, flax and oats with crops of pasture or legumes (pulses, vetch and lupins) to replenish soil nitrogen. European farmers could not explain why these practices worked. A common understanding was that crop paddocks needed 'resting'. In Europe rested paddocks grew European pastures, which included clovers. The resting was in reality a replenishing of nitrogen. In Australia some observers advocated resting wheat land according to European traditions. But in Australia, when selectors rested their land, there were no native pasture legumes that could replenish the nitrogen lost in crops. Instead, a weed fallow developed. This did not regenerate the soil, but merely slowed its exhaustion.

Initially selectors did not embrace the idea of resting the soil with any enthusiasm. Most selectors did not have enough land to afford the luxury.[12] Many farmers disliked the practice because it encouraged weeds.[13] Even in the wetter districts, where European clovers could be grown, most farmers did not understand the importance of pastures or were forced by economic circumstance to continuously crop their land. With continuous cropping many soils were quickly depleted of nutrients and lost their structure. Wheat diseases built up in the soil. In 1879 Horsham farmer John Bawdon explained the decline of yields to a sitting of a Commission of Inquiry into the failure of the Victorian Selection Acts:

> What is your opinion of the durability and fertility of the soil?
> — I think the soil is very good [in the Wimmera district]; my land has been in cultivation 22 years and it is still in a very good condition.
> Do you manure your land — Yes.
> The ordinary selector cannot manure like you?
> — A man does not want any for three years.
> That was your experience when you first came?
> — Thirty three bushels to the acre.
> And you never manured for three years? — Yes.
> And the land continued to return well?
> — Yes, up to three years. I got twenty bushels in the third year.[14]

John Bawdon was an exceptional farmer. He spread stable manure collected from the nearby township of Horsham. Animal manure replaced soil nitrogen. Most farmers rejected this technique because they did not live close to townships and because of a common belief that manure weakened the resistance of the wheat to hot summer winds. Many selectors held this belief so strongly they were afraid to allow livestock to feed on the stubble of a harvested wheat crop for even one night. Instead, farmers burnt the standing stubble. The new Australian stripper encouraged farmers to burn their stubble. In England the stubble was carried off the paddock in sheaves to the threshing machine. The stripper left the stubble standing high in the paddock. The high stubble was difficult to plough but easy to burn. The Welsh immigrant Joseph Jenkins had been a prize-winning farmer in Wales. While travelling in Australia as a farm worker he kept a diary in which he deplored the exploitative practices of the colonial wheat grower:

> There are three characteristics peculiar to the farmers of this colony — exhausting the land, abusing the horses and exploiting the labour . . . The farmers are harrowing and sowing for another crop in the already exhausted land, which is badly infested with weeds of all sorts . . . Some farmers boast they have taken twenty-seven crops from the land without giving it a shovelful of manure in return . . . The land is in an exhausted condition. It is no use talking to farmers about preparing farmyard manure to fertilise the land . . . the farmers should cultivate the land and grow artificial grass as lucerne, cocksfoot grass and clover . . . [Today] I wheeled two tons of manure into old mining shafts 170 feet deep. In Wales I would have 5/- a ton for it . . . Around here they burn stubbles and consider that this provides good manure. I think so too, provided rain follows, otherwise the strong wind will blow it off the surface and render the process useless . . . The exhausted land will simply not produce unless it is heavily, not artificially remunerated. To plough, sow, harrow and reap corn, whilst the straw is either burnt or carted to market, is not cultivation to advantage.[15]

The selectors recognised they were exhausting their lands by continuous cropping. They responded, but not in a manner envisaged by those who drafted the Selection Acts. As the new railways snaked inland, selectors from the settled districts sold their exhausted selections to neighbours and selected again in the virgin lands. In the exploited districts a few remained behind, purchasing exhausted land from the migrating selectors. They became known as 'boss-cockies'. The boss-cockies quit wheat farming and changed to sheep or cattle grazing on their enlarged properties. The wheat fields gradually moved inland with the selectors.

The press and politicians of the day worried that the Selection Acts were a failure. Instead of a society of small holders farming the land, a new farming landscape was being created. It was being taken over by a small number of boss-cockies, squatters in a new form. The exhausted land was described as 'grain sick', depleted of nutrients. Some critics proposed a ban on exports of wheat to conserve the fertility of the land. In Victoria an inquiry was held to determine what had gone wrong. The harsh truth of life on the selections was revealed to the dreamers from Melbourne. The Selection Acts had not created their idealised vision of a yeoman farmer. Instead they had spawned an investment farmer, not emotionally attached to a community or the land. Under financial pressure, the investment farmer could only be interested in making greater profit from the cash-crop wheat. Diverting energy and money into diversified enterprises on a small farm was not sensible. Crop, crop, crop — and when the soil was exhausted, it was time to move on. Soil fertility was not the only reason to move north. The drier climate decreased the chance of loss from the rust fungus. It also made mechanical harvesting easier because the wheat had a better chance of ripening.

Moving inland was not a long-term answer to declining soil fertility; it merely postponed the day of reckoning. There was a limit to how far inland wheat could be grown. With each advance the average annual rainfall decreased. Unreliable rainfall eventually halted the advance. The difficulty for the pioneers was knowing where to stop. Rainfall isohyets appear as neat lines on maps, but there were no neat lines on the ground. There was no easily discernible difference between the rainfall on one property and the rainfall on the next. The temptation was always to develop the next piece of land. The knowledge that the expansion had gone too far would not come until the next drought. Nowhere was this better illustrated than in the development of South Australia's wheatlands.[16]

In 1865 a severe drought gripped the pastoral country beyond the Adelaide plain. The Surveyor-General Goyder mapped the areas of drought, producing his line, to the north of which lay drought-stricken country. Goyder's line followed the boundary of the semi-arid vegetation. In the years that followed, Goyder and others came to argue that the line divided the country between the land that was suitable for wheat growing and the land that was too dry. This interpretation was formalised in the law of the colony. In the late 1860s there was plenty of unploughed land south of the line and no-one took much notice of the restrictions the line imposed. The early 1870s provided good seasons for the South Australian wheat farmers. Land releases south of Goyder's line were rushed by applicants. As available land south of the line ran out, some farmers migrated across the state border to the newly opened lands of the Wimmera in Victoria. Others pressured the South Australian Government for the release of land north of Goyder's line. Vigorous attempts were made to discredit the now infamous line. One impatient would-be settler saw rain was falling on both sides of the line:

> I came to a store at Pekina, on the other side of the 'rainfall', on Tuesday, May 4 about nine o'clock in the forenoon, wet to the skin, and it rained steady all that day and night and part of the next day, and I defy Mr Goyder or any other man to say at which side of the hedge the most rain fell.[17]

Driven by fears of a threatened electoral backlash and the loss of farmers to the Wimmera wheatlands in rival Victoria, the South Australian Government abandoned Goyder's line. The north was opened for wheat farming. The wheat frontier advanced far across Goyder's line. New townships sprang up across the rangelands. But the wet weather that discredited Goyder did not last long enough for the new settlers to get one good crop. An 'old man' drought gripped eastern Australia. After two years of drought the new settlements collapsed and the wheat lands retreated back behind Goyder's line. The press, which had a few years earlier castigated the government for failing to release land beyond the line, now attacked the government for cruelly misleading settlers by releasing obviously unsuitable land.

While critics carped at the South Australian Government and farmers in northern Victoria demanded irrigation, other wheat growers tried to live with the reality of unreliable rainfall. Some farmers imported or selected wheat varieties that seemed better adapted to a short growing season. Most farmed to minimise the potential losses from drought. Despite the admonitions of agricultural scientists, settlers continued to plough as shallowly and as infrequently as possible to minimise their losses when the rains failed. The reasoning was simple — in a good year a crop on poorly ploughed land made almost as much money as a well ploughed crop. It was a lot less work for little or no return in a poor year. A few farmers attempted to maintain fertility. Some cut, chaffed and spread their stubble. Some collected manure from local towns to spread on their fields. Some grazed sheep on their stubble and rested paddocks, but these farmers were the exceptions. To the average selector, driven by debt, these settlers were indulging in a rich man's hobby. The economic facts were stark:

> The wheat-grower, with labour at £2 a week and land at £1/10/1 per acre, has nothing before him but to go on with his double or triple plough from one farm to another while new land is to be obtained. No professor who sat in a chair can ever persuade him from that simple truth.[18]

The result was a gradual decline in the average wheat yields. The southern Australian wheat industry was trapped between the consequences of unsustainable mining of nutrients and the limited area of well watered land.

Fallow, Federation and phosphorus

The dawn of the new century was the dawn of a new age for the wheat industry. The industry was rescued from decline by three new agricultural techniques: dry farming, purposeful wheat breeding, and the application of superphosphate fertiliser. Many years before in the 1840s European scientists had discovered the crucial role played by phosphorus in healthy plant growth. They learned that harvesting wheat rapidly mined the soil of its phosphorus content. Initially rock phosphate was used to replace the harvested phosphorus, but this had not proved very effective. The invention of a soluble form of phosphate, superphosphate, provided a more effective means of replacing depleted soil phosphate. By the 1850s many English wheat farmers were spreading superphosphate on their fields.

The few Australian agricultural scientists were graduates of English colleges. They brought English ideas to Australia. Superphosphate was one of those ideas. The use of superphosphate was recommended to

Australian wheat farmers as early as 1879, but few farmers took any notice. The recommended application rates were heavy and very expensive. Australian wheat farmers, unable to rely on regular rainfall every season, were not accustomed to risking large sums of money on very variable yields. Some farmers who had previously applied rock phosphate remembered how little this had achieved. The agricultural scientists who promoted superphosphate were derided as irrelevant.

Two wheat farmers on the Yorke Peninsula of South Australia were responsible for taking the risk out of using superphosphate on wheat crops. By the 1890s some South Australian farmers were experimenting with seed drills that sowed wheat seeds in straight lines under the soil surface.[19] With the seed drill it was possible to sow wheat seed mixed with small amounts of superphosphate and get yields comparable to those achieved by spreading superphosphate over the whole paddock. This lowered the cost of 'supering' from 40 shillings per acre to 5 shillings an acre.

Improved wheat varieties, particularly the Federation variety developed by William Farrer, were the second important innovation to help save the wheat industry. The English wheat varieties did not bear well in the short growing season of the mainland Australian wheat fields. Some Australian farmers found varieties that grew better in the dry climate. Some imported overseas wheats; some selected wheat from the genetic mix in their paddocks. A few breeders, particularly Farrer and the agricultural scientist Pye, thought more could be achieved by purposeful breeding and selection. They dedicated themselves to the breeding of a superior wheat adapted to the dry conditions of the Australian wheatlands. The new varieties had small leaves and stems and took a shorter period to reach maturity. This made them better adapted to the short growing season. They were also more suitable for mechanical harvesting. These characteristics reduced the risk of crop failure in dry seasons.

The third innovation to transform the Australian wheat industry was 'dry' farming. The heart of dry farming was the bare fallow — ploughing a pasture in autumn and keeping it bare of weeds until the following autumn when a crop was planted. Without weeds to transpire soil moisture, rainfall seeped through the soil to the lower root zone where it could be used by the following season's wheat crop. In the Wimmera the effect of bare fallow was spectacular. It seemed to bear out the old adage of Jethro Tull that 'tillage is manure'. It seemed bare fallow saved the equivalent of 100 millimetres of rainfall. This translated into roughly 15 extra bushels of wheat per acre (or one tonne per hectare). Fallow dramatically reduced the risk of yield loss associated with dry years.

Bare fallow had other advantages. It increased the amount of nitrogen available to the wheat crop. Plants are not able to use the nitrogen in organic matter in the soil. It must first be broken down by bacteria. The ploughed ground of a bare fallow is an excellent medium to facilitate the breakdown of organic nitrogen compounds. Cultivation warms and aerates the soil. With the coming of spring there is a dramatic increase in the amount of available nitrogen in the soil. Because there are no weeds in a bare fallow, the high levels of available nitrogen are not transformed back into the organic matter of weeds before the wheat crop is planted in autumn.

Bare fallow also made weed control easier. Each regular ploughing of the fallow encouraged the germination of weed seeds, depleting the remaining store, which could contaminate the following wheat crop. The fewer weeds in the crop, the better the yield and the less the contamination of the wheat. Bare fallow also provided a break on the proliferation of root diseases of the wheat plant, particularly the root fungus, take-all (*Gaeumannomyces graminis* var. *tritici*). The wheat crop that followed the fallow was relatively free of root disease.[20]

By the 1890s bare fallow was widely used on the black soils of the Victorian Wimmera. Here it could achieve spectacular yield increases while lessening the likelihood of crop failure in a drought. Fallow took a lot of work, but very little capital outlay, so it did not increase the financial risks following crop failure. Adoption of fallow elsewhere was slower as the benefits were not always as spectacular, and working fallow was time consuming.[21] Agricultural societies spread news of fallowing. The Nhill Agricultural Society organised the first Victorian wheat crop competition in 1901. Farmers could see that the best-yielding crops that won the competitions were grown on bare fallow. Fallow and better breeds of wheat made farmers more amenable to investment in superphosphate.[22] By 1910 two-thirds of Victorian wheat farms were sowing superphosphate with their wheat seed and the new Federation wheat was the most widely planted wheat variety in Australia.[23]

Over the next thirty years, as wheat yields rose towards those the first settlers had achieved, the use of dry fallow and superphosphate became accepted practice. Organised crop competitions were concentrated more and more on fallow to the detriment of other aspects of crop growing. At a Wimmera competition in 1906, farms were judged on yield, crop rotation, manuring, systems of saving stable manures, weed control and bare fallow. Ten years later the same competition had two sections: best 100 acres of crop and best 100 acres of fallow.[24] Fallow paddocks were judged by their tilth, stored moisture and absence of weeds. A good bare fallow became an end in itself, as the following typical competition report shows:

> Best Fallowed 100 Acres: Mr R. G. Keam, Woorak, wins in this section with a beautiful bit of fallow, on which two teams were finishing a final touch up. This land had been ploughed once, harrowed twice and cultivated twice, and was in tip top order. Mr O. H. Lienert came second with an excellent piece of work, with not quite as good a mulch as the former and slightly behind in cleanliness. Mr R. Blackwood was third with another fine piece of work, which had it received another stroke of the harrows just previous to inspection, would probably have won.[25]

If some fallow was good, then more was better. By 1920 the advisers were following American trends and recommending longer periods of fallow and a finer tilth. Burning and ploughing for the longer fallow began immediately after the harvest of the crop. The ground would then lie in fallow for four or five additional months with the aim of retaining more moisture. The justification for a fine tilth was ill-founded. Neither farmers nor scientists fully understood the dry fallow. The contemporary scientific explanation was that regular ploughing created a fine mulch of fine soil on the surface, which reduced evaporation. In reality, the crucial role of ploughing was to prevent weed establishment. This misunderstanding was unfortunate, because it led to the worst excesses of the dry fallow enthusiasm — constant ploughing to create a superfine tilth. To farmers and agricultural advisers every ploughing became synonymous with an increase in yield.[26]

By the end of the 1920s, the area allocated to bare fallow in Victoria averaged 70 per cent of the area devoted to wheat crop. There were three common crop rotations. On the black soil plains near Horsham, fallow was followed by as many wheat crops as possible until take-all disease threatened yields. Between each succeeding wheat crop the land was given a short fallow. When the paddock was no longer capable of profitable wheat growing, it was spelled with oats or volunteer weeds. Degraded native pasture offered poor grazing for draught horses and sheep. Oats were a marketable crop, fed the horses and had the added advantage of helping control take-all.

Where the soil could not take this treatment, the most common cropping system was a conservative rotation of fallow, wheat and either oats or pasture. Under this system there was less land in wheat in any year, but the risk of crop loss due to take-all was reduced.[27] Farmers under financial pressure often followed a two-course rotation, alternating wheat and fallow every other year. In each system farmers had learned to burn stubble to help control take-all. The choice of a conservative or intensive rotation by a farmer depended on his immediate need for money to meet debt commitments, his preference for working with sheep in his farming operation and what the soil would allow.

Depression, dustbowls and war

In the first thirty years of the twentieth century, steadily increasing wheat yields justified the faith that farmers and their advisers placed in fallow, Federation and phosphorus. Some scientists and farmers warned of the dangers of fallow, particularly Dr Tepper of the New South Wales Department of Agriculture.[28] But as the area planted to wheat, and the average yield per acre, increased each year, Australian wheat farming seemed successful and sustainable. Australia was able to sell all the wheat it could produce. In the later years of the First World War Great Britain bought all the excess Australian wheat crop. After the First World War stable wheat prices encouraged farmers to invest and expand. But the wheat industry was living on borrowed time. The depression years brought low world prices. Low fertility and soil erosion forced growers and scientists to re-evaluate the basis on which the industry had been built.

In the year following the 1929 Wall Street crash, Australia faced a balance of trade crisis with a surge in imports in the late 1920s. Exports were falling, overseas lenders were calling in loans and few were willing to extend further credit to Australia. The government called on Australia's wheat farmers to help the country trade out of its position. It declared 'Grow More Wheat Year', and led farmers to believe they would receive a minimum price of 40 cents per bushel for all wheat produced in the following season. Australian farmers put their land into intensive rotations and grew more wheat. Unfortunately Australia was not the only wheat producing country with a balance of payments problem. Wheat farmers around the world increased production; the world wheat price slumped to 18 cents and the government was unable to pay the guaranteed price.

Farmers had produced the wheat at an average cost of 30 cents a bushel. Many were bankrupted and others were left in debt. The first priority on most farms was surviving the next season. There was little money to devote to long-term problems. Unfortunately, the long-term ills of heavy cropping rotations, bare fallowing, nutrient depletion, soil structure degradation and increased erosion were now becoming hard to ignore.

In dry lighter soil districts of South Australia and Victoria soil erosion had always been a concern. Agricultural societies complained of drifting dunes in the 1890s.[29] Bare fallow exacerbated the problems and farmers in the Mallee wheatlands had come to accept some wind erosion on their fallowed paddocks. The maxim was 'if it blows, it grows'. But they were not prepared for the blows of the drought years of the 1930s. In the drier parts of the Australian wheat zone bare fallow practices created our own dustbowl.

In wetter areas heavy summer rainstorms could be just as damaging as wind in the Mallee. During the late 1920s the wheatfields of south-eastern Australia experienced particularly heavy summer rainstorms. Bare fallowed ground in some districts was severely eroded. There was enormous soil loss. An agricultural scientist recorded his concern at the erosion in New South Wales:

> At the present time on the South West slopes, deep gullies are to be seen that have been cut by water in the last two years. Land that ten years ago could be cultivated and drilled across is now in many instances cut by gullies seven and eight feet deep and nine and ten feet wide. The damage has to be seen to be believed. Areas of land in these districts were cut out by the heavy rains experienced this season, and some land badly gullied where no sign of gullies had appeared previously.[30]

Bare fallow agriculture was trapping some farmers in a vicious cycle of ploughing to undo the damage of ploughing. The regular pounding of the more fragile cropping soils by ploughing was destroying soil structure. After a shower of rain the powdery surface of many soils set into an impervious crust. The next rainfall was unable to penetrate the crust. Runoff increased dramatically, eroding the soil and defeating the purpose of the dry fallow. The solution adopted was to plough the fallow after each light rainfall. This broke the crust, but further damaged the structure of the soil making the next crust even stronger. After heavy rainfall the farmer would plough to fill in any erosion gullies. Ploughing the small rills and gullies camouflaged the damage, but it did not check the erosion. At the next heavy rainfall all the soil ploughed into the gullies was washed away and the gullies became deeper.

The Second World War ended the years of economic depression and farmers were again exhorted to grow wheat for patriotic reasons. But wheat yields had peaked a decade earlier.[31] The heady days of increasing yields were over; the wheat industry had again reached the limits imposed by declining soil fertility. The war and weather conspired to make even the maintenance of yields an insuperable challenge. Superphosphate was rationed. Farm labour was scarce, with sons and workers joining the army. Only the most pressing tasks were undertaken. Tasks with a longer term payoff, such as controlling rabbits, were often left undone. By the end of the war rabbits infested much of the grain belt. Then the final years of the war brought a very severe drought. The wind blew the bare fallow, creating daily dust storms, moving dunes and sending the top-soil as far as New Zealand. Australia was exporting soil!

Behind the problems of declining yields and erosion was the bare fallow and an exploitative farming regime. Some agricultural scientists searched for physical means of controlling wind and water erosion on bare fallow, importing ideas from the United States. Water erosion was countered by the contour drain, a broad, low ridge that contoured across the property with a definite, but slight, fall. Farmers placed the drains on contours two vertical metres apart. Horses and machines could travel over the banks. Wheat could be grown on the banks. Wheat was drilled in rows parallel to the contour. In the United States the practice was promoted by the slogan 'A deed to the land will not hold the soil, but a contour drain will!'. Australian agricultural scientists promoted the practice to Australian farmers, without the slogan.

Mechanical solutions to wind erosion were more difficult. In the United States farmers developed rod weeders and the blade plough. These implements ploughed under the soil surface to destroy weeds while keeping the surface intact, allowing prairie farmers to keep their fallow weed-free and to control drift. In Australia there were few farmers with tractors strong enough to pull these ploughs and much of the Australian wheatland was unsuited to the plough; the soil was shallow and littered with buried stumps.[32] These ploughs did not appear in any numbers in Australia until the 1980s. United States farmers also practised strip cropping: sowing crops in alternate strips of wheat and fallow across the paddock at right angles to the prevailing wind direction. The crop strips protected the ploughed land on either side from wind and caught any soil that did blow. This form of farming did not mix well with Australian wheat farming practices. Australian farmers expected to feed their sheep on wheat stubbles and the native grasses on their farms. In a strip cropped paddock, sheep could not be grazed without letting them eat the wheat crop as well. American solutions were inappropriate. Home-grown solutions were required.

Australian farmers were advised to try several alternatives: grow another crop instead of fallowing, leave wheat stubble standing, plough deeply to bring up deeper cloddy earth or plough soil into ridges. Each option was supposed to lower the surface speed of the wind and catch moving particles. Growing another crop instead of fallowing controlled erosion, but lowered yields by reducing the availability of water and nitrogen. Controlling weeds in a standing stubble was often impossible, leading to less water and nitrogen in the soil and contaminated crops. The rough ploughing methods only partially controlled drift but crop yields were lower. Farmers could not ignore erosion, though, as it was also causing lower yields.[33]

In the long run, mechanical solutions only controlled the symptoms of a deeper problem inherent in the bare fallow regime: lack of organic material. Bare fallowing had offered an illusory solution to soil exhaustion by increasing the amount of available nitrogen in the soil at planting time. It did this by increasing the rate of decomposition of organic matter and storing the nitrate product until the crop could use it. With each bare fallow the amount of organic matter in the soil was gradually depleted. Agricultural scientists had shown that fifteen years of a fallow–wheat rotation would cause a significant decline in soil fertility with the depletion of most reserves of organic matter. Loss of organic matter prevented the soil from aggregating into a healthy crumb structure. The badly structured soil compacted, formed crusts and plough-pans and was erosion prone. Soil erosion is a result of both cultivation and loss of organic matter in the soil. Solving soil erosion by mechanical means was a classic case of treating the symptom rather than the cause. There would be no long-term control of cropland erosion without a solution to the fertility problem. Some advisers advocated manuring, the application of blood and bone and nitrogen fertilisers when they became available. These proposed solutions were too expensive or too labour intensive. Farmers could not afford them.

Clover ley farming

The answer to the depletion of soil organic matter lay in pasture farming. In 1913 A. E. Richardson, a scientist at the Rutherglen Research Station in north-eastern Victoria, sounded an early alarm at the mining of soil organic matter by wheat farmers:

> The outstanding weakness of our system of wheat culture is that insufficient provision is made for the restoration of organic matter of the soil. It is known that the losses of organic matter due to fallowing in an arid climate are very considerable . . . the value of a soil for agricultural purposes depends in no small matter on its organic content. Deprive the soil of its organic matter and you have rock dust, and what farmer would care to farm a soil made of freshly pulverised bricks![34]

Richardson recommended two management strategies to farmers: the use of a crop rotation that included a year or two of pasture; and growing green manure crops, such as oats, which were ploughed back into the soil. However, he acknowledged that 'it is very questionable whether in these rotations the gain in organic matter is equivalent to the depletion in the period of fallowing and cropping'.[35] The pasture that grew after a long fallow and wheat was mostly disease-hosting grasses: annual ryegrass, barley grass, silver grass and cape weed. It was only in the second-year pasture that small numbers of clovers appeared. Such a pasture restored little fertility to the soil and was of little value for grazing. In search of a better solution he began a series of experiments at the Rutherglen Research Station to measure the effect of pasture, fallow and green manuring on the soil in various rotations. Over the next fifteen years it was established that none of the recommended rotations made much difference to the rate at which bare fallow depleted the soil. The weakness of the rotations was the absence of clover or other legumes adapted to the Mediterranean-like environment of the wheat belt.

In New South Wales scientists urgently advised farmers to integrate the legume lucerne into their wheat rotations, rough plough to retain stubble and to build contour banks across their farms.[36] The staff of the Rutherglen Research Station followed another strategy. In 1931 a depleted wheat paddock at the research station was sown to subterranean clover. After five years of subterranean clover, the paddock was ploughed and wheat was sown. A heavy crop was harvested off the paddock in 1937. The years under the sown pasture had dramatically improved the soil structure.[37] Later experiments showed that a pasture sown with subterranean clover would add between 20 and 80 kilograms of nitrogen to a hectare in one year. The damage done by wheat cropping and bare fallow could be undone.

Rutherglen scientists had developed a new way of growing wheat called 'clover ley' farming. Two years of wheat were grown after a paddock had been under improved subterranean clover pasture for at least four or five years. There was no long fallow before a crop. After the crops, subterranean clover re-established itself without needing to be resown by the farmer.[38] The rotation of crops and subterranean clover pasture was an entirely new way of running a wheat–sheep farm. Previously the wheat was kept permanently on one part of the farm and native pasture on another. This management method arose from a desire to prevent sheep from spreading weeds onto the cropping paddocks of a farm.[39] Under ley crop rotation there was no longer a division of the farm into sheep and cropping paddocks. A farmer rotated his pasture and wheat across the whole of his farm. The system was critically dependent on the application of phosphate to maintain a strong pasture, but was self-sustaining in terms of nitrogen. While European and American wheat farms were developing an appetite for nitrogen fertilisers, the new ley wheat farms did not need nitrogen fertiliser.

Farmers treat most innovations with caution. Ley cropping was no exception. In a twenty-year period from the end of the Second World War the area of improved ley pasture tripled, but large areas of native pasture still remained on farms.[40] Adoption was accelerated by the Korean War, which helped to push wool prices to record levels. The high prices were an incentive for wheat farmers to sow improved ley pastures to increase the number of sheep that could be carried. Otherwise, farmers adopted ley farming gradually through experimentation and adaptation to local conditions. Elsewhere in the wheat belt it was difficult to apply the Rutherglen methods directly. Rutherglen was a comparatively wet district where subterranean clover grew well. Where the rainfall was lower subterranean clover was unreliable. The first farmers to try ley cropping in the drier Wimmera relied instead on the small hairy medic clover, which sprang up naturally in rested paddocks. They met with little success.[41] Ley cropping in these drier areas became practical with the progressive discovery of earlier flowering strains of subterranean clover, and barrel and harbinger medics, which were more productive in the drier environment.

Farmers in drier districts also discovered that ley rotations could return too much fertility to the soil, causing the wheat crop to grow too lushly in the wet of early spring only to 'hay off' (wilt and die before seed set) with the arrival of hot weather. The solution was to reduce the years of pasture in a rotation to one or two. The need for long fallow had to be reconsidered in each district. Wheat farmers in high rainfall areas (such as around the Rutherglen Research Station) no longer needed to plough long fallow to accumulate sufficient nitrogen to grow a crop. Wheat farmers in dry areas found it necessary to continue fallow. The 1950s had been wet. The 1960s brought the first real drought of the post-war period. The Wimmera experience of this drought confirmed that long fallow was still an insurance against low rainfall. Crops grown on long fallow survived. Others failed.[42] Farmers in the Wimmera also discovered that long fallow was still a good protection against root disease. Barley grass grew in the medic ley pasture, hosting take-all until the next wheat crop. The result was that ley pastures gradually spread across the wheat belt, but long fallow remained common in most areas. Wheat farming settled into a new production system in which cultivation continued to play a major role. Around Rutherglen, two years of wheat were grown after five or six years of pasture. There was no long fallow before the crops. In the Wimmera, farmers incorporated ley pastures in a new five-year rotation system: pasture–pasture–long fallow–wheat–short fallow–wheat. The improved fertility of the soil allowed successive wheat crops to be grown, broken by a short fallow between harvest and the next sowing.[43]

The new ley farming overcame the problem of nitrogen depletion on cropping lands. The Australian wheat yield per hectare again began to gradually increase. Ley farming, together with a rejection of ploughing for a fine soil mulch, also brought an end to the worst excesses of the depression dustbowl, but ley farming had not eliminated the practice of bare fallow. While ley farming was making it easier for some farmers not to bare fallow, other changes were making it easier to continue fallowing. Cheaper and more powerful tractors made ploughing fallow easier. After an initial cultivation, it was much less demanding to go out and cultivate after every shower of rain.[44] As tractors became larger, more comfortable and more powerful, cultivation became less of a chore. With the advent of cabin radios, citizens' band radio and air-conditioned cabins in the following decades, ploughing became comfortable; a far cry from the dusty work behind the horse team. Extension officers later coined the term 'recreational ploughing' to describe the satisfaction of working large areas of land in comparative comfort, away from the house and in contact with fellow cultivators on the radio.

While wheat farmers continued to use bare fallow in an unreliable and extreme climate, the risk of serious episodes of soil erosion continued. Some observers realised that the erosion control and productivity promise of ley farming could not be fulfilled unless greater attention was paid to the maintenance of soil structure. In the early part of the century, farmers had been exhorted to value their soil as a living organism. Soil scientists continued to propagate the message in following decades. On farms this biological view of the soil had begun to look like mystical hocus pocus. Wheat farmers needed to return to an appreciation of soil as a living organism, rather than as a repository for water and mineral elements. Regular ploughing was still harming the structure of the soil, limiting the effectiveness of the clover ley system. Some ploughed soil surfaces were crusting after rain. Constant pulverising by implements and compaction by tractors and sheep was creating hardened layers of soil (hard pan) just beneath the maximum depth reached by ploughing implements. The crusted surface hindered both the infiltration of water into the soil and the emergence of seedlings out of the soil, forcing the farmer to plough the land again. The compaction reduced the soil's water storage capacity. Hard pans reduced the root depth available to plants and limited soil drainage, creating waterlogging and flooding in winter and premature dry soil conditions in summer.

Conservation cropping

The obvious mechanical solution to soil hard pans and compaction was to mechanically loosen the soil. In the 1950s P. A. Yeomans advocated soil loosening using a form of ripping. In later decades agricultural scientists also explored the worth of deep ripping. Scientists at the Rutherglen Research Station demonstrated that the soil could be successfully loosened with a deep ripper. However, their work also showed that if conventional cultivation continued after deep ripping, hard pans soon redeveloped. Sustaining soil structure required more than deep ripping. There needed to be an alternative to ploughing. The solution came from an unlikely source: the chemical industry.

The first organic herbicides appeared on Australian farms in the 1930s.[45] Kerosene was used with limited success to control weeds in vegetable row crops. The shortage of farm labour during and immediately after the war stimulated the search for improved herbicides. During the 1960s and 1970s a wide range of new herbicides appeared. Initially the herbicides were expensive and had limited efficiency. With time, the price declined and the effectiveness and specificity of the herbicides improved. Today broad-spectrum contact herbicides kill plant tissue to which they are applied. They can be used to remove annual grasses. The broad-spectrum systemic herbicides are absorbed into plants and kill both the plant top and the roots. They are effective against annual and perennial weeds. Residual herbicides inhibit seed germination. Specific herbicides kill only certain types of plants, leaving others unharmed.

The new herbicides offered a chemical fallow alternative to the ploughed bare fallow. The crucial element of effective fallow was complete weed kill. This could be achieved with systemic and residual herbicides. Not only did this reduce the need for ploughing, but it helped to control wind erosion by leaving a covering of dead grasses to hold the soil surface together. In the United States the opportunities afforded by the new herbicides were developed into a new system of wheat farming called 'conservation cropping'. In the 1970s the high price of fuel, and therefore ploughing, encouraged American farmers to plough less and spray more. In Australia a similar development was encouraged by an alliance of soil scientists, soil conservationists, private consultants and chemical companies, adapting the foreign system to suit local conditions.

The key to conservation cropping is a substitution of herbicides for ploughing. Farmers who decide to use conservation cropping can operate at any of three stages of commitment. The first stage is *minimum tillage* farming. Any fallowing is achieved by use of herbicides. Pasture is killed and left till the autumn rains herald the sowing season. The ground is then cultivated once or perhaps twice to form a seed bed and to control weeds prior to sowing. After harvesting the wheat stubble is burnt. Minimum-tillage farming reduces both the impact of ploughing on soil structure and the period the soil is left exposed to wind or water erosion.

In *direct drill* farming all tillage is eliminated. A farmer burns the wheat stubble before the autumn rains and sows his seed directly into the unploughed paddock after the autumn break. Weed control is achieved entirely with herbicides. Direct drilling aims to eliminate both the risk of erosion and the impact of ploughing on soil structure by reducing the number of times heavy tractors and implements cross paddocks.

The most complex form of conservation farming is *trash* farming. The farmer eliminates the task of burning the stubble and crops are drilled directly into the soil while the stubble is still standing. Sheep may be grazed on the stubble crop before sowing. Trash farming aims to gain all the benefits of direct drilling as well as the added benefit of increasing soil organic matter by retaining the wheat stubble to break down in the soil. Erosion risk is reduced by the standing stubble.[46]

While soil conservationists were worrying about maintaining the *living soil* with conservation cropping, agronomists were pondering a different conundrum. Wheat crops were not achieving the yields they considered were theoretically possible. Casting aside rhetoric about the living soil, one scientist sterilised some soil and doubled the wheat crop in the following year.[47] Obviously not all the organisms in the living soil were beneficial to the wheat crop.

Conservation cropping — drilling seed for a new crop into a standing stubble

Farmers had long known of the fungal root disease take-all or deadheads. Take-all causes blackening of the roots and premature ripening of wheat before the head fills. The result can be major crop losses. Wheat, barley and most grasses are hosts for the disease. Growing wheat crops one after the other leads to an explosion in the population of the fungus. Growing wheat after barley is similarly foolhardy. Over many years farmers learnt the best method of control was to keep paddocks free of grasses and susceptible cereals long enough for the take-all fungus to die. They did this by growing a crop of oats before the bare fallow in a wheat–oats–long fallow rotation. Oats are a poor host for take-all. This rotation included a two-year break without a disease host.

After the Second World War the clover ley pastures took the place of oats in many crop rotations. The early clover ley pastures were often sown in nitrogen-poor land and were dominated by clover. Clover is not a host for take-all. Where fallow remained in the rotation the clover and fallow provided a two-year break. As the clover ley rotations gradually returned the depleted nitrogen reserves in the soil, grasses began to predominate in the ley pastures and the pasture phase of the rotation was transformed from a disease break to a disease host. With the trend away from bare fallow and cultivation, the conditions were set for a re-emergence of root diseases: take-all, cereal cyst nematode (commonly known as eelworm) and *Rhizoctonia*, a fungal root disease of wheat, which colonises grasses (including wheat) grown in alkaline soils.

The threat of the root diseases take-all and *Rhizoctonia* provided a persuasive reason for some farmers to continue with ploughed fallow where they might otherwise have adopted conservation cropping. With fallow there was a disease break of one year. The grass pasture that replaced fallow did not provide a disease break. *Rhizoctonia* is a particular problem for direct drilling. A two-year break is not sufficient to kill the fungus. It must also be disturbed by ploughing. The threat of root diseases has spawned the growth of new farming techniques for pasture management using selective herbicides to kill grasses in clover pastures. Root diseases have also provided an incentive for farmers to grow grain legumes.

Grain legumes are crops that combine the grain productivity of wheat with the nitrogen-fixing abilities of clover. The fertility benefits of grain legumes, such as field peas and lupins, were promoted to farmers in the early 1900s.[48] Some growers tried to integrate them into their rotations. Farmers found them too risky; they were prone to attack by diseases and pests and the markets were unreliable, so few farmers grew them. During the late 1960s Australian wheat production was controlled by a quota system because of low world wheat prices. Most wheat farmers ran more livestock to make up for lost wheat income. Some grain growers grew alternative crops such as oilseeds and grain legumes. The renewed interest was made possible by new chemical herbicides and insecticides. Powerful insecticides were available to control previously uncontrollable pests such as the field pea weevil and heliothis. The greatest advantage of grain legumes was not that they fixed nitrogen, but that they were not grasses. Selective herbicides allowed growers to spray their grain legume crops to kill grasses, something that could not be done in a wheat crop. This allowed grain legumes to be used as a disease break crop to control root

New lupin crop emerging from a wheat stubble

disease and to reduce weeds in following wheat crops. They had the added advantage of deep tap roots, which could break through hard pans and help to improve soil structure.

Because of the opportunities provided by these innovations, the standard clover ley rotations began to change as farmers mixed crops according to the requirements of their paddocks. The simple pasture–pasture–wheat–wheat–oats often became pasture–pasture–wheat–lupins–wheat. Fixed rotations began to disappear. In any year, farmers had more options to consider; wheat farming became a complex management challenge. In making a decision about what to plant in one paddock, in one season, a farmer may need to work through fourteen major decisions.[49] Computer simulations are now used by farmers and consultants to work through the complexities of these decisions and their interactions.

Perplexing problems

Our quick traverse across the history of the Australian wheat industry has revealed a long search for systems of sustainable wheat production. The search has not ended. The wheat industry must continue its quest for a sustainable basis to its production; it faces perplexing problems and exciting possibilities. *Rhizoctonia* is a serious challenge to conservation cropping. One solution may be in simple mechanics. Research has shown that certain types of direct-drill machinery can reduce the risk of *Rhizoctonia* in 'no-till' crops.[50] Another possible solution is in biological control. A sterile red fungus can infect wheat roots without causing any damage, stimulating root growth and secreting a protein that attacks other fungi. It may be possible to inoculate wheat seeds with the fungus to prevent infection by *Rhizoctonia* and take-all. This would be a simple tool to help farmers control root diseases.

Increasing soil acidity in some of the higher rainfall wheatlands has been a growing problem. On a cropping farm, as acidity increases, the variety of crops that can be grown decreases, leaving some farmers with little choice but to grow triticale. Continuing acidification also destroys soil structure. Continued nitrate leaching does not cause acidity to increase indefinitely. Once the limit of acidity is reached, continued nitrate leaching leads to a crystalline transformation of the soil structure and a gradual extension of the acidic layer into the sub-soil. Even more serious is the poor growth of ley pasture in acid soil. Without effective ley pasture, nitrogen levels will not be maintained and soil fertility will decline. Short-term solutions may be found with acid-tolerant legumes like *Medicago murex* but the only long-term solution will be the regular application of lime. The only effective method of applying lime to pastures is by ploughing it into the soil. It seems we cannot avoid the plough.

In drier sandy districts the ley pasture system has exacerbated an entirely different problem. In Western Australia and South Australia sandy soils are being transformed into water-repellent soils by water-repelling

organic compounds created by ley pastures. Farmers faced with this problem will need to plough to incorporate wheat stubble residues and wetting
agents into the soil. Another solution may be to lower the fertility of soils
by reducing the use of legumes in rotations. But where the use of legumes
has been reduced other problems have arisen.

Wheat farmers now face claims of declining protein content of Australian wheat.[51] Protein is a crucial component of wheat. Hard wheats, which
have a high protein level, have good baking qualities. They are used in
white bread. Low protein, or soft wheats, do not have good bread-baking
qualities and are used for biscuit dough. Because bread and pasta are the
major uses for wheat, there is a price premium on high-protein wheats.
Low-protein biscuit wheat commands a lower price and is harder to sell. In
many regions of Australia the protein content of wheat has been declining
over the past two decades. The major cause of this decline is lower soil
fertility caused by intensive cropping and shorter rotations with fewer
years of ley pasture.

The dependence of modern wheat farms on chemical sprays looms as
another potential problem for the wheat industry. By European standards,
Australian wheat farmers have been conservative users of insecticides,
herbicides, fungicides and artificial fertilisers. Extensive wheat farms and
ley farming methods have ensured that the use of nitrogen fertilisers is
almost unknown in the wheat belt.[52] The trend towards reduced-tillage
farming, encouraged by chemical companies, and the adoption of conservation cropping has increased the dependence on herbicides. A wheat crop
may now be produced using two or three sprays: a herbicide to kill weeds,
insecticides to control pasture pests and a nematicide to control eelworm.

The trend towards greater chemical use is more obvious for grain
legume crops. Lupins will often be sown after a contact herbicide, and a
systemic herbicide may be applied to control grass weeds. The seed will be
coated with a fungicide before planting for protection against leaf-spot
disease. Earthmites and fleas, loopers, heliothis or cutworms may have to
be controlled by chemical sprays.

Viral diseases, like cucumber mosaic virus, are spread by aphids. Such
viral diseases are prevented by spraying with an aphicide. Farmers may also
spray selective herbicide to control both weed competition for the lupins
or field peas and root disease in following wheat crops.

Dependence on chemicals raises fears about the long-term sustainability
of this high-input agriculture. Fungicides, herbicides and pesticides do not
offer a permanent solution to pest, weed and disease problems. Often,
resistance to chemicals develops, with the initial response being to increase
the application rate, hastening the development of full resistance. Some
weeds are developing resistance to commonly used herbicides. Chemical
agriculture requires the continuous development of new chemicals before
each cycle of resistance is complete. Wheat breeders have been racing the
resistance cycle since the beginning of the century as their wheats gradually succumb to disease.

A more controversial aspect of chemical agriculture is the question of
chemical residues in either the soil or in food. This is a difficult area in

which to maintain rational and unemotional debate. Ardent supporters on both sides of the debate find it impossible to understand or accept the other side's point of view. In wheat cropping the debate is focused on two methods of farming championed by the self-proclaimed 'conservation cropping' and 'sustainable agriculture' movements. Both groups stress the importance of a healthy living soil, but promote opposing means of achieving this goal. The conservation cropping approach seeks to protect and maintain soil fertility and structure by reducing the use of cultivation. The sustainable agriculture (organic farming) approach seeks to achieve this goal by eliminating artificial chemicals from farming systems. Adherents believe this will protect both soil and consumers. This leads to a more traditional style of grain growing using older rotations such as pasture–fallow–wheat–oats. Crops dependent on insecticides are avoided. Weeds are controlled by ploughing in pastures as green manure crops before they set seed. The paddock is then left unploughed in a fallow for several months through summer until planting. Weeds that germinate after rain are controlled by selective cultivation. Pastures and wheat are fertilised with ground rock phosphate rather than processed superphosphate. The preference for rock phosphate is based on concern that the sulfuric acid used to manufacture superphosphate will damage worms, bacteria and other life in the soil.[53] Stubble is grazed and then lightly cultivated to form a mulch on the surface and protect the seedbed of the next crop. Ploughing is an integral component of this style of farming, but because organic wheat growers run their farms on longer pasture rotations, the extent of ploughing over a number of years may be little more than on a conventional minimum-tillage farm.[54]

Organic wheat farmers claim to have created a profitable and sustainable farm business. They achieve lower yields because of their weed control, phosphorus and pest control choices, but their costs are also lower.[55] Their products are sold at a significant premium, sometimes up to double the standard wheat price. In this light, organic farming has appeared increasingly attractive to many farmers, although the price premiums would not be maintained if there were a large increase in organic production. One of the barriers to entry to organic farming is the period of initial adjustment as chemicals are removed from the farm system. Proponents of sustainable agriculture maintain that it takes time for a farm ecology to adjust to the loss of chemicals. During this weaning period there may be financial difficulties. Conventional farmers with high interest commitments are sceptical of even the short-term financial sustainability of such a change. Others are sceptical of the sustainability of the current high prices for organic wheat.

Both the sustainable agriculture and conservation farming approaches have their ardent spokespersons claiming increased soil organic matter and increased earthworm activity as evidence of the success of their systems. Farmers are suspicious of the mystical elements associated with some organic farming systems and are sceptical of crops supposedly grown without superphosphate on properties where previous fertilising has built up a residual supply in the soil capable of sustaining legumes for five years.

Farmers are also sceptical of the use of chemicals and herbicides associated with conservation farming.[56] Bad experiences with residual herbicides and the discovery of organochlorine residues in Australian beef have sensitised wheat farmers to residue dangers associated with herbicides. There may well be some reason for concern.[57]

Perhaps the greatest cause for concern will come not from insecticides and herbicides, but from the use of superphosphate. The historic opposition of organic farmers to superphosphate may yet be proved correct, but for the wrong reasons. Cadmium is a toxic heavy metal element that accumulates in fatty tissues of the body, much like lead and mercury. Cadmium occurs naturally in some soils. In the United States and Europe, soils naturally high in cadmium have been taken out of agricultural production as a safeguard against contaminated food production. In some countries, where legal limits for cadmium in food have been set, monitoring of cadmium contamination of food has begun. This monitoring may affect future sales of Australian wheat.[58]

The contamination of wheat by cadmium is linked to the use of superphosphate and the pasture revolution in post-war farming. In native unimproved soils, cadmium is generally found at levels between 0.01 and 0.05 parts per million. Higher levels occur in some areas. In alkaline soils, high cadmium levels do not contaminate wheat because the cadmium is unavailable. Increasing soil acidity makes cadmium in soil more available to plants. Besides helping to develop high organic matter in soils, which leads to increasing acidity, superphosphate has added cadmium to the soil. This is not because the superphosphate is an 'unnatural' compound that has been processed with sulfuric acid, but because it is drawn from guano phosphate. Cadmium salts are present in sea water in greater concentrations than in most soils. Guano phosphate deposits are the residues of thousands of years of sea-bird droppings. Sea birds concentrate cadmium in their droppings, so the guano contains concentrated levels of cadmium. Most of Australia's supplies of phosphate have come from the ocean islands of Nauru and Christmas Island, which were once covered with guano phosphate.

It is unclear to what extent cadmium contamination of wheat is due to contaminated soil or to contaminated wheat because of transport in vehicles that have been previously used to transport superphosphate. As the supplies of island guano phosphate are running out, Australia is now importing low-cadmium rock phosphate and may have to develop its own rock phosphate deposits in the next decade. Superphosphate manufactured from this base material will be low in cadmium. In the meantime, we may have been left with a legacy of cadmium in the soil. This provides another reason to control increasing soil acidity.

On the basis of these problems we can conclude that the aim of a sustainable wheat industry has not yet been achieved. What are the prospects for sustaining the wheat crop? Judging by the complexity of the problems it is reasonable to question whether permanently sustainable cropping systems can ever be achieved. Wheat farmers must balance on a narrow path between the many competing constraints. The relative importance of the competing influences will change with each district, between farms and possibly even between paddocks on the same farm. Permanently sustainable crop production is likely to be a goal that is never quite achieved. It is more realistic to view sustainable cropping systems as a series of temporarily sustainable states.

7

Selling sustainable cropping

The small town of St James in north-eastern Victoria is set in rolling plains dotted with the remnants of eucalypt and native pine woodlands. St James has one major claim to cultural significance. The first of the G. J. Coles chain of stores was established there. The St James skyline is dominated by a wheat silo, as are most railway sidings in this wheat producing district. St James does not have the low rainfall of many wheat growing areas to the west. Dry years elsewhere in the southern wheat belt are often bumper seasons in north-eastern Victoria. Because it rains more often, the soils are not alkaline but tend to be naturally acid. This acidity confers some advantages. Subterranean clover has grown well in the pasture phase of crop rotations. The root diseases take-all and *Rhizoctonia* are not as well adapted to the local acid soils as they are to alkaline soils farther west. This makes direct drilling easier to sustain in the long term. Lupins grow well, enabling farmers to gain the benefits of a legume to break the root disease cycle; they also improve the soil structure and add soil nitrogen. The soils are not sandy, so there is little risk of wind erosion.

Few silver clouds are without a grey lining. In the north-east of Victoria too much rain is more often a problem than too little. Heavy rain early in autumn waterlogs the land and prevents farmers driving their tractors on to their fields to sow wheat. The acid nature of the soils has been exacerbated by the subterranean clover pasture system, so that for some paddocks the limit of safety has been passed, and only acid-tolerant crops such as triticale may be grown successfully. The soil particles, far from being too large as in the sandy Mallee, are sometimes far too small to cope with traditional agricultural practice. Depending on the locality, they form hard pans, crust or set solid if ploughed when too wet. In this chapter we follow the story of scientists, advisers, commercial companies and farmers seeking to develop and promote sustainable cropping on these soils. While the methods of sustainable cropping appropriate to this area may have limited application elsewhere in Australia, the lessons learned in trying to promote these practices have relevance far beyond this part of Australia.

Developing scientific credibility

Launching a major campaign to increase application of lime in 1912, the Director of the Victorian Department of Agriculture observed: 'It has become almost common place to say, in respect of a large proportion of land in this State, that it would be better for a dose of lime'.[1] Sustainable agriculture was very much on the minds of the scientists of the department. Wheat farming was emerging from a period of extreme exploitation when farmers had continuously cropped land without the replenishment of any nutrients. The scientists believed the widespread use of superphosphate to replace the minimal reserves of soil phosphate that were carted away with every bag of wheat was a step towards a more sustainable cropping system. They also believed wheat growing would become more sustainable if farmers also replaced the lime removed with every harvest.

Farmers in Europe regularly applied lime to their land without looking to reap an immediate profit. Their practice was based on a tradition of maintaining the soil for following generations. Lime pits were common in Europe's rural landscape. Lime pits were not common in Australia, partly because there are few lime deposits close to agricultural areas and because the application of lime was almost universally ignored. The Department of Agriculture scientists believed that continued harvesting of crops would remove alkaline products, making the soils more acid, 'souring' and 'stiffening' the soil.[2] Measurement of the 'lime content' of the soils of north-eastern Victoria had shown these were among the most lime-deficient, or acid, soils in south-eastern Australia. The scientific advisers believed regular applications of lime would help to maintain the fertility of these soils by improving soil structure and liberating potash and phosphorus. The government advisers embarked on a major publicity campaign to promote liming as a regular agricultural practice.

The first rule of marketing is to understand the consumers. One contemporary scientific adviser described how the farming community was unconvinced of the value of lime:

> Hitherto the use of lime in this State has oftentimes been regarded by many practical agriculturalists as an expensive luxury . . . incredulity has been expressed concerning its benefits as a regular means of maintaining or increasing soil fertility.[2]

The scientists believed the farming community's inaction was because of confusion arising from conflicting information about liming. Their strategy to overcome this problem was simple: provide a definitive statement of the value of lime and rely on the credibility of scientific authority. Reading the writings of the advisers we can see that the scientists assumed that the farming community shared their own values, particularly a respect for the scientific adviser:

> The application of lime has been advocated in a light and airy way by many advisers, but all sorts of confusing advice as to the quantities, periodicity of

application, and the like have been given. Farmers, however, with that conservative wisdom which is sometimes charged to their detriment, but which is really their abiding safeguard against irresponsible advisers and wasteful expenditure, have been loth to act on the exhortation of other than dependable investigators who can advance sound research and scientific proof for their guidance.[3]

The Department of Agriculture produced a monthly publication, the *Journal of Agriculture*, for farmers and their families; it offered advice for a wide range of agricultural industries. In 1912, the October issue of the *Journal* was committed entirely to the question of liming. One article described the chemistry of lime, another the relationship between lime and soil fertility. Other articles considered liming for vineyards, tobacco fields, orchards and potato farms. The *Journal* gave the locations of all significant lime deposits in the state. To underline the gravity of the lime-depletion problems farmers were creating for the future, the issue was introduced by a five-page appeal written by the Director of Agriculture, who promised to initiate long-term research into the value of lime. An experimental farm was to be established on unused land on the State Viticultural School near the town of Rutherglen in north-eastern Victoria and the possibility of subsidising the cost of freighting lime on the railways was canvassed.

The high hopes held for the 'Lime in Agriculture' campaign were not realised. We have no measure of the use of lime by farmers in the years that followed, but it is clear from the then current traditions that there was little change in farm management. The high hopes for the research program were also unrealised. The results showed that earlier confident assertions on the value of lime applications were not founded on scientific understanding, but on hope and emotion. Six years of heavy liming of Rutherglen pastures produced only minor increases in yield.[4] In 1923 farmers were told that with yield increases of the order of only one bushel per acre, there was no incentive to cart lime any distance from a pit. The early 1912 campaign had failed because the basic scientific research had not been done. At that time there was no economic return from uncritically applying lime across the district. The marketing strategy in the campaign was also flawed. The campaign was based on the assumption that scientific advice by itself would be sufficient to encourage farmers to change the way they managed their farms. The concern about acid soils would become a more substantial problem for a future generation of researchers.

The campaign may have failed, but it left a valuable legacy: the Rutherglen Experimental Farm. The first director of the new Rutherglen farm made clear his intention to help local farmers develop more sustainable farming practices:

... there are systems of farming in practice which have already depleted many soils of their virgin richness and which threaten to further deplete them below the limits of profitable production. It must be obvious that with constantly increasing population and diminishing productive power of the soil, the time must soon come when our national welfare will be threatened unless provident methods of cultivation are followed. The conduct of systematic, rationally conceived, permanent experimental plots will be a most invaluable medium in unfolding those systems of farming which not only provide for maximum crops, but also for the maintenance and increasing of soil fertility.[5]

The key to testing the sustainability of farming systems was a permanent site on which to conduct permanent experiments. Until that time nearly all agricultural research in the state had been taking place in farmers' paddocks — sixty sites across the state. In an age when cars were rare and telephones were unknown, the logistics were complex. Most experiments were limited to simple tests of fertilisers and seeds. Investigation into other matters of farm management relied on the good will and commitment of the property owner. Experiments that strayed from the normal methods of farm management were impossible. Long-term experiments depended on long-term cooperation and had little chance of survival.[6] With a new permanent site at Rutherglen the three resident scientists unleashed a period of feverish activity, making up for all they had been unable to achieve on farm plots. Amid the experiments, pride of place went to a series of long-term crop rotation experiments to test the sustainability of the contemporary crop and pasture systems. These trials were the first steps along a path that eventually led to the development of improved subterranean clover pastures and ley farming. Today those first plots offer clues as to the causes of increasing soil acidity.

The experimental farm offered more than a venue for permanent experiments. It offered a better means of promoting improved farm management. By gathering experiments together in one place, the farm became a showcase to display new techniques and innovations to farmers. The farm was opened to the public for an annual field day and visitors could see all the work of the department scientists without the need to travel between numerous farms hosting single experimental plots. From the start the field days were well attended and high hopes were held that they would prove an improvement over the promotional strategies used in the 'Lime in Agriculture campaign:

> The publication of articles on the growing of cereals and the results of researches carried out give farmers theoretical knowledge of their industry. On the other hand, the bringing of three or four hundred tillers of the soil together, men who have had practical knowledge of farming under the old system — what father did is right for me to do — and showing them the great advantages that are to be gained by bringing scientific work together side by side with practical knowledge, is doing something that in the future will have beneficial results.[7]

Scientists published a guide to the experimental farm for each field day. The extensive guide was reprinted in the *Journal of Agriculture*. On the field day the advisers led a guided buggy convoy tour of the experimental sites. The hope was that this 'seeing is believing' strategy would lead inevitably to changes in farming practices across the district.

Publicising farmers' achievements

Ten years after the 'Lime in Agriculture' campaign, the Department of Agriculture had a new system of soil management for the grain growers of north-eastern Victoria. The department sought to increase the annual wheat yield by convincing farmers to use the dry mulch fallowing methods popular on the Wimmera plains. Farms of the Wimmera with a similar rainfall were achieving higher yields than farms in the north-east:

> While the Wimmera wheat yields have been increased by 169 per cent [in the past 25 years], the yields in the Northern district, including the Goulburn Valley have increased . . . from 10.4 to 13.3 bushels per acre, or by 28 per cent. Do the present averages constitute the limit for the Goulburn Valley? If not, how far are they to be increased?[8]

The higher Wimmera yields were indeed due to the use of dry fallowing techniques, though the comparison conveniently overlooked the fact that the fallowed yields should have been averaged over two years to take account of the unproductive year of fallow. Crop experiments at the Rutherglen Experimental Farm showed improved yield from following Wimmera fallowing methods. The task was to convince north-eastern farmers to follow the methods of their Wimmera compatriots. One method of encouragement had been crucial in fostering the Wimmera interest in fallow: the 'Crop and Fallow Competition'.

The 'Better Farming Train' during the 1920s

In 1923 crop and fallow competitions were introduced to north-eastern Victoria. The best farmers were encouraged to enter a paddock into their local district competition. Departmental advisers judged the crops and the results of each competition were reported annually in the *Journal of Agriculture*, together with a description of the management techniques used. The role of fallow was given prominent place. Competing farmers benefited from the consequent discussion and exchange of opinion with other competitors. The competitions improved the district knowledge of successful farming practices by publicising the successes of the better farmers.

The competitions were an admission that the withdrawal of all research work to the fields of the experimental farm had not been a total promotional success. In an aside in one of his competition reports, H. A. Mullett explained why journal articles and experimental field days had not been as successful as expected:

> ... better wheat growing practices are not often discovered and put into practice unless the rural community co-operates in the work, and the experience generally is that it is easier to make the discoveries than to get the community to believe that they are worthwhile searching for.[9]

The reason for Mullet's diagnosis was the farming community's response to the experimental farm's pasture top-dressing trials. The response of pasture to superphosphate had been impressive, yet the farming community had shown little interest.[10] Clearly the research scientists had not solved a problem about which the farming community were concerned.

The Shepparton, Dookie and Elmore agricultural societies were the first to hold crop and fallow competitions. In a few years other Victorian societies, Rutherglen, Yarrawonga, Numurkah and Wangaratta, followed. High hopes were held for the competitions. In some districts the hopes were realised. In other places the competitions were a failure. The Dookie district was one of the successes. Parts of Dookie are covered by rich grey soil similar to the best soils of the Wimmera. Despite the similarity of soils, Wimmera farms consistently outyielded Dookie farms. Unflattering comparisons convinced a group of farmers from Dookie and nearby areas to take a trip to the Wimmera to see the Wimmera method of farming. The travellers returned and put into practice what they had seen. They began ploughing fallow in March instead of August or September, a full fourteen months before sowing. Using the new Wimmera scarifier, they cultivated after every major rain to maintain a dust mulch and kill each successive generation of germinating weeds. Sowing was delayed as late as possible, with a heavy rate of seeding and dramatically increased superphosphate applications. Local farmers were sceptical at first. The winter was wet and the late-sown crops were 'pugged' into mud rather than soil. While the experiment at first looked to be a failure, the first impressions were misleading.

It will be remembered that the seeding season opened particularly late, and once the rains began they continued almost without intermission until August. Consequently, the crops sown in mid June and July [according to the Wimmera method] had to face exceptionally cold and wet conditions. Nevertheless the pioneers persisted with their efforts. In August the crops were backward but clean. Early-sown wheat at that time appeared as if it would be superior. The Wimmera style crops were the subject of derisive comment. However, they continued to make headway, though the spring was not all favourable to them. By harvest time they were seen to be free from excessive straw development, very dense and well headed, and quite free from wild oats and weeds. The harvester has since demonstrated how well they filled the bag.[11]

The resulting harvest silenced the many critics. The Wimmera-style crops were so heavy the judges had trouble estimating the yield. These crops easily won the crop competitions. In three years Wimmera fallow became the norm in the Dookie area. Paddocks were fallowed for over a year. Seven, eight or even nine cultivations before seeding became commonplace. The farmers who had the highest yielding crop in the 1923 competition somehow managed to fit thirteen cultivations into the year:

> The winning crop of Messrs. Bennett Bros. was . . . treated as follows. The paddock was summer fallowed in March with a disc cultivator to three inches deep. In August the land was reploughed to three inches. In September it was harrowed, in October spring toothed and harrowed. After the harvest a stroke of the harrows was given in the first week in January, and another in March to fill up the cracks that appeared. After rain in May the harrows were again used, then the paddock was scarified in June and twice harrowed. Seeding commenced in the second week in July. After . . . two strokes with the harrows were given.[12]

All this was done by horses and man as tractors were yet to be used on farms.

Around Shepparton, Wimmera fallowing was less successful. In local parlance, the fragile red-brown earths of the district were described as 'cementy' — they set like cement when mistreated. After the first Shepparton crop and fallow competition, the judge warned of the dangers of dry ploughing, which could lead to soil compaction. Despite this, he was optimistic that 'far better results would have been obtained if all farmers had used superior methods' — 'superior methods' meant better fallow.[13] A year later the advice to farmers was much more cautious. The Wimmera methods could not be applied to these soils without considerable modification. No one knew what these modifications should be. This wisdom had been won at some cost to the adventurous local farmers who had taken the fact-finding trip to the Wimmera with their Dookie neighbours. Fired with enthusiasm, they, like the Bennetts, must have spent half the year behind the plough. Their efforts would have earned them the same initial derision, but without the final vindication of higher wheat yields:

> Some Goulburn Valley farmers, in a burst of enthusiasm following a visit to the Wimmera, worked their fallowed land over so often that it was reduced to a dust

like powder, which in a season like the present was apt to be puddled down flat with the heavy rain. Consequently the germination has not been as good as it should have been, while the tillering has been unsatisfactory and the growth of the plants stunted. On the other hand, farmers who merely left the land in a comparatively rough state had more satisfactory results. Indeed, in a few instances in comparatively new land fairly good crops have been grown on rough ploughed stubble land.[14]

Rough ploughing, which today would be called minimal tillage, proved superior. The winner in the 1922 competition cultivated four times in the season, and the second place-getter cultivated only twice. Both competitors avoided any form of spring cultivation.

A short distance north in Numurkah the newly instigated crop and fallow competition highlighted different problems. The light silty soil did not set like cement bricks. Instead, ploughing helped to form a solid surface crust, limiting the soil's capacity to absorb water. Local farmers continued to cultivate and fallow, but modified their ploughing practices, rejecting disc ploughs, which turned the soil and brought fine soil particles to the surface, opting instead for scarifiers, which killed weeds but left the soil surface relatively undisturbed. There were good reasons for continuing with fallowing and burning stubble, even if it did risk crusting or cementing the soil. Farmers needed to plough at some time to prepare a weed-free seed bed in which to sow the crop. Early fallowing allowed farmers to plough when they had the time, rather than immediately before planting when timing was crucial to getting the crop in before the soil became too wet. Ploughing was also the only short-term remedy where land had cemented or crusted. Farmers also believed ploughing helped to control far more serious problems. Smut and take-all fungi raged in the local crops. All crops were affected to some degree, and some crops were damaged considerably. The recommended control was a combination of stubble burning and fallow:

> In combating [fungal disease] the object is . . . the destruction of seeds or spores of the fungus . . . A thorough burning of the stubble will destroy many of them. This should be followed by a very thorough cultivation, ensuring all unburnt stubble is turned to rot, and that the soil is kept moist and mellow. This condition encourages the germination of seeds of the fungus, and having no host plants in which to live, they are starved out. Where possible a rotation of crops is also useful in suppressing the disease . . . The methods recommended for the suppression of flag smut also apply to [take all]. The fungus carrying the disease can live on several of the grasses prevalent in wheat country, and thus the disease is carried on. The fallacy of giving land affected with take all a spell under grass is thus understood.[15]

Fallowing had its dangers. Advisers changed their rhetoric from enthusiastic support of Wimmera fallow to gentler exhortations to farmers to find a balance between the needs of soil and the advantages of fallow. The balance varied from area to area. Often it meant minimising spring cultivation or using sheep to control weeds. Disc ploughs, which turned the soil over, were replaced by tyned cultivators and Wimmera scarifiers. Pasture rotation to control 'plough sickness' was balanced against the risk of take-all. A consensus developed among farmers that fallowing increased yields for the loss of a little grazing, but the loss was at a time of year when there was plenty of grazing anyway. The crop and fallow competitions helped to develop this consensus and to identify the least destructive method of fallowing in each district. As a tool for changing farming practices, they proved a great advance on the earlier liming campaign. However, the competitions were not the perfect tool. Their greatest fault was failing to identify farming practices that did not work. Farmers were more willing to share their successes than failures. If a crop was less than satisfactory, farmers would withdraw it from the competition rather than display a failure to their peers and read of it in the *Journal of Agriculture*.[16] Judges pleaded for all crops to be entered to allow others to share the valuable lessons of failure, but few farmers were willing. While some farmers may have learned by experience that gypsum was not effective in protecting a soil from the damage of fallow, the lesson was only widely publicised after experiments with gypsum at the experimental farm. The competitions did not reveal this problem.

Fallow did nothing to increase the sustainability of wheat farming. At Dookie, where the fallowing had seemed so successful, the practice soon led to some of the worst erosion in the state. Many decades later fallowing has also left a legacy of soil salting. The continuous mining of soil nitrogen remained a concern of scientists of the Department of Agriculture:

> Over a large area of Australia, soil fertility is declining owing to the effects of exhausting crop rotations and faulty cultivation . . . Even where cultivation has been carried out to the best advantage and the crops adequately manured, this deterioration goes on . . .[17]

The incorporation of legumes into wheat rotations was the only long-term answer. In 1937 Rutherglen scientists cultivated and sowed wheat on land that had been a subterranean clover pasture plot in an experiment to compare the yield of crops grown on land that had been under clover pasture with the yields on continuously cropped wheat land. In one season it was clear the subterranean clover had dramatically increased the wheat yield on the new plot.[18] In later research it was shown that fallowing made no difference to the yield of a crop grown after a period of clover.[19] On the basis of these spectacularly successful experiments the experimental station staff advocated two revolutionary changes to the accepted methods of wheat farming. One was ley farming — the abandonment of the strict division of farms into cropping and grazing paddocks in favour of the rotation of cropping and pasture across the farm. The other was short fallow — the abandonment of long fallow in favour of a shallower cultivation in the late summer or autumn immediately before the sowing.[20]

The first step in promoting this new style of farming was to show the results of the successful experiment at the popular, and locally influential, Rutherglen Field Day. The practice of ley rotation took a remarkably short time to spread from the Rutherglen Experimental Farm to farms around

Rutherglen.[21] Though the field days were obviously very influential around Rutherglen, they could not be relied on to promote ley rotations and short fallow throughout the region. Those who did not see the crop at the field day may have read of it in the 1940 *Journal of Agriculture*.[22] The Rutherglen scientists knew, on past experience, this would not be sufficient to promote the practice. Two organisations tackled this task, again using farm competitions. The Agriculture Department initiated a new District Crop Championship. The newly formed Soil Conservation Board and the State Rivers and Water Supply Commission initiated another competition, the Hanslow Cup.

Harold Hanslow donated a perpetual trophy to the Soil Conservation Board in 1941 to support a competition in the Goulburn River catchment. The aim of the competition was idealistic: 'putting the farm on a practical basis of permanency' or 'maintaining the land in profitable and permanent production'.[23] Today it would be called sustainable agriculture. The priority was erosion control. In the north-eastern cropping areas the summers were dry, but heavy summer thunderstorms often caused erosion on fallowed land with depleted organic matter. The State Rivers and Water Supply Commission wished to encourage soil conservation in the catchment of the Goulburn River to protect reservoirs from silting and to reduce flood damage in the lower river irrigation areas.[23] The commission believed that erosion would be dramatically reduced if farmers built grassed waterways and contour banks and used contour ploughing and adopted ley rotations, as well as eliminating long fallow in favour of autumn fallow. The judges for the cup awarded points for the adoption of various erosion control measures on the farm.[24]

The Hanslow Cup was not a complete success in encouraging erosion control in the catchment. It encouraged individual effort, but did little to encourage group action. Erosion control based on contour banks, grassed waterways and gully stabilisation needed to be implemented across property boundaries.[25] The competition was not popular with cropping farmers because it was perceived as promoting the erosion control goals of the State Rivers and Water Supply Commission rather than the goal of improved production shared by most cropping farmers.

In contrast to the Hanslow Cup, judging of the District Crop Championship was based on the more popular goal of growing a good crop. Promoters of the competition were concerned about the unsustainable nature of current cropping practices, but they realised that concern about soil fertility and erosion alone was not enough to encourage most farmers to adopt ley farming and autumn fallow. Two advisers put the matter succinctly:

> Methods are usually dictated by the immediate need for revenue rather than considerations of fertility maintenance leading to a larger revenue over a period of years.[26]

Ley pastures allowed farmers both to meet the immediate financial need and improve the soil, by maintaining soil fertility and by increasing production with higher crop yields as well as dramatic increases in fat lamb production and the control of wild oats. These benefits were used to promote ley pastures.[27] Farmers who adopted ley pastures for these reasons averted or delayed some of the dangers of long-term depletion of soil fertility and organic matter. In 1946 the advisers reactivated the moribund crop competition as the Northern Region Crop Championship. Agricultural societies across the north-east ran their own crop competitions and the winning crop was entered in the district championship. Crops were judged on yield, purity and freedom from disease. The advisers correctly anticipated that crops grown on ley ground would win most local competitions. Five agricultural societies entered their winning local crop in the second championship. Four crops had been grown following a ley pasture.[28] The judge of the competition commented that ley rotations were becoming a standard feature of district cropping. The championship was an advertisement for ley rotations. The exception was the Yarrawonga Competition in the north of the North-East Region. Here rainfall was lower and subterranean clover did not grow well. Ley paddocks grew only volunteer clovers and the results were not rewarding.[29] The success of ley rotation would not come until the selection of earlier maturing subterranean clover strains, which were able to set seed in the drier climate.

There were major short-term and long-term benefits from abandoning long fallow. The short-term benefit was the elimination of the unproductive year of fallow. It took four years to harvest two wheat crops in the fallow–wheat–fallow–wheat rotation. Under the Rutherglen system the two wheat crops were grown in successive years and then the paddock was returned to a productive pasture. The long-term benefit was the reduced risk of erosion.

Despite these advantages, convincing farmers to eliminate long fallow was a more difficult task than promoting ley cropping. All but one of the winning crops in the 1947 crop competition were sown after a period of ley pasture followed by long fallow beginning in September. These crops would receive between five and six cultivations before sowing.[30] The exception was the crop that won the local Wangaratta competition. It was grown without the normal long fallow, the first of five cultivations delayed till January. The judge commented: 'the fact the crop was grown without normal fallow was of interest'.[31] Over the next decade short fallow became more common here and around Rutherglen.[32] It did not spread further. To the west and south-west of Rutherglen, in the Nathalia, Yarrawonga and Dookie districts, long fallow was practised throughout the 1950s.

Although the crop championship was a good vehicle for promoting ley rotations it was not for promoting short fallow. Its emphasis on a single year's yield gave a flattering view of the success of long fallow crops. Long fallow crops won the championship seven years in succession between 1948 and 1954. A competition based on yield or profitability over a whole rotation may have shown very different results. One advantage of the crop competitions was to take the scientists into the fields to see the problems faced by farmers at first hand. The judges could see that farmers were wary of abandoning long fallow and deep cultivation because this seemed to increase the short-term risks of farming. Fallow was still seen as an insurance against drought. Also, research at Rutherglen showed that in dry years deeper ploughing gave approximately double the yield of a paddock given shallow cultivation.[33] Short fallow also had its risks in a wet year. Farmers who delayed their preparation until the autumn 'break' risked missing the chance of sowing the crop if heavy rain came early and the soil became too wet. The long fallow spread out the work of ploughing over the season and ensured the soil would be prepared for sowing whenever the 'break' arrived.

Demonstrating cropping practices on farms

Farmers in the Yarrawonga district found the string of local crop championships won with long fallow techniques more reassuring than the results of experiments conducted on the more distant Rutherglen Research Station. In 1955 the agricultural research scientists decided to take their experiments and demonstrations back to the farms. They began with a large clover ley experiment and demonstration on a Yarrawonga farm.[34] Unlike experimenters working on farms at the turn of the century, the scientists were not limited to simple experimental work. This time they planned to demonstrate a complete farming system.[35] This was made possible by the improving transport infrastructure of the country: better roads and better trucks. The logistics were still far from simple. Instead of merely supplying a farmer with new seed or fertilisers and advice, the research scientists had to transport farm machinery to the experimental block whenever work was required.[36]

The difficulties with farm demonstration blocks helped to precipitate important changes in the organisation of research. From 1912 the setting of Rutherglen research priorities had been the responsibility of scientists in Melbourne. By the early 1960s farmer organisations had formed research committees. The Wheat Industry Research Committee provided funds for the Rutherglen Research Institute (previously named the Rutherglen Research Station) to purchase equipment and a truck to continue its work on farms, thus ensuring paddock demonstrations remained part of the research methods of the region.[37]

An enduring difficulty of the short fallow was the risk of wet seasons, which prevented timely paddock preparation. During the 1960s and early 1970s scientists at Rutherglen worked to develop direct drilling as an alternative to short fallow. Initially direct drilling did not offer a yield advantage over short-fallow or minimum-tillage cropping. Any financial advantage came from savings in time and costs. Improved soil structure was gained over a number of years. These advantages had not been highlighted in the competitions that had been used to successfully promote Wimmera fallow and ley cropping. Profitability from lower costs and reduced risk could not be seen by walking through a paddock. New strategies were needed.

The scientists initially relied on journal articles and field days, extolling the message of reduced costs, timely preparation and eventually improved soil structure.[38] Direct drilling was promoted as part of a wider package, which included the use of herbicides for weed control and lupins as a legume crop. While many farmers began using other parts of the package, direct drilling remained unpopular. Market research showed farmers knew of the technology, but did not consider it relevant to their own farm.[39] This was ironic given that the growers were funding the work through levies paid to the Wheat Research Council, whose members obviously believed the research was relevant. Farmers did not trust the results of small de-monstration plots on a research farm and had not been involved in the evaluation themselves. They were particularly concerned about the effec-tiveness of weed control under direct drilling. The answer was to involve farmers in evaluations in their own districts. At this stage new institutions sought to influence farmers' cropping practices. Chemical companies that sold the necessary herbicides brought their own marketing techniques to the promotional task.

The British firm Imperial Chemical Industries (ICI) had developed two important broad-spectrum herbicides, paraquat and diquat. Despite the effectiveness of these herbicides, sales growth was slow. The marketing division attempted to establish the chemicals as being easily integrated into traditional farming practices, marketing a premixed package of paraquat and diquat under the tradename Sprayseed.' This eliminated the task of mixing the chemicals on the farm and the tradename established a link between the chemicals and sowing. Some farmers integrated the new product into their conventional management system. They used Sprayseed as a replacement for cultivating in wet seasons when the ground was too wet to be cultivated without bogging the tractor. Using the conventional autumn fallow system farmers had to wait until the ground dried after the autumn break before cultivating to kill the weeds. In a wet year the chance to cultivate might not come until well after the best date for sowing, thus forfeiting a crop. With Sprayseed, farmers could use a light tractor and boom spray unit to kill weeds. Used this way Sprayseed reduced the risk associated with conventional cropping. The resulting pattern of sales was not particularly satisfactory for ICI. Sales of Sprayseed were good in a wet season but in normal or dry years there were few sales.[40]

ICI decided to link Sprayseed with direct drilling of crop seed as a better marketing strategy for maintaining a constant demand, irrespective of the season. Sprayseed was launched as a tool for direct drilling with the slogan 'It's all so simple'. The supposed simplicity was based on the observation that the task of spraying was simpler than the task of ploughing. The strategy was not successful. Ploughing was part of the farming culture, and any farmer who did not plough was seen as being lazy. To market direct drilling as being simple reinforced this unfortunate stereotype. Those who did try direct drilling soon found that it was not so simple. There was more to direct drilling than simply mixing a spray and driving over a paddock. The difficulties farmers discovered were beyond the experience and know-ledge of the local chemical agents. The successful establishment of direct drilling, which implied a radical change in farm culture, was to take more than slick advertising. ICI had to choose between giving up the direct-drilling campaign and devoting more resources to foster the changes.[41]

To provide greater personal support to farmers interested in trying direct drilling ICI established teams of consultants in areas where there was potential for direct drilling. The consultants were not chemical salesmen; they worked with individual farmers to ease the difficulties of learning to direct drill. The ICI consultants in the north-east collaborated with the research scientists at Rutherglen Research Institute and departmental advisers who were promoting direct drilling and better soil structure.

The campaign was more successful than previous strategies. As interest in direct drilling expanded, the ICI direct-drill teams were unable to keep up with the demand for their advice. A new approach was needed — to encourage the direct drilling farmers to help each other. ICI sponsored marketing surveys to identify the existing social groups of direct-drill farmers and the most influential farmers in these groups. They next approached the influential farmers with a proposal to form 'cell groups'. Group members met regularly and swapped ideas and experience of direct drilling. The ICI teams gave technical support to the groups rather than individuals. The cell groups flourished. In a sense, the groups were an informal adaptation of the old crop and fallow competitions. The groups provided a forum for both informal cooperative learning and also informal competition.[42]

One of the major concerns of the cell group members was the lack of good financial data about the profitability of direct drilling compared with traditional practices. Farmers knew the yield differences between the two cropping systems from the work of the Rutherglen Research Institute, but

A farmers' group examining soil structure

there was very little information about the profitability. ICI formed select groups of the best conventional and direct-drill farmers in various areas. These groups were called crop production groups. The groups' main task was to collect financial data for the company to use in the cell groups; in return, the group members received personal support from the ICI consultants and from the other leading farmers in the groups.

Promoting concern for soil structure as part of direct drilling helped to increase sales of Sprayseed. The work of the groups also provided new opportunities for other competitors. A later entrant to the marketplace was Monsanto with their systemic herbicide glyphosate, marketed under the tradename Roundup. Monsanto promoted Roundup as the 'plough in a drum' by linking it with the term 'conservation tillage'. This was a shrewd piece of advertising. It staked a claim for the product in the subsequent expanding direct-drill market and took advantage of the work done by ICI.

Many of the cell and crop production groups are still running in the north-east of Victoria and in southern New South Wales. Direct drilling has been established as a method of farming.[43] The initial difficulties of direct drilling have not been totally overcome. It still is not 'so simple'; in a more recent marketing campaign Monsanto repositioned Roundup as a generic weed killer, moving away from conservation tillage and direct drilling.

While ICI had limited itself to the task of promoting the technique of direct drilling, the Rutherglen Research Institute and the local Soil Conservation Authority officers had a broader objective of promoting conservation cropping, not just conservation tillage. For the north-east, conservation tillage included minimum tillage and direct drilling. Conservation cropping encompassed direct drilling, minimum tillage, the retention of crop stubble, liming to reduce soil acidity, deep ripping to remove compacted layers of sub-soil, and increasing crop water use efficiency to control acidity, watertables and waterlogging. These priorities had been set in a meeting with local farmers in 1980.[44]

These farm practices required complex and difficult changes in farm management. Many of these changes were not directly linked to the use of a commercial product the way direct drilling was linked to Sprayseed, so there was no reason to expect commercial companies would promote them. If conservation cropping was to be promoted, then it required government support. Early in the 1980s several well tried strategies were being used to promote conservation cropping. The Soil Conservation Authority Officers used a combination of direct, personal extension with farmers and a newsletter in which farmers purportedly talked to other farmers about solving their conservation cropping problems. While Rutherglen Research Station field days were well attended, conservation cropping techniques did not spread quickly beyond the immediate Rutherglen district. Many farmers did not believe the Rutherglen solutions would work beyond the Rutherglen district. This was particularly the case on the cementy red-brown earths to the west, where farmers had integrated chemicals into their farming systems but saw no reason to change to minimum tillage.[45] Demonstrations away from Rutherglen were required.

Advisers set up four demonstration sites in different localities. A co-operative farmer in each district provided a paddock. The paddock was divided in two, half being farmed as before by the owner, and the other half by government extension officers using conservation cropping techniques. Local farmers were able to drive past and see the differences. There were some unintended consequences when the participating farmers, allowing their farming skills to be put on display against those of a government officer, perceived their farming credibility was at stake. Some of the demonstrations were taken as a competition by the host farmers. Extra effort was put into the conventional paddock to make sure that face was not lost. In other cases, the host farmers were impressed with conservation cropping at the start of the exercise, and a little way into the demonstration started using the conservation cropping techniques on the conventional paddock. In either case, the value of the demonstration as a comparison for the local farmers was destroyed. Within two years the extension officers closed half the demonstration blocks and continued the others knowing they were of no value for comparative purposes.

The demonstration blocks provided some lessons for the promotion of conservation cropping. The blocks obviously encouraged competition when competition was not the objective. It was also clear in some areas that the time was not ripe for demonstration. The results had not been as good as the advisers had expected. The traditional practices had won the competition against the conservation cropping techniques. Not enough was known about conservation cropping on some soil types.

SoilCare

A new approach was developed using the best of the demonstrations, the best of the ICI strategy and of the Landcare group approach. Demonstrations on paddocks owned by local farmers became a focus for groups of local farmers. The group members discussed their concerns about conservation cropping and, with advisory and research officers, decided on the experiments to be run on the demonstration blocks with the aim of solving cropping problems specified by the group. A small experimental plot in the corner of the demonstration block helped farmers and advisers to understand how much each individual technique was contributing to the success or failure of the package being demonstrated on the whole paddock.[46] A catchy title, 'SoilCare', was used to appeal to funding agencies and the farming community. The shared decision-making of SoilCare avoided the problems of 'loss of face' that beset earlier demonstrations. The responsibility for the success or failure of management was shared, with no individual farmer feeling his management skills were on display. SoilCare was another step in the devolution of responsibility for research direction to the farming community.

SoilCare has been a campaign of the 1990s. Early in the campaign, market position research was commissioned to explore what farmers thought about the conservation cropping message. The research confirmed that there had been a steady increase in the use of direct drilling and minimum-tillage cropping through the 1980s.[47] Direct drilling was well established in the wetter south-east of the region, but rare in the drier areas to the north-west around Nathalia and Picola.[48] Forty per cent of farmers were using direct drilling on their farms, though only a minority of these farmers were using it exclusively. Most farmers chose between minimum tillage and direct drilling depending on the season and the state of the paddock.[49]

Farmers were not direct drilling to achieve increased yields. Rutherglen research scientists had measured increased yields after five years of direct drilling. Very few farmers believed direct drilling would give increased yields, despite the widespread belief that it improved soil structure. This implies a rather surprising conclusion: many farmers did not believe improved soil structure increased production. There is further evidence to support this conclusion. While many farmers were aware of soil compaction on their farms, a farmer who saw soil compaction or crusting on his farm was no more likely to direct drill than a farmer who believed he had neither problem on his farm.[50] The crucial advantages that convinced farmers to direct drill were savings in time and fuel. Farmers who did not direct drill were much less likely to believe direct drilling offered these advantages.[51]

Not everyone who tried direct drilling found the technology satisfactory. For every two farmers who successfully adopted direct drilling there has been one farmer who has abandoned it.[52] Most who gave up the practice believed their soil was unsuited because, with direct drilling, it crusted over or set hard in the upper layers. Because fewer wheat seedlings break through, early growth is poor and yields are lower; farmers with these soil problems believed cultivation was necessary to prepare a seed-bed.[53] Despite these difficulties, many of those who abandoned direct drilling still held positive attitudes to the technology, believing the advantages outweighed the disadvantages, but not on their own farm. Farmers who had never tried direct drilling had a much more negative attitude to the technique, acknowledging few of the advantages but being aware of most of the disadvantages.[54]

The continuing positive attitude of the relapsed direct drillers indicates the need to concentrate on helping solve technical problems rather than promoting attitude change and awareness of soil degradation. The link between conservation attitudes and farm management behaviour is weak. This has been found not only in north-eastern Victoria, but also in New South Wales, Queensland and other parts of Victoria.[55] There is little point in trying to change the negative attitudes of the non-adopters until the technical problems experienced by those who have adopted and rejected direct drilling are solved.[56]

The weak link between conservation attitude and behaviour can be seen in beliefs about chemicals. Farmers considering the choice between direct drill, minimum tillage and conventional cultivation say the safety of chemical herbicides and the possibility of residual damage to the soil is a major concern. But north-eastern farmers are also very impressed with the weed-killing advantages of herbicides and the advantages for soil structure.[57]

Most farmers see cropping as a trade-off between one environmental danger and another. The nature of this ambivalence can be seen in a map of north-eastern farmers' beliefs about chemical weed control and cultivation (Figure 7.1).[58] *Safe for the environment* is a long way from *chemical weed control* (and a moderate distance from *weed control by cultivation*). *Better soil structure* is a long way from *weed control by cultivation* (and a moderate distance from *chemical weed control*). Like Buridan's ass, north-eastern farmers are caught in the middle. Concerns about environmental safety are an attitudinal barrier to the use of herbicides. Concerns about soil structure are a reason to reject cultivation. The practical advantages of herbicides have won out over fears about their safety. Nearly all north-eastern croppers have used herbicides. Very few farmers have rejected herbicides after using them on their farms.[59]

The weak link between conservation attitudes and behavior is also obvious when we consider the adoption of stubble retention. Attitudes to stubble retention are positive. Half the farmers in the north-east said they thought the advantages of stubble retention outweighed its disadvantages. Many farmers believed stubble retention improved soil structure and the soil's moisture-holding capacity. But as with direct drilling, very few farmers believed the soil improvements from stubble retention would result in yield improvements. Unlike direct drilling, few farmers mentioned other financial benefits from stubble retention. It did not save time or money. Not surprisingly, because there is no immediate financial attraction, positive attitudes have not been translated into changes in farm practices. There was a gradual increase in the use of stubble retention over the latter half of the 1980s, but burning crop residues is still a common practice on north-eastern Victorian farms. Only 10 per cent of cropping farmers had

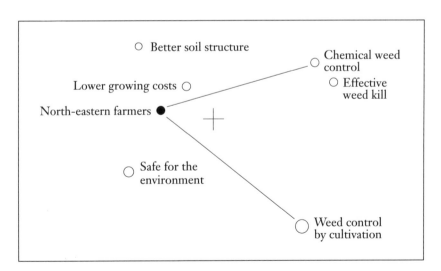

Figure 7.1 Cropping farmers' beliefs about chemical weed control in north-eastern Victoria

eliminated burning from their farm management and a further 10 per cent had made stubble retention a regular, though not exclusive, feature of their cropping program.[60]

For every farmer who has successfully adopted stubble retention, there is a farmer who has abandoned the practice.[61] Again, many of those who have abandoned the practice still have a positive attitude to stubble retention, but they perceive it as impractical with existing technology.[62] The main problem is that stubble blocks the machinery used for sowing the next crop. A proposed solution is specially designed machinery. Trash clearance combines can sow seed into paddocks with standing dry stubble. However, many farmers believe no machine will cope with wet stubble:[63]

> In an experimental plot they [the scientists] had no trouble. They get to the end and kick the stubble out. Its that kick that destroys us when we buy the machine and 500 yards down the paddock there's a great lump stuck in it . . . He [a farmer] spent weeks, every spare moment he had, converting his combine to a trash machine. When the season broke he went out and there wasn't anything it wouldn't go through. Then, when it rained a second time, it just wasn't working. Also, the equipment he bought to put under the combine was having problems because the tynes were breaking. He just got sick of it and sold it [the machinery] at the end of the season.[64]

Where a trash seeder can sow through stubble, other problems emerge. Standing stubble harbours some diseases.[65] Standing stubble also makes it harder to get an efficient coverage with herbicides. In some seasons the stubble harbours increased populations of slugs, which attack the emerging wheat seedlings.

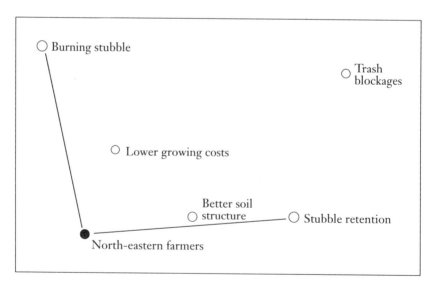

Figure 7.2 Cropping farmers' beliefs about crop stubble management in north-eastern Victoria

A belief map of north-eastern Victorian farmers' attitudes to stubble management (Figure 7.2) shows, as with the question of herbicides or cultivation, that farmers were caught between two unpleasant options, burning stubble and trash blockages. Burning stubble is a very unpleasant and sometime dangerous job. Despite these drawbacks, farmers choose to burn because they see it as the more practical option. Trash clearance combines are expensive and sometimes ineffective. No amount of promotion of the long-term improvements to soil will be of use if farmers do not believe the machines to be effective. There is a remaining challenge for SoilCare demonstrations to help farmers solve these problems.

If the SoilCare program is to help to achieve a more sustainable agriculture in north-eastern Victoria, it will need to halt the gradual increase in soil acidity. Much experimental work has been done since the 1912 liming study at Rutherglen. The mechanisms of increasing soil acidity and the advantages and limitations of liming are now much better understood.[66] Despite this work, promotion of acidity control still faces many of the problems revealed by the 1912 Lime in Agriculture campaign. The mechanism of soil acidity is complex and is not well understood by many farmers. Those farmers with a grasp of the mechanics of acidity often express frustration at their situation. Farming practices, which were promoted thirty years ago as part of the solution for permanent agriculture, are now seen as part of the problem of unsustainable agriculture. Their paddocks are suffering for their responsible farming.

Liming is commonly accepted by farmers as the solution to increasing acidity. In 1989 half the farmers in the district had limed some of their land; but of these, half were not liming on a regular program. Many had given up the practice.[67] Research has shown that some acid soils in the east of the region are very responsive to lime applications. Elsewhere the disappointing experimental results achieved in the 1920s are still valid where there is an acidic subsoil. For farmers on this land it does not pay to use lime. The only financially viable solution is to adapt to an increasingly acidic environment. Where crops do respond to lime, farmers deciding whether to apply lime to soil face a difficult decision. Because lime is bulky, transport costs make lime very expensive by the time it is delivered to the farm. Liming soil becomes a major capital investment in the soil from which returns will not be realised for many years. Faced with a significant initial cost, a farmer will ask himself 'What will I lose by not liming this year?' The answer has been, almost invariably, 'very little'.[68] Yet if the decision is delayed year after year top-soils, and eventually sub-soils, will become increasingly acidic. Eventually the process may become economically irreversible as sub-soils acidify. When farmers have decided to apply lime, economic considerations have encouraged liming at a less than optimal level. Lower rates of lime application will merely delay or slow the increasing acidification of the soil. Learning to live with acid soils may be necessary even on land that responds to lime. This means farmers will need to integrate deeply rooted crops and pastures and acid-tolerant species into their rotations. Helping farmers sort out their liming options is another important task for the SoilCare project.

We end this consideration of attempts to change cropping land management by returning to the town of St James and its wheat silo. The G. J. Coles emporium has long ago moved from the town. The future of St James and the surrounding district is linked to the grazing and wheat industries. If the wheat industry is not to move on like G. J. Coles, if farmers are to be able to deliver wheat to the St James silo fifty years hence, then farmers, research scientists and advisers will need to continue to work together to find practical solutions to the physical and financial dilemmas posed by acidity, compaction and soil fertility. Practical solutions will be those that offer a financial advantage to farmers while solving these problems. Impractical solutions will be those that need to be promoted with a hand on the heart and a call to farmers to invest for the good of the soils or the nation. The challenge of impractical technology is to transform it into practicality. If that goal cannot be achieved in north-eastern Victoria, the wheat silo at St James will face an uncertain future. If the goal of practical conservation farming cannot be achieved across Australia's wheat fields, many Australian wheat farmers will share an uncertain future.

8

Potatoes and processors

On the north coast of Tasmania, the farmland has soils that most Australian farmers can only dream about. In the hinterlands of the Mersey Estuary there are fields of rich, deep soil, which support some of the most intensive cropping in Australia. Despite this advantage all has not been well with this apparently prosperous and productive rural industry. Local community groups protest against the flights of the crop-dusting aircraft that are a necessary part of farming systems in the area. When it rains heavily in spring, soil washes off the paddocks, collects against the fences, fills dams

Vegetable cropping land in northern Tasmania. The steep slopes make this land prone to soil erosion after heavy rainfall.

167

and covers roads. Giant red crescents spread in the sea as the local rivers empty silt into Bass Strait.

Northern Tasmania was once the grain bowl of early Australia. Nowhere else in Australia did soils and rainfall so closely resemble the English homeland of the first European settlers. The traditional mould-board ploughs worked well in the deep soils. The English wheat varieties grew better here than elsewhere in the new land. The Mersey Estuary was a good port so transport was not the economic obstacle it was elsewhere. It was cheaper to cart the local wheat the short distance to port, and ship it to Sydney, than it was to grow wheat 160 kilometres from Sydney and haul it across the Blue Mountains. The history of Australian wheat farming is the story of developing better adapted wheat varieties and providing cheaper land transport. As progress was made on each of these fronts, Tasmania gradually lost its competitive advantage as a wheat producer. First, South Australia developed its own wheat industry close to the port of Adelaide. Then Victoria built a rail network through the north of that state. Improved wheat varieties led to higher yields in the drier mainland colonies. By the turn of the century Tasmania's wheat industry had lost its pre-eminence. In Tasmania, wheat's place on the farm was replaced by potato growing.

Unlike wheat, potatoes had an enduring advantage as a crop in northern Tasmania. Potatoes needed deep, fertile, friable soils and regular rainfall in the growing season. Potatoes were first grown commercially in the district in 1827. The potato variety was called 'redskin', and it arrived with the first convict settlers at Port Arthur. In northern Tasmania it grew well in the fertile soils. In some fields the first crops yielded 30 tons to the acre (75 tonnes per hectare). The quality of the north Tasmanian potato was high. During the gold rush years of the 1850s the arrival of Tasmanian potatoes was always welcomed at the Melbourne wharf across Bass Strait.

Tasmania's potato famine

The high early potato yields were a chimera. While a few growers could expect 30 tons per acre in virgin paddocks, no one could have anticipated that the average yield would decline to two-and-a-half tons per acre within a few decades of the first potatoes being sown.[1] The original yields were not sustainable. Irish blight, the mosaic virus and soil decline had taken their toll.

A modern potato is nothing like its wild relative that grew in the Andes. The original wild potato was toxic, containing dangerous nerve poisons such as chaconine and solanine. The indians of South America soaked and washed the wild potato to leach out the poison before eating the remaining flour. European breeders selected plants to breed out the high poison levels in the wild potato. They produced a less toxic plant, but one that had lost much of its natural resistance to disease. A single, low-toxicity potato plant could be simply cloned, over a number of generations, into thousands

Plate I   Thomas Clark **Muntham Station** circa 1860
            City of Hamilton Art Gallery

*Edward Henty established Muntham Station near Casterton in south-western
Victoria soon after Major Mitchell's visit in 1836. Clark's depiction
corresponds with early descriptions of grass-covered hills with occasional
clumps of trees.*

Plate II   Duncan Elphinstone Cooper   **Panorama of Challicum, Nos 7 & 8**
            circa 1848
            National Library of Australia

The earliest pastoralist settlers made their mark on the landscape but did not radically change the treescapes. They occupied a relatively empty land of open plains and lightly timbered hills.

Plate III   Eugene von Guérard   **View from the Bald Hills between Ballarat
and Creswick Creek** 1858
National Gallery of Victoria (Felton Bequest 1960)

*An early view of the Pyrenees and the woodlands and grassy hills in the upper
Loddon–Avoca catchment. The demise of the Aborigines and the cessation of the
periodic burning of the grass and young saplings produced thicker forests of
young trees in the open woodlands.*

Plate IV   Thomas Mitchell   **Mammeloid Hills from Mount Greenock** 1836
Mitchell Library, State Library of New South Wales

The west Victorian uplands when first seen by a European in 1836.
The symmetrical, treeless hills are interspersed with woods.

Plate V   Kerrie Youngs   **View from Mount Greenock** 1991

In this scene very little has changed since when Mitchell viewed it.

Plate VI   G.F. Angas   **Eaglehawk Gully, Bendigo**
La Trobe Collection, State Library of Victoria

Turning up the soil in the search for gold in the early 1850s

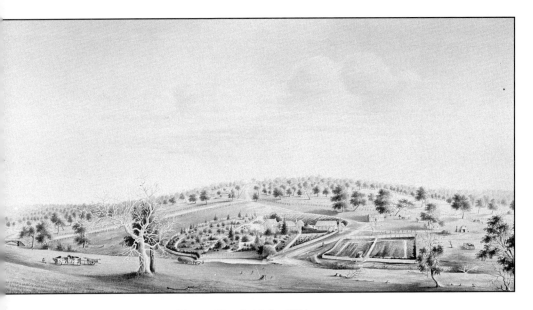

Plate VII   William Tibbits   **Wando Dale** 1876
City of Hamilton Art Gallery·

The Wando Dale homestead, situated between Hamilton and Casterton,
near present-day Wando Vale Potter Farmland Plan farms

Plate VIII Ludwig Becker **Crossing the Terrick-Terrick Plains** 1860
La Trobe Collection, State Library of Victoria

The Burke and Wills expedition travelled across the dry, flat Terrick-Terrick Plains east of Pyramid Hill in northern Victoria in August 1860. Land to the west of Pyramid Hill was irrigated from the Loddon River in the 1880s.

of individual plants by planting pieces of potato and letting the eyes sprout. The result was acres of genetically identical potato plants with equal resistance to disease. For some varieties the common resistance was almost non-existent. Societies that grew to depend on this potato monoculture were very vulnerable. Disease could devastate a community's potato fields, destroying whole crops. In Ireland the late blight of the staple potato brought about the starvation of whole village populations and precipitated a great wave of Irish immigration to the new worlds of North America and Australia. The blight fungus migrated with the Irish.

Blight arrived in northern Tasmania about 1900. Until then potato farmers in Tasmania had had a relatively simple time with their crops. In 1906 a strange new breakdown was noticed among the seed potatoes. In the following year, several late crops suffered a strange fungal attack in the wet autumn. The local growers called the disease brown rust. The next year, 1908, was again wetter than usual and the new disease spread to many crops. Sydney potato merchants complained that the consignments from Tasmania were rotting in their warehouses. The government introduced inspections at the port to protect the mainland markets. Infected districts were quarantined. Scientists from Victoria and New Zealand identified brown rust as Irish blight. The following year was again wet and few growers made a profit. In one week, 10 000 bags of potatoes were destroyed at Devonport wharf. They had deteriorated after packing. Preparations for the 1911 crop were feverish. Clean seed potato was selected and growers sprayed as best they could with Bordeaux mixture, but the precautions made little difference. Blight swept across northern Tasmanian crops in a matter of a few days. Only a quarter of the crop was saved, by leaving it in the ground until late in the season.[2] The potato industry appeared doomed.

The reason for the devastation was not disease but a lack of genetic diversity. Almost the whole Tasmanian potato industry was based on the genes of a single potato variety, the redskin, which was introduced to the settlement with the first settlers. Unfortunately for the settlers this potato had no natural resistance to blight. In following years the variety was abandoned in favour of others with some resistance to the disease. Further control came with a better understanding of the disease. Originally, growers believed the blight spread down the stems into the potatoes. But the fungus infected the leaves and stem of the plant, and released spores that dripped down onto the ground with rain, infecting exposed potatoes. With this knowledge, growers planted rows farther apart and hilled more soil onto the rows of growing potatoes. A greater depth of soil gave greater protection to the tuber. Nature helped as well. With dry seasons the fungus retreated, causing minor problems for only the most genetically vulnerable potato varieties. Vigilance could not be relaxed because blight was always a threat in wet years. A smaller disaster occurred in 1930, when half the crops were lost in some areas and a quarter of a million bags of potatoes were destroyed.[3] Another outbreak occurred in the wet years after the Second World War.

Irish blight caused spectacular losses in bad years and evoked great concern when it appeared. In dry years blight was not a problem and caused less damage than the mosaic diseases. These stunted the potatoes and reduced yields in wet and dry seasons. Each year more plants on a farm would succumb to the progressing mosaic symptoms. To the growers there was no observable cause of the symptoms, which were unrelated to any disease: it appeared the potato variety was losing vigour. The problem was caused by a virus spread by aphids. Once infected, a plant passed the disease to the next generation through the seed potato. The solution was to plant next year's crop using uninfected seed potatoes. This explained the gradual spread of the symptoms over a number of seasons. A new industry grew up, supplying virus-free seed potatoes to commercial growers. The seed potatoes were grown on high altitude land where aphids were less active. Seed farmers were required by government legislation to destroy any infected potatoes. In return for rigorous attention to virus and aphid control, the government registered seed potato growers, enabling them to charge a higher price for their potatoes.

The third reason for declining yields was declining soil fertility and structure. Despite their depth and richness, the soils of northern Tasmania were still fragile; continuous cropping had depleted humus. The result was a cloddy soil, which did not suit potato growing or potato harvesting. As potatoes were being grown on steeply sloped land, erosion was also a problem; bare, ploughed soil was left exposed, to be washed away by heavy rain in early spring after planting. The use of bare fallow hastened humus decline and erosion. In 1930 potato growers were using a short rotation, with almost two hectares of potato crop for every hectare of fertility-restoring legume crop or ley pasture.[4] This tight rotation regime exhausted the soil of nitrogen, phosphorus and trace elements. The government encouraged farmers to plant temporary clover pastures and green manure crops to restore nitrogen fertility. Superphosphate was also promoted; but trace element deficiencies were undiagnosed. Another new disease attacked potatoes at flowering time, mottling leaves and eventually killing whole plants within three weeks of its first appearance. It was given the name fire blight, but it was not like the Irish blight. It was not a disease but a deficiency in soil potassium. The potato crops had depleted the soil of available potassium.[5] After the Second World War fire blight was eliminated by adding potassium fertilisers.

After the war: rebuilding yields

Blight-resistant potatoes and certified virus-free seed potatoes established the potato industry on a more secure footing. By 1945 average yields had reached 4 tons per acre, almost doubling the yields of a few decades earlier.[6] Attention to the soil had played little part in this improvement. To help the war effort potato growers had pushed their land with ever tighter rotations to feed a beleaguered England. The continuous cropping depleted the soil of nutrients and built up soil disease.

After the war government advisers encouraged farmers to pay more attention to the health of their soil. They advocated whole-farm planning to husband the land:

> If the actual position in many farms could be assessed, it would probably be seen that actual fertility losses through soil erosion far exceed losses due to overcropping. Soil erosion is undoubtedly taking place . . . so gradually that the danger is unsuspected. It is a disturbing thought . . . that some of these landholders are literally, if slowly, being washed off their properties without realising the fact. A plan of urgent reconstruction for many farms is now an urgent necessity and, as the first step, each landholder is counselled to make a careful appraisal of the property and its productive resources, both actual and potential. The plan should envisage as one of its primary objectives the restoration and conservation of soil fertility, the most important of all capital assets to the farmer.[7]

Today trees seem to play an integral part in most people's idea of a whole-farm plan. In the post-war period trees did not have a part in farm planning. The contemporary view was of sustainable development based on pasture and animals. Farmers were encouraged to contour map their land and divide it into areas that could be sustainably developed and areas to be left undeveloped. The undeveloped land was left under tree cover because it was unprofitable to develop. The rest of the land was to be renovated with clover pasture and superphosphate to restore lost fertility. This was the era of the sub-clover and superphosphate revolution and livestock farming was profitable. No more than 5 per cent of the land was recommended to be in crop at any time. Farmers were advised not to crop steep land, and to plough on the contour, across the fall of the land.[8]

Farmers with large farms increased their numbers of livestock and decreased their cropping. The work was easier and livestock prices were high. Farmers with smaller farms could not make enough money running cattle, so they increased the amount of their land under crop. By the mid-1950s Tasmanian potatoes were being produced by a large number of farmers with small properties. In commercial agriculture, many small farmers in an industry is rarely economically sustainable. Between 1945 and 1975 four out of five potato farms in northern Tasmania disappeared. To understand such a radical change, and its implications for the conservation of the soil, we need to look at how potatoes were grown in 1950.[9]

A potato farmer began the season with a fallow ploughing of the potato paddock (to conserve water in the soil), or by ploughing in a green manure crop or the residues of the last potato crop (to improve soil structure). Virus-free seed potatoes were cut into pieces, dipped in a fungicide and hand planted in lines that were then disced and harrowed to cover the seed potatoes. When the potato plants showed above the soil, the farmer ploughed soil from the sides of the rows towards the plant to create ridges of soil in which the new potatoes could grow. There were good reasons for all this ploughing. These ridges protected the new potatoes against blight, potato greening and the potato moth: the deeper the ridges, the better the protection. The potato moth found it more difficult to burrow through the soil to the tubers, the blight spore was less likely to drip onto exposed potatoes and there was less likelihood of any potato greening through exposure to the sunlight. The farmer ploughed between the rows whenever necessary to keep the weeds under control. Harvesting was done by hand with a fork and shovel. A friable, well ploughed soil without clods made harvesting easier and meant fewer potatoes were damaged.

For most growers the marketing of potatoes was full of uncertainty. There were large fluctuations in production and prices. With only a small domestic market and a reliance on interstate markets to absorb three-quarters of its potatoes, the Tasmanian growers were in a vulnerable position. The growers tried to limit the price fluctuations by controlling the Tasmanian market through a Potato Marketing Board; but they had less control over the interstate markets, and so they remained vulnerable. In the 1950s and 1960s, with increased potato production in the rest of Australia, Tasmanian potatoes were gradually squeezed out of the mainland markets.

Farming with the factories

The small number of potato farmers who remained were saved by the arrival of three processing companies. The companies built three potato processing factories in the district surrounding Devonport and created a new market for Tasmanian potatoes. By the mid-1970s only 5 per cent of the Tasmanian potato crop crossed Bass Strait as unprocessed potatoes; most potatoes were processed directly by companies. Over the next decade potato growing concentrated around the factories. There was little point growing potatoes elsewhere.

Unlike the greengrocer, the factory managers did not buy their potatoes on the open market. The managers needed a reliable and continuous supply of vegetables for their processing lines. They ensured this by entering into contracts with growers. The terms of the contract were set in negotiations with a government appointed Growers' Board. The contracts not only gave the processors security of supply, they gave new security to the grower — security that allowed the growers to make large investments in mechanical harvesters, planters and travelling irrigators. In the short run the contracts made investment in these machines possible; in the long run they made it obligatory. The price of potatoes was soon set on the assumption that these labour-saving machines were used. Farmers without these machines were unable to compete. The contracts dramatically changed the way potatoes were farmed.

Mechanised planting and harvesting machines required good soil structure. Mechanical harvesters working in poorly structured or weed-infested soils damaged the potatoes, which were then not acceptable to the factories. This gave farmers an incentive to maintain good soil structure. But the new machines and mechanised methods made maintaining good soil structure more difficult. The harvesters were heavy and so compacted the soil. Their tyres were deliberately narrow to minimise the damage to the buried potatoes, but this increased the compaction of the soil under the tyres. Heavy trucks now crossed the paddocks to collect the harvested potatoes. Farmers 'solved' the problem with more ploughing. In the 1960s and 1970s farmers using mechanical planters and pickers were advised to plough their soil to a depth of 30 centimetres to achieve a good tilth. This meant four cultivations before planting, followed by cultivations to heap up the rows and control weeds after the potato plants had appeared — a total of six or eight cultivations.

The deeper cultivation did little to help control erosion. The new machines were awkward, unstable and heavy. It was dangerous to drive them along the contour of the hill. It was far easier and safer to drive them up and down hills; thus cultivation and potato rows ran up and down the hills. The deep cultivation up and down the hills caused serious erosion. Heavy spring rains washed the soil against fences; soil flowed into dams and covered roads. After heavy rain farmers would use a mechanical scoop to pick up the soil at the bottom of the hill and carry it to the top of the hill again. After very wet weather some roads were covered by topsoil.

Because there was no need to plant potatoes to meet the needs of the interstate market, growers were free to plant at the time of year that maximised the yield of potatoes. Potatoes were grown during spring and summer, when temperatures were highest, but rainfall was most unreliable. Irrigation was needed. In the 1950s very few properties were irrigated. By the 1970s all had irrigation. Fixed sprinklers in turn gave way to travelling sprinklers, machines that crawled across the paddock.

The travelling sprinklers presented another soil problem; they had to water the soil quickly as they passed over it. The 'rain' that falls from them is more intense than the heaviest tropical downpour, giving the crop a drenching equivalent to 500 millimetres per hour. Travelling irrigators can only travel up and down hills, another reason to run rows up and down the hill. Travelling across ploughed ground, the wheels and the water pipe dragged along behind compact the soil. If the travelling pipe breaks away from the irrigator, or the irrigator stops, the rush of water can quickly turn a small gully formed by the wheels of the travelling irrigator into a massive washout. Irrigator failure can move a lot of soil. These technical problems remain a challenge for the agricultural engineers. Irrigation has also meant that potatoes are planted at the time of year when there is a great risk of rain storms. For two or three weeks until plant cover is achieved, the soil is at a high risk of erosion.

During the period of mechanisation there was a gradual decline in the area planted to potatoes. This was not matched by a decline in total potato yields. By 1970 the average yields per hectare were three times greater than thirty years earlier. Farmers needed to replace depleted soil nutrients to maintain the yields. The potato farm of the 1940s had used pasture to maintain fertility. With the increasing specialisation of potato growing, farmers were less likely to mix cropping and livestock. The potato growers used green manure crops such as peas and beans, which were not as efficient as clover pasture in replacing mined nitrogen. The soils needed to be supplemented with nitrogen fertilisers. Ploughing between the rows to control weeds was gradually replaced by the use of herbicides. This reduced the number of cultivations, but it still involved the tractor wheels straddling the rows of potatoes. Each pass by the tractor damaged the potato plants above and below the ground, reducing crop yields. A similar problem arose with new fungicides and insecticides, which gave the grower greater control over disease, but at the cost of more tractor passes over the maturing plants and tubers. There were two solutions: widen the space between the rows and widen tractor axles to pass between the rows, or spray from the air. Many growers chose aerial spraying.

The processing factories offered markets for more than just potatoes. They processed other crops such as peas, brassicas, onions and beans. Onions could be harvested with potato harvesters, giving farmers a reliable crop to use as a disease break after potatoes. Onions have a very small seed, which needs to be planted in very fine soil. Working the soil up to a suitably fine tilth takes many cultivations, sometimes as many as six or seven before sowing. Soil in this state is at high risk of erosion until a crop can cover the ground.

The processing industry changed northern Tasmanian potato farming. It became more productive, more profitable and more secure than in the years after the war. But the intensified cropping, the decline of pasture and tighter rotations meant that the magnificent soils were under pressure and, without careful management, the long-term prospect was again declining fertility. Potato growers in southern Tasmania and elsewhere continued with pasture rotation, mixing potatoes with livestock enterprises, particularly dairying. This maintained fertility, and a more diversified income, but was not without its problems.

For all its advantages, pasture had some drawbacks. One was a small native insect, known as the click beetle. In its larval form, when it was called the wireworm, it fed on the roots of native trees and grasses. The wireworm adapted very well to the soil conditions under sown pastures. In some areas the wireworm under the pasture multiplied to such an extent that it became a threat to the following potato crop. When the grass was destroyed, the wireworm burrowed its way into the growing potato, destroying the potato. Wireworms did not survive to infect a following potato crop.

Wireworm in potatoes could be avoided, at the expense of exhausting the soil, by avoiding pasture and continuously cropping potatoes. The insecticide dieldrin also provided a means of control. When pasture was ploughed to plant potatoes, and there was a high population of wireworms in the soil, fertiliser mixed with dieldrin was ploughed into the soil before planting to kill the wireworms. As dieldrin did not break down quickly in the soil, government regulations required that the crop was not harvested for 90 days after dieldrin was applied.

In the late 1960s the United States reduced the level of organochlorines it would accept in food products. Many Tasmanian dairy products regularly tested positive for residues and dairyfarmers were advised to stop spraying pastures with organochlorines, which were used to control pasture pests such as grassgrubs, armyworms and cockchafers.[10] The prohibition was not extended to wireworm control. Testing showed dieldrin, an organochlorine, was not absorbed by plants, so there was little concern about the residues in either the potatoes or the cattle that grazed the land in the next pasture phase. This logic was exposed as flawed in the 1980s. When residue levels were further lowered in the United States, Australian beef carcases from potato farms were identified as being contaminated with dieldrin. The grazing cattle had ingested the chemical because they swallow large amounts of soil each day, often attached to the roots of uprooted grasses. The government quarantined contaminated potato farms. No cattle were able to be sold from the properties, and the basis of a pasture and cropping rotation was destroyed.

Green lines and processing lines

Soil conservationists claimed that the erosion and compaction of the potato soils could be controlled. Their idea of an environmentally friendly potato farm looks like a patchwork quilt of varying shades of green. Steep hills and gullies are left permanently in pasture, thus minimising the risk of erosion. Pathways for travelling irrigators are kept grass-covered to minimise the risk of washout from mechanical failure of the supply pipe. In the paddock, diversion hills and ditches are built up across the paddock to divert runoff into the grassed waterways. Lower down the hill, sub-surface drains are laid in soils that are prone to waterlogging and instability after heavy rain.

Between the lines of grassed waterways and irrigator pathways are the paddocks. A vegetative cover is kept on them during winter by growing green manure crops, such as beans, which protect the soil and reduce the need for nitrogen fertilisers. Ploughing is kept to a minimum, and pulverising implements like the rotary hoe are avoided in favour of tyned implements. Ploughing is not done when the soil is too wet or too dry. During harvesting or planting, the tractor wheels are confined to permanent wheel tracks to minimise the compaction damage of heavy machinery.

The solutions seem reasonably simple. In 1988 northern Tasmania farmers acknowledged the erosion in the district and most growers thought it was at least a moderate problem. But when farmers considered erosion on their own farms, one-third of the farmers thought their own erosion problem was insignificant. Only one in eleven farmers thought they had a substantial problem.[11] These soil problems are of only moderate concern to most potato farmers. Crop diseases and crop pests are of much more concern, as are the usual problems of marketing and low income. Oversupplied markets, costs and high land prices all similarly outranked soil erosion and soil structure as issues of concern.[12] Erosion was not seen as a precursor to fertility decline and unsustainable agriculture. The prevailing view was that erosion had been around for generations. Farmers had learnt to judge the rate of erosion by all manner of techniques, even to the extent of noticing the weight difference between a load of onions leaving the paddock and the weight of the onions after washing.[13] Most farmers believed erosion was a result of high rainfall, steep slopes or the travelling irrigators, all factors beyond their control. Few farmers saw farming practices as the cause of soil erosion and showed little interest in planning for erosion control. They saw no alternative to the use of travelling irrigators, ploughing vertically up and down hills and cultivation to prepare a seed bed. Conservation farming was perceived as making management of the farm harder. Potato farmers generally doubted the effectiveness of grassed water and irrigator pathways.[14]

Farmers believed soil compaction and crusting were caused by poor farming practices such as harvesting and cultivating with heavy machinery in wet weather, continuous cropping and rotary hoeing. Farmers believed they had little choice in the use of these practices. The critical management decisions for protecting potato soils are the timing of ploughing and harvesting. In northern Tasmania these decisions were not fully within farmers' control. The contracts with processing companies had given potato farmers some income stability, at the cost of management autonomy. The processing companies were able to determine the varieties to be planted, the timing of planting, the nature of the spraying program, and the timing of harvest. Processors often required a farmer to harvest on a certain day, irrespective of the weather. A call from the company saying 'We need 50 tonnes by Wednesday' could mean heavy potato harvesters being used on saturated soil.[15] The imperative for the processing companies was to maintain supply to the processing lines in the factories. Gaps in supply meant laying off the local casual workforce.

Many farmers blamed the processing companies for the damage to the soil. They disliked leaving the topsoil exposed and harvesting in wet weather; but decisions about harvesting were determined by the requirements of the factories. These perceptions are depicted in a belief map of Tasmanian onion growers in 1988 (see Figure 8.1).[16] *Maintaining good soil*

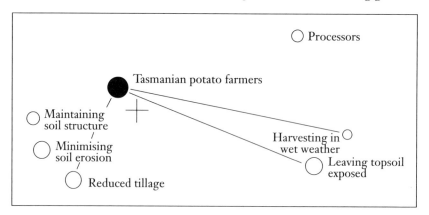

Figure 8.1 Vegetable farmers' beliefs about soil erosion in northern Tasmania

structure, minimising soil erosion, and *reduced tillage* were closely associated. *Harvesting in wet weather* and *leaving the soil exposed* were seen as far removed from practices of good soil care; these practices were more closely associated with the processing companies.

The solution to the problem of wet weather harvesting was to harvest when weather conditions were appropriate and to store the potatoes until the processing companies required them. The companies already stored potatoes, in refrigerated and ventilated sheds, to keep the processing lines going after the finish of the potato harvest. The use of less expensive, unrefrigerated storage on farms did not provide protection against the potatoes sprouting; sprouted potatoes were of no use to the processing companies. Neither the processors nor the farmers were willing to build more of the expensive sheds. Building more sheds to protect the soil resource on which the processing lines depended was not an investment that provided returns to the processors in the short term. The farmers were unwilling to make major investments in storage sheds because the storage was only required for such a short time. In 1989 the processors and farmers forced change on each other.

One processing company decided to expand its potato processing operations. It offered potato growing contracts to farmers farther south in Tasmania. The landowners wanted contracts, but were unwilling to accept the demand of wet weather harvesting. To attract suppliers the company agreed to build storage sheds. Growers elsewhere then asked why their land should be treated differently. In the climate of current public concern for land degradation a refusal by the processors to extend their storage could result in bad publicity. It seems likely that wet weather harvesting will be replaced by dry weather harvesting and storage.

Pressure from growers resulted in a more soil-friendly harvesting policy by one processing company; but the processors were about to play a role in improving growers' soil management. A severe rainstorm passed over the Forth and Kindred districts in 1989, soon after the onion crop had been planted. One processing company, worried by the 10 per cent crop loss from the resulting erosion, sought to have farmers improve their soil management practices. At a meeting of growers the company executives expressed their concern; they encouraged their growers to adopt soil conservation planning, and employed a soil conservation consultant to advise growers how to implement better soil conservation practices.

A recently arrived processing company, CIG, took the further step of including a clause in growing contracts that required growers to use recommended soil conservation measures. The company wanted farmers to grow a new crop, pyrethrum, to be processed into organic insecticides for household use. The new product meant new contracts, in which farmers were required to have certification from soil conservation professionals that the pyrethrum crop was grown with the most appropriate soil conservation practices. There were sensible reasons to guard the soil under a pyrethrum crop. Pyrethrum takes three years to reach harvestable size; during this time it provides little protection for the soil. The company required soil conservation practices to protect its investment in developing the new market. The certification was also useful for another purpose. Pyrethrum-based insecticides will be vying for 'green spot' labelling. Certification of soil conservation practices provides evidence that the product is not only 'green', but also it is produced with minimal damage to the soil. It would appear the pyrethrum company is anticipating consumer pressure.[17] Consumer preferences have brought about a favourable change in soil management.

9

Beyond the silent spring

In 1987 the United States tightened its procedures for detecting organochlorine residues in imported foodstuffs. Organochlorine residues were soon detected in beef imported from Australia. With a major market threatened, Australian authorities tightened controls on the use of organochlorine chemicals and instituted 'traceback' procedures to detect the sources of contamination. Fifteen hundred properties with contaminated soil were quarantined.[1] No cattle could be sold from the properties until the contamination was removed or biodegraded. Many of these properties will be quarantined into the next century. How had this happened?

Australian agricultural exports had been threatened once before by organochlorine residues. In the early 1960s the United States reduced the level of organochlorines it would accept in food products. Their tests showed some Australian dairy products contained organochlorine residues above the acceptable limit. Dairy farmers were spraying organochlorines on pastures to control grass grubs, armyworms and cockchafers. Cattle eating the sprayed grass were concentrating the organochlorines in their milk. The Australian regulatory authorities banned pasture spraying with organochlorines and exports were, for the time being, safeguarded.

Throughout the following decade the United States progressively banned the agricultural use of organochlorines. Regulatory authorities outside the United States disagreed with the US interpretation of toxicological evidence on the safety of organochlorines. In Australia the limited agricultural use of organochlorines continued on farms. Though no longer allowed to spray pastures with organochlorines, Australian graziers could still use organochlorines to protect fence posts from termites. On potato farms dieldrin was mixed with fertilisers to control wireworm, a small native worm that eats potatoes. Because dieldrin was not absorbed by potato plants or grasses that grew following a potato rotation, there was no expectation that this agricultural practice would lead to significant organochlorine residues in either potatoes or meat. However, some contamination was inevitable. Cattle eat soil attached to the roots of grasses they uproot. A beast can swallow 1 or 2 kilograms of soil in a day.

Although by the 1980s organochlorine use was banned in US agriculture, the US Customs Service tolerated a very small percentage of imported carcases with organochlorine contamination.[2] In 1986 this tolerance to residues evaporated. Protestations that the United States was responding to political pressure rather than scientific evidence were of little practical value. The Australian export beef industry was threatened unless drastic action was taken to eliminate all residues from beef exported to the United States. Farms with organochlorine contaminated soil were quarantined.

From a wider perspective, this story of the politics of the regulatory process is not a satisfying explanation of the reasons for the quarantining of 1500 farms. A deeper question remains. How had organochlorine chemicals become accepted as a tool of trade on Australian farms? For some understanding of this question we need to explore the development of what the organic agriculture movement calls 'chemical agriculture'. In this chapter we will trace this development in the horticultural industries and explore some of the dilemmas that farmers face in making decisions about how to use agricultural chemicals.

The early orchards

When Europeans arrived in Australia, they brought their European orchard trees with them. Arthur Phillip planted fig, orange and pear trees and grape vines on the shores of Sydney Cove. These trees became Australia's first orchard. The Hentys, the first permanent European settlers on the south-east coast, established Victoria's first orchard at Portland in 1834, planting raspberries, strawberries, apples, gooseberries and grapes.[3] In contrast, many of the squatters who moved inland from the coast had few dreams of permanency. Their intention was a passing occupation, sufficient to make a fortune and return home. They built homestead huts of rough bark. They planned to be gone by the time any fruit trees began to bear. But where the English set down their roots in permanent homes, fruit trees were an essential part of the migrants' re-creation of England in the new land. These early orchards were not for profit, but for luxury. In a new country with poor roads and a small population, fresh fruit provided one of the few culinary delicacies in an otherwise uninspiring diet. Growing fruit trees ensured seasonal luxuries: berries in spring, plums in summer, apples and pears in autumn and maybe oranges in winter.

The first commercial orchards grew on the edges of the larger towns or near the goldmining settlements. William Howitt visited an early orchard at Plenty, now a leafy Melbourne suburb:

> I was surprised to see the flats of the garden planted with vines . . . The apples, pears and plums here flourish and bear immensely. They have plenty of gooseberries which do well in places shady and not too dry for them . . . All other fruits flourish beautifully. They have the finest and most abundant peaches,

where they are cultivated, but this is yet but rarely. People are in too great a hurry to make fortunes, with the favourite and universal idea of going home. Therefore fruit is very rare and dear. Their apples and plums are superb, and of a large size and flavour. Plums, apricots, melon, grapes, and almost any kind of fruit, are as fine as they can be grown, if they are grown. Almonds and figs abound, the latter producing two crops a year. The quinces are gigantic in size, and make the most admirable marmalade. They have oranges and lemons in the open air . . . English cherries are splendid . . . and they are obliged to prop the branches of their apple trees, the crops are so heavy.[4]

The early orchardists were able to produce a crop without the protection from the chemicals available to the modern grower. They did not have the myriad of diseases and pests that are the bane of the modern grower. The difficulties of transport that kept the commercial orchards on the outskirts of towns acted as a quarantine. There was no fruit coming from overseas, and trees grown from imported seeds were often free of overseas diseases and pests. The major pest problems for the early orchardists were cherry borer and American blight.[5] It was just as well there were few pests, because the controls used at the time were limited and time consuming: orchard cleanliness, and nicotine, pine resin and soap sprays.

The gold rush brought people, expansion and wealth. Immigrants brought fruit trees and seeds and new pests. The greatest danger to apple growers was the arrival of the codlin moth and the black spot fungus. The codlin moth struck Tasmanian orchards in 1857. From there it quickly spread to South Australia and then the remaining eastern states.[6] The codlin moth lays its eggs on apple fruit and leaves. After hatching, the codlin grub burrows into apples and eats until it is ready to pupate. There may be three generations of moths in a season. To survive winter, grubs will crawl down the trunk of the tree to find a dark cranny in which to hide and weave their cocoon. The moths re-emerge next spring to lay the next generation of eggs. The moth causes extensive damage, sometimes destroying three-quarters of the crop.

Orchardists tried to keep the codlin moth under some control by cleaning the loose bark from the trunks of trees and the ground underneath, picking up damaged or fallen fruit and by wrapping bandages around the trunks of the trees. Instead of hiding in the crevices and loose bark, where it laid its eggs, the moth hid under the folds of the cloth. Growers removed the bandages once a week during summer and autumn and boiled them to kill the hiding moths.[7] On some orchards pigs or poultry patrolled the ground, eating up dropped apples, which could act as a safe haven for moths.[8] Keeping a clean orchard was time consuming. One South Australian grower complained that despite picking 6000 caterpillars from trunk bandages and picking 13 000 infected apples, the codlin moth numbers were as bad as ever.[9] The best fruit often came from the orchards behind the mansions of the wealthy owners who had enough money to pay others to keep their orchards clean.

Western Australia, protected by its remoteness, remained free of codlin moth and jealously guarded this unique status. The first codlin moth did not appear in this state until 1902. This signalled the beginning of a running battle to suppress sporadic outbreaks and regular finger waving at the other states, which were the source of these infections.[10]

The black spot fungus appeared in eastern Australia a few years after codlin moth. The fungus attacked the leaves of apple trees early in the season, later moving to the fruit, leaving large black marks, cracks and deformed, unsaleable fruit. Infected leaves dropped to the ground, where they released spores to continue the infection in the next season. Black spot completed the destruction wrought by codlin moth. It devastated some regional apple industries. Growers could not make a living and pulled out their trees.[11]

Similar pests arrived to infest peach trees. In 1856 peach leaf curl appeared. This fungus destroys the newly emerging leaves of the peach tree in spring. Early peach crops fail. Trees bearing later crops recover, but the trees produce immature fruiting wood leading to crop loss in the following season. Repeated attacks over several years destroy the trees. Peach production became a matter of luck, dependent on the season. Oriental fruit moth appeared in mainland peach orchards sometime before 1900. The moth has a life cycle similar to that of the codlin moth, but instead of burrowing into apples, it burrows into growing shoots of peach trees and into peaches. Continued attacks destroy fruit and weaken the tree. Worse than its direct damage, the oriental fruit moth helps to spread the fungal disease brown rot.

Like the fruit moth, brown rot appeared around the turn of the century. This fungus attacks the blossoms of peach trees and other stone fruit, infecting the embryo fruit. The infection reactivates in the fruit when it ripens. By the time the fruit is ripe, it is rotten. At the height of the rot, the fungus sprouts grey hairy fruiting bodies. The spore overwinters on dried rotten fruit that hang on the tree and spread the spore to start a new generation of rot infection the following season. The spread of brown rot can be spectacular. In the right conditions a small brown spot on a peach in the morning can spoil the fruit in the course of a day. Growers could sell their fruit in apparently wholesome condition, only to be faced with demands for refunds from grocers on the next market day because the fruit had rotted on the shelves.

Immigrant insects like codlin moth and oriental fruit moth quickly became pests because they were set free in a new land without their homeland predators to keep them in check. This was not a problem unique to Australia. Just as exotic insect pests caused havoc in Australian orchards, Australian insect pests caused havoc in overseas orchards. The cottony cushion scale was hardly noticed in Australia because natural predators kept it under control. But when it reached America it became a major pest in Californian citrus groves, where it had no natural enemies.

There was another cause for the quick spread of fungal diseases and pests: the gradual commercialisation of fruit production. In the artificial world of the commercial orchard, where trees of the one type are planted close together, there was little to stop an infestation spreading from one tree to the next, quickly infesting a whole orchard district. It was easy enough to spread infection between districts. Apples stored in fruit sheds introduced codlin moth to a ready made site for pupation. Clean orchards could be reinfected in spring by moths emerging from thousands of nooks and crannies in the dark packing sheds.[12] Growers supplied the local market with fruit packed in recycled pine boxes. The recycling saved money and timber, but it also recycled fungal spores and insect pupae between farms. Peaches that rotted in the boxes covered the walls with fungal spore. The peaches fell apart and soaked the box in their juices. The fungus then attacked the juice, appearing to grow out of the wood of the boxes.

Early in the life of the Australian colonies these diseases and the other pestilent imports were of little economic importance to the community. But by 1900 the fruit business had changed. In Victoria orchard growing was no longer a cottage industry. The emphasis had turned from home orchard production to closer settlement. Fruit was being grown for income, not for seasonal luxury. Growers produced too much fruit for the local market. The Victorian government was heavily committed to closer settlement policies. The surplus of fruit had to be exported. The government built coolstores in the major fruit-producing areas and government advisory officers encouraged production to meet export standards. The advisers tried to encourage the growers to produce what the market demanded:

> Nothing is more conservative in a man than his stomach. He will change his country, his king, his politics, his religion, and still adhere to the food of his fathers. Hence we may not hope to win people over to consume our surplus foodstuffs, because they are consumed and beloved by us. We must cease planting without an objective for the fruit when grown. Our first concern should be to increase the productivity and quality of existing orchards, and when this is done plant steadily those areas which are distinctly more favourable than many which have been employed in the past.[13]

The overseas markets required high quality, blemish free, uninfected apples that would travel, without deteriorating, to the distant European markets. But apple orchards were under siege from codlin moth, black spot, woolly aphid and mildew. Peach orchards were attacked by rot, leaf curl, peach aphid and oriental fruit moth. Scales threatened pear and citrus trees.

The first step in traditional disease and pest control was a clean orchard. Cleanliness was believed to help control the codlin moth by destroying pupating moths.[14] It helped control the spread of black spot by the burying of infected leaves. Removing loose bark and mummified fruit, pruning infected branches and ploughing beneath the trees to bury rotted fruit or infected leaves helped to limit the sources of infection that could threaten the following year's crop. Attempts were made to clean up stray codlin moths with lures, lights and traps.[15]

For effective disease control everyone in a district has to keep their orchard clean, but not everyone did. The South Australian government passed laws requiring growers to inspect, pick and destroy infected fruit once a week.[16] The regulations were not well enforced due to lack of funding. This led to heated conflict between growers who took differing opinions on the need for levies and stronger enforcement.[17] In Victoria the government attempted to ensure cooperation by passing the *Vegetative Diseases Act*, prohibiting growers from selling diseased fruit or trees. Inspectors were appointed to enforce cleanliness on those who did not maintain clean orchards. The inspectors razed old, abandoned orchards left to waste by speculators. The inspectors could destroy infected orchards or nurseries. Disinfecting steam baths operated at the central fruit and vegetable market to clean fruit boxes. The *Vegetative Diseases Act* helped to create a clean orchard ethos that persisted in following generations. An orchardist was judged by his peers on how his orchard looked, not on how much profit he made. An untidy orchard was a bad orchard. A good orchard was neatly ploughed, without a sign of weeds, trees neatly pruned in perfectly straight rows.

Despite advisers' claims that 'cleanliness is the first step to prosperity',[18] cleanliness was not a panacea for the orchardists' problems.[19] In South Australia scientific advisers expressed alarm at the increasing ravages of disease.[20] Other solutions were needed; at the turn of the century there were high hopes for biological control. Australia had helped to solve California's cottony cushion scale problem by exporting the scale's natural predators, native ladybirds and a parasitic fly. The predators had flourished in the California citrus groves and quickly tamed the scale, providing an impressive example of the value of biological control. Perhaps somewhere overseas there was a bug that would control codlin moth, fruit moth or the myriad of scales worrying the Australian orchardists? In each state economic entomologists advocated the use of natural controls for insect pests.[21] One government entomologist printed 9000 copies of a guide to natural controls.[22] The entomologists were constantly on the look-out for economically useful predators or parasites. One wet summer Mr Thiele of Doncaster, on the outskirts of Melbourne, noticed a fungus attacking the codlin moths on his apple trees. He and some of the state's agricultural scientists tried to make a spray out of the fungal spores, but the spray was ineffective.[23] Government economic entomologists tried to convince growers that the insect-eating birds living around their orchards should be protected from the sporting shooters and tree clearing. The Victorian Department of Agriculture produced a series of colour plates of birds to help growers distinguish between insectivorous and fruit-eating birds. The growers were ambivalent. They saw as many fruit-eating birds as insect-eating birds flying over their properties. Like the fungus that attacked codlin moth, the impact of bird conservation on orchard pests was not obvious. At one agricultural conference South Australian government scientists called on growers to plant trees to harbour insectivorous birds. Growers responded with tales of extreme losses of fruit because of fruit-eating birds and with calls for their destruction.[24] The disillusionment was summed up by one of the government scientists:

> There is no doubt that the method of pitting nature against itself is a most economical one, and that most satisfactory results might be expected from it,

but it would appear we have not yet learned how to apply these remedies to best advantage.[25]

The remaining weapon in the armoury was spraying. The tobacco-derived, nicotine-based sprays offered limited control of a few pests, but the success of these sprays seemed to be diminishing. Some growers had been forced to gradually increase the frequency of spraying until constant spraying had little impact on pest numbers.[26] Perhaps the aphids had grown resistant to the nicotine, or the nicotine was destroying predators. Growers looked for new ways to control the pests and diseases in their orchards. The answers came from overseas, particularly from France and the United States.

Blue Bordeaux, Paris green and red oil

In the 1880s the vineyards of Bordeaux faced attack from a particularly persistent pest. The local children took great delight raiding the grape-vines before the pickers arrived. The vignerons painted a mixture of blue copper sulfate and white lime as a paste on the grape bunches nearest the roads. The unsightly and ill tasting paste scared off the children, who believed they would be poisoned by what was a harmless concoction. In 1884 Bordeaux grape vines had a particularly bad infestation of mildew. Many grapes rotted, but the bunches near the fences were comparatively unscathed. The growers had serendipitously discovered the susceptibility of fungi to copper. In the next few year Bordeaux vignerons used the paste over large areas of vines. News of the mixture spread across the world, and the simple mixture of lime and copper sulfate became known as 'Bordeaux'. The effect of the spray on black spot, mildew and peach leaf curl was impressive and growers quickly accepted this new treatment.[27] By the turn of the century Bordeaux was commonly used in Australian orchards. Each grower prepared his own spray in an arduous and messy process, mixing separate suspensions of copper sulfate and lime, and then gradually combining the two while continuously stirring the mixture. Bordeaux was not effective in controlling brown rot, so the search for fungicides continued.

A few years after Bordeaux's dramatic arrival, European growers discovered that arsenic-based dyes were extremely effective in controlling leaf-eating insects. In solutions innocuously called 'London purple' and 'Paris green', arsenic became as popular as Bordeaux in apple orchards around Australia.[28] Initially orchardists used Paris green in addition to the usual cleanliness and bandage controls because it was not totally effective against codlin moth.[29] By 1910 Paris green was replaced by lead arsenate. Lead arsenate was particularly effective for controlling codlin moth because it remained potent for longer on the apple.[30]

After fungi and leaf eaters, the other major orchard pests were leaf suckers: aphids and scale. Scale did not eat leaves, so could not be poisoned by arsenic. The oil industry provided a partial solution. Spraying unrefined

Citrus trees prepared for fumigation with hydrogen cyanide

light petroleum oil on dormant deciduous fruit trees killed the aphid eggs. When combined with nicotine in summer, this 'red' oil controlled peach aphid. Citrus trees did not have a dormant period when growers could spray red oil on their evergreen orange and lemon trees, without damaging the growing leaves. They resorted to the most dangerous of the inorganic controls: cyanide gas. Cyanide gas was produced from the highly poisonous cyanide of potassium and sulfuric acid. This American idea involved putting a tent over an orange tree and fumigating it with hydrogen cyanide. The production of the gas was dangerous work. Growers had to set up the tent, bury its walls to make it airtight, mix the chemicals inside the tent and retreat quickly before breathing any of the fumes.[31]

The growers relied increasingly on these inorganic chemical sprays. Traditional pest control work was time consuming. Growers saw no reason to continue bandaging apple trees to trap codlin moths, so the practice lapsed.[32] The use of these chemicals strengthened the concept of calendar spraying, spraying at a particular time, rather than when indicated by the pest populations. Bordeaux was a preventative spray and farmers had to learn to use it before the fungal infection occurred.[33] The spraying ritual was reinforced by the experience of the 1895–1901 drought; the dry conditions effectively controlled the fungus, so many orchardists stopped spraying to control black spot. When the drought ended, the wet weather brought record fungal infection to orchards. Many trees were destroyed. Scientists derided the orchardists for slackening off the task of spraying during the good conditions in the drought.[34] Spraying was seen as an insurance, which demanded regular premium payment.

After the First World War, Victorian government enthusiasm for rewarding the returning soldiers with land led to the planting of more apple and pear orchards in the Goulburn Valley.[35] The domestic market could not absorb the increased production, so much of the fruit was exported. Profitable returns from canned fruit allowed peach trees to be planted for the export market. But returned soldiers and farmers were planting new orchards in many other countries. World production increased, prices fell and the pressure for financial survival became more intense. A farm adviser of the day described the growers' predicament:

> Whilst our early growers had their worries and troubles . . . I am of the considered opinion that the present day grower has still more problems. For instance he has more competition to meet in the world's markets; he has more pests and diseases to fight; the cost of production is about 100 per cent higher than it was ten to fifteen years ago. Today's grower, in order to provide himself with today's living requirements, really has to produce three cases of apples for every one produced fifteen years ago.[36]

The export markets were increasingly selective regarding fruit quality, but it was becoming harder to produce quality fruit. The insect pests were becoming less susceptible to the chemicals used for their control. Lead arsenate did not kill codlin moth as readily as previously. Each season's population of moths became more resistant to the chemical. As lead arsenate lost its power, growers sprayed more often, accelerating the process. Even cyanide seemed to be less effective. Scale was taking less time to reinfest fumigated trees. Growers blamed the lax farming methods of their neighbours who did not control scale, and they demanded that the government enforce pest control on recalcitrant farmers.[37] But the recalcitrant neighbours may have actually been helping to control the scale. The cyanide was killing off the scale's predators faster than it was killing the scale, and after each successive fumigation the scale took less time to infest the tree. The neighbours' unfumigated trees were the only refuge for the predators.

The response of most farmers to increased resistance was to increase the strength or frequency of spraying. Overseas consumers were unimpressed with apples shipped with the lead arsenate still clinging to the skin of the fruit. Britain threatened to ban Australian apples if the residue levels did not decrease. To protect the export market, scientists advised farmers to change their management practices to decrease the use of lead arsenate. This was not as simple as it seemed. Growers were advised not to spray within a month of harvest, but to use nicotine instead in this period. The scientists claimed it would work, although twenty years earlier they had agreed with farmers that nicotine was no longer effective.[38] Later, arsenic was mixed with oil to dilute the arsenic and reduce the amount deposited on the fruit. It also weather-proofed the arsenic against rain, allowing growers to spray less often.[39] This did not decrease the residues. Traditionally arsenic was removed by hand wiping each apple before packing. Hand wiping did not remove the oil and arsenic mixture.

Orchardists and scientists looked to other methods to control codlin moth. Molasses or apple juice mixed with lead arsenate was used as a lure to catch and kill flying moths. The lures, by indicating when infestations occurred, also allowed growers to spray the fruit more selectively.[40] The government established demonstration farms to show growers how apples could be produced using less lead arsenate.

Understanding of biological control had advanced since the earlier disappointments. Bird protection was still encouraged, but there were more impressive strategies.[41] Entomologists had learned that the oriental fruit moth was parasitised by several small wasps; the yellow fronted ladybird and the *Leis* ladybird were recognised as major predators of peach aphis. Growers were encouraged to use anti-aphid sprays less often to protect these predators.[42] The aphelinus wasp, a natural predator of the woolly aphid, was introduced into orchards to protect apple trees. The biological controls did not eradicate the pests, but merely limited their numbers. At the beginning of each season the grower who relied on natural controls had to watch the pest insect population increase until the predators could multiply sufficiently to effect control. In suitable seasons the early pest population could explode before any appearance of the predators and advocates of biological control had to convince farmers to wait for the predators, to be patient, and to accept some infestations.[43]

For fungi such as brown rot and insects such as codlin moth there was no biological control and farmers depended on chemical control measures. Other advisers recommended against moderation in the pursuit of higher yields of clean fruit:

> At times, it is doubted whether the additional five or ten per cent of clean fruit obtained justifies the extra cost obtained in procuring it. This argument is often advanced in a heavy cropping season. It should be remembered however that the effects of efficient control are cumulative. A control which gives five per cent more clean fruit in the first year probably will give double that increase in the following year. Then again, a heavy crop is usually followed by a light crop. The result is that a much higher percentage of fruit is lost, although, because of prospective high prices, greater care is usually taken by the orchardist. In this regard it may be said that the effects of neglect are cumulative.[44]

To the orchardist these were conflicting messages. For biological control, entomologists advised short-term loss to avoid long-term risks. For chemical control, growers were advised to avoid any short-term risk for long-term advantage. In the economic climate of the time many orchardists did not have the luxury of worrying about long-term risk. In the mid-1930s the apple market collapsed. In the next ten years one in every three apple trees in south-eastern Australia was abandoned or grubbed out.

The organic pesticides

One of the darker aspects of the Second World War and the cold war that followed was the acceleration of research into toxic chemical weapons.

From this morbid pursuit came a new family of pesticides based on complex hydrocarbon molecules. The beginning of the organic pesticide revolution came with the improvement of the simple red oil spray. Chemists discovered that the destructive effects of red oil on green vegetation were caused by unsaturated oil molecules. Removing these from the oil made it safe to spray oil on citrus trees in summer and winter. Chemists next discovered that refined summer oil contained ring molecules and long-chain molecules. The long-chain molecules blocked the insects' breathing holes. The ring molecules, which did very little, were removed, producing a more effective pesticide. The new oil was mixed with an emulsifier allowing growers to dilute the spray with water. The resulting white mixture was named summer oil, an innocuous compound, still used today on orchards and in home gardens.

Chemists added new foreign atoms to the carbon rings and chains. Some of the most spectacular compounds were made by adding chlorine. Some of these organochlorine chemicals proved to be extremely efficient insecticides; the most famous was DDT. It was nowhere near as poisonous to humans as lead arsenate, and yet small doses killed insects such as flies, mosquitos and fleas with great efficiency. The first spectacular coup with

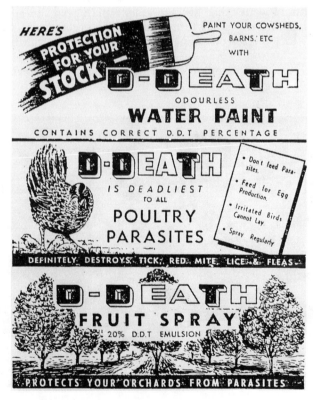

An advertisement for DDT about 1950

DDT was the defeat of a typhoid epidemic in Naples during the American occupation in 1944. Because DDT was a long lasting chemical, it proved a boon for the control of malarial mosquitos and helped to eradicate malaria in Brazil and Ceylon (now Sri Lanka). On orchards it was far safer to use DDT than lead arsenate to exterminate leaf-eating insects. Sold under trade names like DDeaTh, DDT gained acceptance for many farm applications.

Another group of chemicals, the organophosphates, were created by adding phosphate atoms to the carbon rings and chains. These chemicals were not as long lasting as the organochlorines, but some of them offered a spectacular advantage. Plants absorbed them into their leaves and distributed them throughout the plant. The chemicals flowing in the plants were toxic to leaf-sucking insects such as aphids and scale. These systemic chemicals were very poisonous to humans and governments banned several before their release. Others, such as parathion, were judged to be safe enough for orchard use, but not for use by home gardeners. The less toxic malathion and rogor were released for home and orchard use.

The chemical revolution in insect control was not initially matched by a similar revolution in fungus control. Organic fungicides appeared in the 1950s, but they offered little to peach producers. In the Goulburn Valley a six-year run of wetter than average years waterlogged and killed many peach trees. Brown rot took a firm hold in many orchards, causing heavy fruit losses. In some seasons some growers lost all their fruit because of brown rot.

A special research committee searched for a control for brown rot. The early organic fungicides did not perform well in the committee's experi-

Severe brown-rot damage in a case of peaches prepared for market

ments. Spraying at blossom time seemed to offer the best result, but the fungicides did little to save orchards where the infection was taking hold of ripening fruit. Strangely, the organochlorine insecticides offered more control than the fungicides. Oriental fruit moth carried the brown rot spore from fruit to fruit, establishing new infections as they pierced the skins of fruit. Complete control of the fruit moth with DDT reduced the losses from brown rot. Research also confirmed the importance of two existing strategies: winter Bordeaux sprays and orchard hygiene. Ploughing was still one of the most effective ways a grower could limit the spread of the infection from one year to the next.[45]

The ethos of the clean, cultivated orchard was still very strong in the 1950s. Peach producers had been advised for many years that 'frequent stirrings of the soil by cultivators, by hand hoeing and by ploughing, are essential to successful peach growing'.[46] Apple orchardists were told ploughing aerated the soil, conserved moisture, killed weeds and controlled black spot. An orchard might be ploughed up to ten times in the season. The danger of frequent ploughing was the depletion of organic matter and erosion of the soil. In the early part of the century orchardists replaced the lost organic matter by spreading manure on their farms. The manure was mainly dung and urine-soaked stable straw or droppings from poultry farms. The city fringe orchards were living on the nitrogen mined from the inland wheat farms. The manure came from horses that ate oats grown inland. The horse and fowl manure was supplemented by green manure: legume crops of peas and beans grown between the rows in winter and ploughed into the soil in spring. For the farmer who conscientiously followed these practices, the risks of ploughing were minimised.

The post-war urban enthusiasm for the motor car spelt an end to the easily available horse stable manure, which was replaced by newly available, manufactured nitrogen fertilisers. The nitrogen fertiliser did not give the organic bulk the manure had supplied, which had preserved the soil structure against the constant ploughing. By 1950 soil conservation advisers were warning orchardists of the dangers of erosion caused by ploughing.[47] Orchardists near Melbourne were advised to plant orchards on the contour rather than up and down the hill.[48] The few who followed this advice found their trees died of waterlogging and the collar rot fungus.[49]

The typical orchardist kept ploughing and planting green manure crops. When the soil eroded it was scooped from the bottom of the hills and carried back to the top of the orchard rows. For orchardists on the suburban fringe, the long-term problem of erosion hardly mattered. The orchards their family might have husbanded for several generations would soon be replaced by urban housing. The greatest financial reward for these orchardists came from the orchard's proximity to the metropolis, not from the fertility of the soil.[50]

The grassed orchard became a possibility in the 1960s with the introduction of systemic herbicides and fungicides. The systemic fungicide benomyl offered the first effective chemical control for brown rot. Once absorbed by the plant, benomyl could kill brown rot within the fruit. The suite of chemical controls was complete. With this combination, farmers could control both weeds and brown rot without ploughing. Within ten years the grassed orchard was a common sight.

The smell of sex

By the 1960s the number of agricultural chemicals had proliferated. There were over 300 agricultural chemicals registered in about 10 000 formulations and trade names across the world. As with the inorganic chemical revolution, twenty years of widespread use of the new organic chemicals was revealing familiar problems.

Pests were becoming resistant to the new chemicals. Some farmers had anticipated this problem.[51] In areas where farmers enthusiastically adopted the calendar method of spraying at a particular time, resistance to the chemicals appeared in several pests. After initial spectacular success, DDT resistance developed among codlin moths. The problems of resistance were not limited to insecticides. Unlike inorganic copper, the systemic fungicide benomyl is specifically toxic to a single fungal metabolic process. This specificity allowed the development of resistant strains of brown rot.

As serious as the growing resistance were the new pests that the pesticides encouraged. Two-spotted mites were one of these pests. These mites live on the leaves of stone fruit and pome fruit trees. They are very small spiders, which suck juice from leaves. One mite can lay 200 eggs, and a generation can hatch and lay within seven days in favourable conditions. In most circumstances the mites and their predators maintain a stable population balance, with mites predominant in spring and predators predominant in autumn. The short generation span of the mites, and their apparent genetic plasticity, gave the mites an advantage over other insects in the race to develop resistance to chemical sprays. Lead arsenate made the mite into a pest by differentially killing the predators and not the resistant mites. Mite-infested trees grew poorly and produced low yields. DDT not only killed the predators, but also stimulated mites to lay their eggs. Orchardists using DDT or other insecticides to control leaf-eating pests included a mite control spray in the spray program. Over the years the mites gained resistance to miticides.[52] Currently there are no effective registered chemical controls for mites.

A third problem was farmers' growing sense of familiarity with the new chemicals. Many farmers were apparently indifferent to the danger of the chemicals they handled. Some of the new organophosphates were extremely toxic. The anti-cholinesterase chemicals were closely related to the nerve gases developed in the Second World War. They disable the action of cholinesterase, a messenger chemical in nerve cells. The human body takes up to three months to recover the cholinesterase disabled by the chemical. Repeated exposure to the chemical has a cumulative effect unless there is sufficient time for the body to recover. The anti-cholinesterase chemicals were easily absorbed through the skin or by breathing the droplets. Orchardists who employed others to spray the chemicals were required by law to have antidote chemicals on hand for immediate use. But most orchardists did their own spraying, and the law did not protect them from themselves. Many did not know the toxicity of the chemicals they used, did not understand the labels, did not have antidotes and were indifferent to safety. Sprays were used with little regard to others in the vicinity. Many growers ignored the warnings and sprayed without full protective clothing.[53] Cases of fatal poisoning were fortunately few, but free blood testing at agricultural field days showed many cases of sub-clinical poisoning.[54]

The initial post-war enthusiasm for organochlorines and other pesticides was gradually replaced by a search for less chemically dependent ways of controlling the non-fungal insect pests and for ways to reduce the need to use sprays in orchards. One method was to encourage farmers to limit spraying to periods of greatest disease risk by issuing warnings of periods of high disease risk. In the 1960s the Department of Agriculture started a black spot warning service for apple growers in the Red Hill district in Victoria. The services failed because growers were loath to change from spraying as an insurance to spraying strategically according to the conditions.[55]

In the late 1960s a predator mite was discovered that had grown resistant to sprays in the same way as the normal mites on which it fed. The predatory mite was tested by the CSIRO and introduced to Australian orchards in the mid-1970s. In the Goulburn Valley nearly half the farmers released the predator mite into their orchards in the first year of its introduction.[56] These farmers had to learn to spray to control, not eliminate, the mite. If they eliminated the normal mite, they eliminated the predator mite. They had to learn to spray strategically, only when the mite populations justified it. When combined with the brown rot warning service, which advised growers when climatic conditions were conducive to brown rot infection, orchardists had the beginnings of an integrated system of pest management to minimise chemical spraying.

The missing component of the whole system was control of oriental fruit moth. Past attempts at biological control had been only marginally successful. Farmers still needed to spray against it, risking destruction of natural predators of mites. The best that could be done was to count the populations to establish when the infestation of moths was sufficiently high to warrant spraying. As the moths only came out at night they were difficult to count. The solution was to place female moths in cages inside traps. The scent of the female acted as a lure to males and a count of the contents of the trap the next morning gave a reasonable indication of moth populations. Scientists realised that if it was possible to artificially manufacture the chemical which attracted the male moths to the females, a lure could be made for moth control. By the 1980s chemists had synthesised the chemical attractor. When orchardists hung strips of material impregnated with this pheromone in their orchards, the smell of female moth confused the males and interfered with mating. When all the orchardists in a district used the pheromone strips, the oriental fruit moth population was conspicuously reduced. Both brown rot control and biological control of mites was made simpler. Using this system, spray usage in orchards has been cut by two-thirds, while fruit quality has not been affected.

This system of pest control is an example of what has become known as integrated pest management (IPM). IPM strategies are designed to minimise pesticide use on the farm while maintaining the marketable quality of produce. IPM is based on four principles: spray only when the level of pest infestation reaches a threshold of economic significance, or when the climatic conditions threaten a disease outbreak; use non-chemical controls wherever possible;[57] avoid using sprays in a way that limits the effectiveness of natural controls; and avoid spraying regularly with the one chemical to slow the development of pest resistance to chemical controls.

Natural farming

The development of IPM strategies in the 1970s and 1980s coincided with greater restrictions on chemical availability. Pest resistance, predator imbalance and farmer health are important to agriculture, but they are not things that normally stir public opinion. It was another matter that brought the use of the new pesticides to public attention: the cumulative effects of DDT and other organochlorines on wildlife and the wider community. In 1962 a book, *Silent Spring*, written by the US author Rachel Carson, gave an account of the environmental impact of organochlorines.[58] The book caught the public's attention in its first pages with an apocryphal scene of destruction:

> . . . a strange blight crept over the area and everything began to change. Some evil spell had settled on the community: mysterious maladies swept the flocks of chickens; the cattle and sheep sickened and died . . . The farmers spoke of much illness among their families. In the town the doctors had become more puzzled by new kinds of sickness appearing among their patients. There had been several sudden and unexplained deaths, not only among adults, but also among children, who would be stricken suddenly while at play and die within a few hours.[59]

This was an extreme scenario created to publicise the dangers of pesticide residues. DDT and similar chemicals used in agriculture were entering the natural food chain. They accumulated in body fat and concentrated in animals at the end of a food chain. One of the most publicised ecological issues of the time was the concentration of DDT in eagles, falcons and other birds of prey. The pesticides were weakening the birds' egg shells and threatening their survival. More ominously, organochlorine residues were measured in human breast milk. *Silent Spring* was the beginning of the inevitable end for the use of organochlorine pesticides, the last chapter of which is only now being played out in Australia.

Carson's tales of indiscriminate agricultural chemical use and the unintended effects of organochlorines rallied the environmental movement and urban consumer concern in the United States and then elsewhere in the world. *Silent Spring* precipitated the United States' progressive prohibition of agricultural uses of organochlorines over the next fifteen years, a process that culminated in the quarantining of 1500 Australian grazing properties in the late 1980s.

The progressive prohibition of organochlorine chemicals had little effect on pome and stone fruit orchardists. Other chemicals replaced DDT and Endrin Oil in spray programs to control codlin moths, oriental fruit moths and aphids. The publicity surrounding organochlorine residues in food presented another challenge to the orchardist. In the late 1980s, public concern about organochlorine contamination of beef presented product quality and marketing challenges for orchardists. The use of pesticides was publicly debated and the demand for produce grown without pesticides increased.[60] The alternative farming movement was not new, but it had been little more than a fringe group for decades. Suddenly a larger group of consumers was demanding its products.

The organic farming movement started in England in the 1920s and 1930s, amidst the concern over arsenic residues in food. The movement was based on a belief that the inorganic pesticides and fertilisers of the day were bad for the health of humans and the soil. The members of the movement eschewed the use of artificial fertilisers and sprays, and advocated a farming strategy aimed at indirectly nourishing plants by feeding and maintaining a healthy soil, returning to traditional cultural practices such as ploughing, composting, crop rotation and companion planting, and the use of natural chemicals and biological control. The movement believed the best way to fight pests was with healthy plants, and the healthiest plants were produced by the nutrition created by the action of soil microbes. Soil deficiencies are made good with ground mineral rocks rather than artificial fertilisers. The Soil Association of England today runs a farmer advisory and accreditation service. Similar organisations exist in the United States and European countries.

The biodynamic movement, based on Rudolph Steiner's philosophy of social and natural relationships, started in the same period as English organic agriculture. Biodynamics shared the same concern for soil health as organic agriculture, but its scope widened to include concern for the relationship between the farmer, his soil and society. The farm was viewed as part of a whole web of relationships, both organic and social. Ideally, a farmer made a life-long commitment to his land with the aim of healing the damaged earth. In contrast, chemical farmers were seen as seeking the maximum yield from their land with little regard for the ecological or social consequences. There are major differences between biodynamic and organic farming. The biodynamic philosophy has not always been as strongly anti-chemical as the organic movement. Some Australian biodynamic advocates have contended that judicious use of artificial fertilisers can be necessary to achieve the appropriate *balance* on a farm.[61] The biodynamic movement is more willing to assist natural processes than the organic movement, believing in human intervention that is in harmony with nature. Many of the recommended practices have a mystical dimension that stretches traditional scientific credibility.[62] Adherents of biodynamic agriculture have their own regulatory system, education program and marketing brand for product verification and differentiation, and conduct research programs in Sweden, Switzerland and Germany. Many in the biodynamic movement have a general distrust of artificial fertilisers and chemicals, and a strong distaste for corporate agriculture and agricultural science. The views of Alex Podolinski, a leader of the Australian biodynamic movement, illustrate this position:

> What is agri-culture actually about? Is it to establish scientific theories and to prove these by costly and time consuming statistical evidence? A history of this evidence would disclose how little notice has often been taken of the results and then been contradicted in later years. Furthermore can this current science method be so entirely trusted? It did, along with many others, pronounce DDT and Thalidomide as safe. Of course, such a system exerts a powerful hold and has to continue increasing — to justify itself.[63]

The United States has its own alternative agriculture movement. Increasing concern over pesticides in the 1960s and 1970s led to a tightening of restrictions on the use of pesticides and a transfer of some of the costs of chemical use from the general community to farmers. Increased understanding of the costs of chemicals and the realisation of the advantages the chemicals gave to large-scale vertically integrated agriculture aroused suspicion in the minds of small farm communities, giving rise to a new interest in organic agriculture. This 'eco-agriculture' movement has more in common with the biodynamic movement than the organic movement.[64] Its members are distrustful of large corporate agriculture, scientists and university experts and resent the impact of commercial innovations on the social structure of rural America.

The various alternative agriculture movements have their advocates and adherents in Australia. As well as these, Australia has its own pioneers of alternative agricultures, people who have developed systems applicable to the peculiar problems of Australian agriculture. In the 1950s P. A. Yeomans published his book on keyline agriculture, discussed in Chapter 2. The keyline approach was developed in response to a particularly Australian problem: shortage of water during part of the year. The system is based around an integrated farm water storage plan, with dams constructed in gullies along the keyline, the break of slope between the hills and the valleys of the farm. Contour channels are built to distribute the water from dams for flood irrigation of the lower parts of the farm. Keyline philosophy, in common with organic agriculture, emphasised the building up of the living soil, achieved by gradually deeper chisel plough cultivations and the establishment of subterranean clover pastures. Yeomans criticised the application of fertilisers to the surface of the soil without building up the lower soil to improve the availability of mineral nutrients. By improving soil structure and absorption capacity, and by building dams and contour channels, Yeomans offered a solution to the problem of erosion that bedevilled many hill farms in the 1950s.

While Yeomans publicised his practices on changing the water balance on farms, in Tasmania Bill Mollison was developing a new style of agriculture based on eliminating the monocultural design of commercial agriculture, which contributed to the development of pest and disease problems. In his permanent agriculture or permaculture system, trees are planted in a plan to recreate something resembling a natural ecosystem, with micro-environments created to benefit particular species groups. The aim was the establishment of a permanent, self-sustaining, regenerating productive system. The initial emphasis of permaculture was towards:

> . . . small groups, living on marginal land available cheaply, where the ethics of farming are aimed at a future, and different, lifestyle, and where regional self sufficiency is more important than cash cropping for export, or monoculture for commercial gain.[65]

Permaculture has had strong appeal to refugees from urban life who have settled on country properties. It has little appeal for commercial broadacre Australian farmers. Mollison believes most of the land needed to feed Australia's population can be found within existing towns and that the only long-term future for the large agricultural property will be either water supply, forest or restricted meat production. The implications of this scenario for Australia and its economic system are extreme.[66]

Since the dieldrin residue controversy farmers have been increasingly courted by advocates of chemical-free agriculture and by advocates of IPM promising a cleaner agriculture. The chemical-free advocates run regular seminars, field days and conferences, mostly on members' farms. The seminars offer a promise of profitability on organic farms, together with an increased sense of satisfaction gained from looking after the soil. Chemical companies run sophisticated advertising campaigns, and employ consultants to run groups to help promote the effectiveness of their products, including high yields, greater profits and the satisfaction of looking after the soil.[67] As the costs of developing and registering agricultural chemicals have increased, companies have an increasing financial incentive to promote moderate chemical usage to slow the development of pest resistance and prolong the commercial life of their products.[68]

Playing a middle role between the chemical companies and the opponents of agricultural chemicals are governments. It is clear that government departments of agriculture believe any dramatic regulatory reduction in the use of agricultural chemicals would lead to significantly lower agricultural production and a lower standard of living for the country.[69] Governments are also concerned about the impact of pesticides on the environment and public health, and the advantages of being seen by overseas markets as a source of 'clean' produce.[70] The outcome is a government policy to promote 'clean agriculture'. A 'clean agriculture' is an agriculture that uses enough agricultural chemicals to maintain production yet minimise the risks associated with agricultural chemical usage. For most governments the components of a 'clean agriculture' program are research and promotion of integrated pest management, education programs to encourage responsible chemical use and the regulation of those involved in selling, promoting and using agricultural chemicals.

One of the motivations for 'clean agriculture' programs is to maintain public confidence in current patterns of agricultural chemical use. Market research shows that the public believe that chemical sprays, even when used within regulations, are a danger to public health, that it is practical to grow our food without pesticides and that chemical residues in food are increasing.[71] Though government policy makers can marshall convincing evidence that each of these statements is untrue, their fear is that a public loss of confidence in current agricultural practices could lead to increased regulation and decreases in production. The basis for dismissing consumers' concern about agricultural chemicals is a faith in the current system of chemical regulation. Government departments believe the results of residue monitoring justify their faith.[72]

Poisons in our food

Regulatory authorities seek to protect the public by setting a maximum residue level for each pesticide. The level is based on experimental testing to infer the 'no observable adverse effect' threshold. This threshold is the highest dose level that will produce no observable adverse effect in the most sensitive test animal. The 'no observable adverse effect' threshold is used to calculate an 'acceptable daily intake', the average daily dose that will produce no appreciable effect during an entire lifetime. The 'acceptable daily intake', together with information about dietary habits, is used to set a 'maximum residue level' that will ensure that 'acceptable daily intake' levels will not be exceeded. The 'maximum residue level' may be set lower than indicated by toxicological work if a lower residue level can be achieved with good conventional agricultural practice.

The setting of maximum residue levels depends crucially on the setting of the 'no observable adverse effect' threshold. There is a great difficulty in defining this threshold when searching for sub-lethal and carcinogenic effects. This threshold is often determined by experimentation with animals. The ability to detect a carcinogenic effect depends critically on the size of an experiment. A small experiment with only a few animals will only detect carcinogenic effects caused by large doses. With increasing experimental size it is possible to detect toxic effects from smaller doses. Thus, no matter how large the size of an experiment, scientists who claim to have detected a 'no observable adverse effect' threshold are left with the question of whether a still larger experiment would have detected a toxicological effect at a lower dose. One of the largest recent toxicological studies used 24 000 mice over five years at a cost of US$7 million, only to leave this question of low-dose responses unresolved.[73] To claim a scientific proof of a 'no observable adverse effect' threshold is misleading. It could equally be argued that only a few molecules could be carcinogenic and the failure to find a 'no observable adverse effect' threshold is merely a reflection of an inadequate experiment. The existence of a 'no observable effect' threshold cannot be *proven* by experimental techniques. It can only be inferred. That is the nature of science. Likewise, it is difficult to establish epidemiological evidence that there are no chronic effects from long-term low-dose exposure to chemicals.[74]

It is superficially attractive to conclude that because science cannot show the synthetic chemicals to be absolutely safe, they should be eliminated. But the same arguments can be applied to natural chemicals. Is a cancer risk from a natural chemical any better than a risk from an artificial chemical? Until recently natural chemicals were rarely tested for carcinogenicity. However, new tests of carcinogenicity and mutagenicity have revolutionised the science of toxicology. Using cultured bacteria, chemical risk can be quickly evaluated. The new tests have revealed research fakery by some companies and revealed weakness in old data. Some artificial chemicals once thought safe have been banned.[75] The new tests, used on natural chemicals, have shown that nature is not benign. Many of the natural flavour constituents of common foodstuffs are potentially carcinogenic or mutagenic. Peanuts contain highly carcinogenic aflatoxins; alfalfa sprouts contain the natural pesticide canavanine; celery, parsnip, parsley and figs contain carcinogenic furocoumarins; pepper, mustard and horseradish contain carcinogenic compounds such as safrole, pipperine, isothiocyanates and phorbol esters.[76] The toxicologist Bruce Ames estimated that in our daily diet we consume between 1 and 2 grams of these natural carcinogens each day. This is 10 000 times the average intake of artificial pesticides. If these chemicals were manufactured synthetically and presented as new food additives, they would not be registered. The natural chemicals are quite capable of producing their own 'silent spring'. In California, a sudden run of birth deformities in one family involving a child, a litter of goat kids and a litter of puppies was blamed on herbicide spraying. Later investigation showed that the pregnant mother and dog had been drinking milk from the family's goats, which had been feeding on lupins. Alkaloids in the lupins had been transferred into the goat's milk, causing the birth defects.[77]

Plants manufacture these natural pesticides as defence against insects and fungi. While wholesome healthy looking vegetables or fruit may have these hidden risks, damaged or 'off' food can be much worse. Many plants increase their toxin concentrations when damaged or under attack. The attackers, in turn, manufacture their own array of toxins to overcome their prey. Moulds generate aflatoxins and sterigmatocystins. These are extremely carcinogenic compounds. The ergot fungus that attacks rye is a potent psycho-active drug. Earlier this century infected rye found its way into the flour used by a village bakery in France. Most of the villagers suffered terrifying hallucinogenic trips; some suffered permanent brain damage. Consumers' historical preference for undamaged food can be explained by damaged food being more likely to be poisonous. In a sense, quality is safety.

It is clear natural carcinogens cannot be eliminated from the human diet. Breeding for disease resistance, a strategy of non-chemical pest control, is likely to increase the intake of natural pesticides. It will create plants with higher concentrations of these chemicals. Breeders have been

attempting to increase the disease resistance of domestic lettuce by crossing with wild lettuce to transfer the gene that produces a mutagenic natural
pesticide from the sesquiterpene lactone family. In the United States pest-
resistant varieties of potato and celery have been taken off the market
because they exceeded maximum desirable levels for glycoalkaloids and
psoralin derivatives. Applying the conservative low dose argument used for
artificial pesticides to the natural pesticides suggests the counter-intuitive
conclusions of supporting plant breeding to reduce the level of natural
toxins in food.[78] This has occurred for centuries. Potatoes were once much
more toxic than they are today. The Incas processed them to remove the
poisonous chaconine and solanine to make them safe to eat. Today's
potato is the product of a long period of selection to reduce toxicity. In this
sense, the safe potato is an unnatural plant, as are many of the staple foods
in our diet. Yet, if we were to reduce the level of natural pesticides in our
food, we would need to use many more artificial chemicals for pest control.

We have to accept some risk, whether it be from natural or synthetic
chemicals. The risk attached to any chemical should not be judged in
isolation, but in the context of the wider range of risks that we bear as part
of living. Natural chemicals should not be treated differently from artificial
chemicals. The dangers from each should be assessed and decisions made
on the best available information.

The key to a decision about risk is information. Consumers can choose
to accept the assessments of the regulatory boards that evaluate and register chemicals for agricultural use. For those who choose to use chemical-
free products, there needs to be a credible certification of the status of
organic produce. On the farm there is a need for reliable and credible
information about the basis of organic food production. Advocates of
various persuasions are apt to make sweeping claims for the sustainability
and profitability of their chosen farming system. Some claims are credible,
some are incredible and others are somewhere between. Both organic and
conventional growers want information about the many unanswered questions. The basis of a change to organic food production is a change in farm
management techniques. There is little incentive for private firms to undertake private research. This is a task that requires government support.
Properly designed independent research could answer many of the concerns of conventional farmers about claims made of organic farming for
practices such as companion planting. Australian governments are just
beginning the task, a decade behind their overseas counterparts.

With respect to plant health, a most important role of the national
government is the operation of the quarantine service. It is interesting to
speculate on Australia's usage of chemicals if an effective quarantine service had been operating since 1850. Farm management would be simpler
and involve fewer chemicals if brown rot, peach moth and black spot had
not yet reached Australia.

In the late 1980s a fierce debate raged between orchardists and the
Federal Government over fireblight and fruit imports from New Zealand.
Fireblight is a virulent bacterial disease that attacks pear and apple trees. In
a wet cold season it can devastate a pear orchard, destroying the growing

tips of the trees. The disease is present in New Zealand, where growers are forced to inject antibiotics into their trees as a defence against the disease. For many years Australia had banned the importation of pears and apples from New Zealand to protect the Australian industry from this disease. The Closer Economic Relations Treaty will end this ban. Orchardists believed that the Australian apple and pear industry was at risk, arguing that conditions in southern Australia would suit the blight and large numbers of trees could be destroyed.[79] To the orchardists it seemed strange that the Federal Government was happy to explore the distantly possible dire consequences of the greenhouse effect on their industry, yet remained comparatively unmoved by the threat of the industry's vulnerability to fireblight.

There is some wisdom to be gained from reflecting on Rachel Carson's apocalyptic vision of death and destruction caused by agricultural chemicals. In the developing world this vision has proved at times depressingly accurate. The World Health Organisation has estimated that each year across the world some 3 million people suffer acute poisoning from agricultural chemicals, leading to 220 000 deaths. WHO also estimates that in each year there are a further 735 000 cases of chronic poisoning and 37 000 cancer deaths caused by agricultural chemicals.[80] Most of these cases occur in the developing world. We in the developed world hear of the most gruesome mass poisoning tragedies such as the chemical leak at Bhopal or the mass poisoning in Iran caused by the milling of pesticide-treated grain. By far the greatest cause of fatal poisoning is suicide, estimated at being over 90 per cent of all pesticide fatalities.

In Australia we have been spared these tragedies. The 'evil spell' predicted by Rachel Carson has not 'settled on the community'. With hindsight Carson's vision appears exaggerated and overstated. However, her vision played an important role in changing the emphasis of conventional pest control towards integrated pest management. Overstatement was a useful rhetorical tool in arousing public concern. The question we face now, thirty years later, is how far we travel down the path of alternative agriculture. If we are to accept the evidence of most farmers' behaviour, it seems *Silent Spring* also overstated the viability of the non-chemical alternative. Unlike consumers, most farmers concerned about agricultural chemicals see no clear alternative, only difficult choices and ambivalent decisions.

In the early days of IPM, orchardists who abandoned calendar spraying and introduced predatory mites were attracted by the possibility of reduced spray usage and lower costs.[81] The same advantages are available today. Experience with IPM today indicates that orchardists may be able to reduce their chemical costs by up to 80 per cent while maintaining or increasing fruit quality. The increased publicity of the disadvantages of agricultural chemicals has changed orchardists' opinions of chemicals. Many orchardists' have broadened their 1970s concern about the high cost of chemicals to embrace concerns about their health, their soil and especially their markets.[82] But it is generally accepted that this concern and interest has not yet been sufficient to cause a major shift to chemical-free fruit production or significant increases in the adoption of IPM in orchards.[83] The adjustments required of farmers are not simple to achieve. IPM programs and organic agriculture are more complex than calendar spraying. Greater understanding of farm ecosystems is required. Systems of biological control are slow and risky in the short term. Pests develop into large populations early in the season before the predators have time to catch up and achieve control. This is easier to bear if the pest is a mite that does not directly attack fruit. Where a pest directly attacks fruit, large losses can ensue. The use of chemical controls achieves a guaranteed, immediate short-term effect, even if the long-term effect is unclear. Orchardists are naturally reluctant to risk major crop losses waiting for the slower responding biological controls. There is hope though that this may change. Recent research in the Goulburn Valley showed that many are interested in discovering more about IPM and testing it on their orchards.[84]

Many farmers also know of, and express interest in, organic farming.[85] However, like IPM, adoption of organic agriculture does not yet match this interest. Despite claims that organic agriculture is one of the fastest growing forms of agriculture, it is in reality a very small industry, producing significantly less than 1 per cent of domestic consumption.[86] As with IPM, organic farming advocates promise farmers financial advantages. Net returns have been found to be similar for chemical-free and conventional farming in non-horticultural industries.[87] Lower costs and lower yields are balanced by higher unit returns. The entry to true organic farming is perhaps even more difficult than a transfer to IPM. One of the barriers to entry to organic production is that the period of initial adjustment as chemicals are removed from the farm system is difficult. Advocates of organic cropping maintain that it takes time for a farm ecology to adjust to the loss of chemicals. During this weaning period there may be financial difficulties.[88] These financial difficulties are exacerbated by the waiting period until produce may be certified as organic. Conventional farmers with high interest commitments are sceptical of even the short-term financial sustainability of such a change. The Australian Conservation Foundation is advocating direct government financial support for farmers trying to make the transition.[89]

Conventional horticulturalists are also doubtful of the basic claims of supporters of organic farming, let alone the mystical elements associated with some organic farming systems. Many do not believe it is possible to profitably produce fruit or vegetables without using chemicals.[90] Despite the professed interest of many consumers in chemical-free produce, most consumers demand well presented, large, blemish-free produce. Blemished produce sells poorly and, in oversupplied markets, it does not sell without an *organic* label. This has encouraged less than scrupulous producers to dishonestly label produce, leading to pressure for a government supervised certification scheme to replace the plethora of voluntary schemes.

There are also marketing doubts about the long-term viability of the organic produce market. The relative profitability of conventional or chemical-free farming will depend on the demand for, and supply of, the respective products. Despite its rapid growth, the organic produce market has remained quite small, and demand and supply have been finely balanced. Will price premiums disappear as production increases?

The differences between orchardists committed to IPM and orchardists growing produce organically are not as great in practice as they are in rhetoric. The certification schemes for both the National Australian Association for Sustainable Agriculture and biodynamic farming have three levels of chemical-free production, the lowest level of which certifies produce sprayed with a minimum of artificial pesticides or fungicides. This is to make allowance for the unavoidable situation where a farmer faces a major crop loss because of some fungal or pest disaster. Production at this level is similar to the IPM approach of spraying only when deemed necessary. Farmers attempting to achieve either method of production face similar challenges to their management skill. The key to their success is maintaining adaptability in their farm ecosystem and flexibility in their management systems. Agriculture will always demand flexibility.

A lesson from any historical assessment of pest control is that reliance on any one control measure is ultimately doomed to failure, whether the control be artificial or natural. Past generations have believed they had sorted out the problems of pest control, only to be proven wrong. Pests develop resistance to natural or artificial chemicals. They adapt to plants bred for insect resistance. Natural selection usually adapts to subvert controls aimed at a pest's life cycle or behaviour. Flexibility and adaptation are the keys to sustainability. It is hard to see any pest management system being permanently sustainable.

Part Three

A green vision

10

The thirst for water

Over a period much longer than the white settlement of Australia the Aborigines from New South Wales and Victoria met annually in the Snowy Mountains. Having harvested the Bogong moth, they would return to their respective localities. In 1949 the Federal Government decided to harvest the water of the Snowy Mountains to provide hydroelectricity and irrigation for New South Wales and Victoria. This was one of the last major irrigation projects in Australia. It was widely supported as an expression of national development; very few questioned the advantages that the expansion of irrigation farming would bring. Other major irrigation projects on the tributaries of the Murray River, the expansion of the Hume and Eildon Dams in the 1950s and Dartmouth in 1978, completed a lemming-like rush to water for the inhabitants of the world's driest continent. The enthusiasm was even reflected in children's story books. (See box: *Murray, Bidgee and Snowy*!) Between 1950 and 1980 the area of irrigated land doubled and water storages for irrigation quadrupled.[1] The nemesis of the Snowy and these other dams has been a legacy of rising water tables and salinity in many of Australia's irrigation areas. The salinisation of irrigated lands seeped into the public consciousness in the 1980s. Bureaucrats, engineers, agricultural scientists and farmers knew of it earlier, there was plenty of evidence of failures of irrigated agriculture and horticulture in the 1950s when the last great irrigation engineering feats were being wrought.[2]

Early antecedents

The thirst for irrigation water in Australia has long established antecedents. Farmers in irrigation areas of Australia have their own culture, their own way of seeing themselves, possibly even their own political party. Although it is the irrigators' culture today, it is not a culture that irrigators developed. Irrigators have merely maintained and adapted beliefs that had their roots in the stand of the Eureka miners, the selection movement, the

For the children: Murray, Bidgee and Snowy!

About the same time that Snowy was born two neighbouring snow fields, just across the mountain, had children too. Like Snowy they quickly grew strong and venturesome and before they were a week old they went dancing down the Western side of the mountain.

When they reached the plain they were wide rivers, ready to help all they could. And this was just as well, for the plain was hot and dry. So they did their best with Man's help to grow things wherever they went, grass and trees, cotton and rice, oranges, lemons and grapes. And Man built his home near their banks.

But however hard they worked, they never seemed to be able to finish their task. There was always more to do. Man became sad and worried because there was not enough food for his kind who were coming to live in Australia from all over the world. And there was not enough food because there was not enough water to grow the food.

'How we wish he would come over here and join us where he is really needed,' Murray and Bidgee said to one another. 'Over there he is wasting his strength.' So Murray and Bidgee spoke to Man about it. 'Can something be done to get Snowy to come over and give us a hand?' they asked, looking tired and discouraged after a hard day's work.

From *Snowy: The Story of A River*, by Howard Guinness, Anzea Books.

squalor and depression of eighteenth century urban life and the patriotism of the First World War. Understanding the complexities of the degradation of irrigated lands means understanding the politics of irrigation, politics that grew from these beliefs. The initial settlement of Mitchell's Australia Felix was by a rush of fortune hunters. They were graziers who, acting in defiance of the colonial government, sought to steal land from the decimated Aboriginal tribes. Within fourteen years of Mitchell's glowing account of the northern plains of Victoria in 1836, all land suitable for grazing was occupied by a small number of squatters. Squatting was a rich man's game. It required significant capital and it involved risks that the undercapitalised could not afford. Most ordinary men were excluded and viewed the squatters' position with envy. Poor immigrants had come to Australia with visions of freely available land.

Social activists such as Caroline Chisholm promoted Australia as the land of opportunity. They believed settlement in Australia offered a new hope for the agricultural labourers of Britain and Ireland where labourers lived in poverty and squalor under the oppression of a landed class. Immigration schemes lured agricultural workers with the promise of new prosperity based on available work and land. The schemes were supported by the squatters, who desperately needed labour, and English landowners, who wanted to get rid of troublesome and unwanted labour. When these immigrants arrived, they found the land in the hands of a small group of opportunists unable to manage their runs by themselves and now seeking labour. Many of the immigrants became agricultural workers again. The severe shortage of labour meant wages were very high and employers had to take any worker available. The early annals of Victoria record many instances of poor relations between masters and servants. The masters complained that servants were failing to show due respect. The shortage of labour in Australia turned the English deference to employer by employee on its head.

To formalise the land occupation, which the colonial government had not controlled, squatters were given recognition in the form of a lease, renewable at yearly intervals. In the late 1840s, after a depression and the occupation of all suitable land, the squatters engaged in a campaign to obtain greater security of tenure. Using their political muscle in Sydney and London, they succeeded in having the government pass the 1847 Orders in Council. These orders allowed the governor to grant fourteen-year leases that recognised ownership of improvements and gave the squatter an option to purchase at the end of the lease. Individual squatters needed then to convince the governor to approve applications for their individual leases under the Order in Council. Unlike the governor in New South Wales, who quickly formalised the position of many squatters, Governor La Trobe in Victoria delayed, siding with the interests opposed to squatters.

The sudden influx of a new fortune seeker, the gold digger, changed the political landscape to the squatters' disadvantage. The diggers' initial interest was gold. Alluvial and surface gold only lasted for a few years. The average digger did not have the capital to work the deeper seams. Only a minority made their fortune on the fields. For the others there were only two choices: either join the next rushes to New Zealand, California and Queensland or stay in the colony of Victoria and seek employment. For many diggers, neither of these options was attractive. The squatters' land was more attractive. It was vast, and it appeared to be under utilised and unfairly held. In Victoria the settled land was held by fewer than 700 men. The travelling diggers had seen the few small farmers selling their crops at a good profit to the hungry diggers.[3] Why could they not do the same?

An alliance of middle-class storekeepers, workers and diggers formed. They had a dream — the creation of a yeomanry in the countryside. This dream, in many guises, was to dominate Australian political life until the late 1960s as selection, closer settlement, irrigation development and soldier settlement. To the middle class of the 1850s the creation of a yeomanry meant a return to the values of rural England before the industrial revolution, and the independent rural dweller, the yeoman. It meant keeping some of the population gains of the gold rush to maintain the profitability of the businesses of the middle class. The middle classes also feared the political power of the squatters, who had gained influence in Sydney and London. Their champion was Edward Wakefield, who advocated the opening up of land for purchase by those with sufficient means to undertake development. Wakefield's theories about the need for a proper balance between capital, land and labour to encourage balanced development appealed to the English Government because they would constrain settlement within the effective limits of the colonial administration. Wakefield advocated that the price of land be fixed at a sufficient price to preclude the immediate purchase by those without capital, but not so high that they would not be able to purchase land after working as labourers for a few years.[4] Those without capital would remain farm labourers until they had accumulated some capital.

The diggers had a different vision from Wakefield. Many were once Irish tenants or English working-class reformists, known as Chartists. To the Irish, Wakefield's balanced development looked much like the British land tenure system that spawned the potato famine. The Chartists remembered the enclosures of the commons in England, the precursor to the industrial revolution. They believed the loss of the commons forced farm workers into the city to become wage slaves. The promise of land was a protection against the exploitation by the nineteenth century capitalists. Influenced by revolutionary movements and social reform movements in England, the diggers formed land leagues and agitated for land reform that would open up the squatters' runs for selection and not exclude any man from taking up land, merely for the want of money. The chartists dreamed of independently based small holdings of 1 to 5 acres. The squatters naturally opposed these ideas. They pointed out that far from monopolising the public lands, they were the only producers of note and that there was no other way to make money from the land. To them the urban agitators seemed extremely ill-informed.[5] There was little sympathy for the squatters and their demands for compensation. Their claims of developing the land were dismissed by William Howitt:

> These squatters, who give 10 shillings for a run equal to an English county, sell hay at 60 shillings a ton to the government . . . I have already spoken of that famous contract, by which a squatter . . . gives 10 shillings a year for his station and lets to his landlord, the government, one paddock of it for 500 shillings a year. Yet these are the gentlemen clamorous on the score of compensation . . . 'But', say they, 'our runs are grown so much more valuable in our hands; and, in proportion to their present value we ought to be compensated, if they are taken away' . . . The answer is 'They have grown valuable, but not by your improvements, but by the influx of the public, and it is the public which demands, and has a right to enjoy, the advantage'.[6]

Ten years earlier the squatters in Victoria would have had little trouble in overcoming a challenge from this alliance. But now Victoria was a self-governing colony with its own parliament. All men voted for the Lower House. Reflecting the wishes of the majority of the population and concern about social unrest from urban unemployment, the Lower House put forward a number of Bills to introduce land selection. The Upper House was elected only by men with property. Reflecting the views of this privileged class, it defeated each Bill. The deadlock between the Houses was broken in 1860 by a riot on the steps of parliament by those wanting the land 'unlocked'. A compromise Bill was quickly passed.

The compromise Bill was a failure for the land-reform movement. The first Victorian land opened up was in the Western District; it was made available for 1 pound an acre. Squatters managed to purchase most of this by either honest or dishonest means. The local townspeople saw little future in being small farmers and entered into a frantic lottery to extract payment from the squatters in return for complicity in undermining the intent of the Selection laws. Squatters paid local townspeople to select land, which they then transferred to the squatter for a payment. The Selection laws raised money for the government but did little to promote closer settlement. There followed a series of Bills, each attempting to overcome anomalies effectively exploited by the squatters. The first effective Selection Act was passed in 1868. This allowed 320-acre (130-hectare) selections with stringent residential and development provisions, but less stringent purchase conditions. This Bill opened up selection to the ex-diggers of little means. The best land in the Western District was now in the hands of the squatters, so attention turned north of the divide.

The Goulburn Valley was settled in the early 1870s. Previously, the land north of the Great Dividing Range had been considered less desirable.[7] Now it was fit for an army of small farmers. Unlike their Western District compatriots, the northern squatters soon lost their runs, replaced by small wheat farmers on 320 acres. Life as a selector was hard. The squatters had been right. Wool was the only commodity with a suitable market; but the selectors' holdings were too small to allow them to be profitable wool producers. Almost half the selectors of the Rodney district in northern Victoria left their selections before formalising their selection leases.

Despite the hardship, fortune smiled on the new settlers in the first few years of selection. The early 1870s were years of high rainfall. Land north of the Great Dividing Range received almost double the long-term average rainfall over a number of years. Then the latter half of the 1870s turned to continuous drought. The selectors had no experience of these new conditions and were ill prepared. They had by then exhausted the natural fertility of their soil by continuous cropping. Survival seemed impossible. The Welsh swagman Jenkins described the general depression:

> The drought continues. Cattle are falling for the want of water and fodder. The losses are general. The bankers exert pressure on the farmers who have borrowed money, with their crops as surety. The crops have failed. The farmers are selling their crops and their land, so they cannot employ labour. Consequently, the labourers are unemployed in their hundreds. In that the labourers earn no wages, they cannot buy even the bare necessities of life such as food and clothing from the stores. Thus, the tradesmen suffer . . . The nation is depressed.[8]

The 320-acre limit to farm size may have been appropriate for the more fertile Western District, but it was too small for the drier northern plains. It allowed neither spelling of crop land, nor a buffer for drought. Many farmers sold their exhausted and parched land to neighbours and migrated to the new lands being selected farther out. Those who remained amalgamated properties. Within twenty years only one in ten of the original selectors remained. The dream of closer settlement seemed a failure.

Pressure for irrigation

While the selectors had been enjoying their early wet years, a Scotsman, Benjamin Dods, had presented to the government a grandiose scheme for turning the waters of Victoria's Goulburn River westward in a series of canals to irrigate vast areas of the northern plains. His dream was the Grand North-Western Canal Project. Dods made no headway in selling his scheme to the government. His request for a lease of 3 million acres of unalienated Crown Land did not excite a government that was committed to breaking up the remaining leases of the squatters. However, his enthusiasm for irrigation was taken up by Hugh McColl who was agitating for the construction of a permanent water supply for Bendigo.[9] McColl joined with Dods to promote the Grand North-Western Canal Project. The Victorian Government would not support this private development. Dods withdrew but McColl was not so easily dissuaded. He transformed his arguments from support of private irrigation development into arguments for public development, arguing the national interest would be served by the state constructing a weir on the Goulburn to irrigate the Waranga and Rodney plains by a system of canals. McColl, inspired by the irrigation development in California, was an ardent publicist for irrigation in Australia; he found fertile ground for his scheme with the remaining selectors of the Goulburn who were battling drought. McColl and the selectors formed the Central Irrigation League to lobby government. They argued that the reason for the failure of the selections was lack of water. In 1880 McColl stood for parliament and the clamour for irrigation saw him elected. He continued his enthusiastic campaign from within the parliament.

The popular clamour for irrigation could not be ignored. The government responded to the pressure from McColl by appointing George Gordon, an engineer, and Alexander Black, Assistant Surveyor General, as a Water Conservancy Board to inquire into the provision of a stock and domestic supply for the northern plains at reasonable expense. Gordon and Black recommended the development of a modest water supply for stock and domestic needs, with water being distributed from existing drainage lines. It would not be transferred across catchment boundaries and farmers would be responsible for pumping the water onto their property. The government accepted the recommendations and planning began.

McColl and his supporters were not satisfied with water only for stock. The government had merely skirted the issue. The Water Conservancy Board was then asked to inquire into the feasibility of providing an irrigation supply. The board rejected government involvement in irrigation as uneconomic. Foremost of the arguments against irrigation was a conviction that irrigation profitability was greatly exaggerated by its proponents.

Gordon and Black pointed out that the successful Californian irrigation schemes, so lauded by McColl, relied on snow-fed rivers that flowed throughout summer. The Californians had no need for expensive dams. Australia lacked this advantage. Irrigation would also require the construction of an expensive drainage network. The cost of the required works could not be borne by the existing rural population. The state was not large enough to support these costs. Even if the costs could be met, Gordon and Black doubted that farmers had the skills to effectively use irrigation water. Finally, Australia lacked the rail access to large markets, which ensured the profitability of irrigation in California.[10]

History has since borne out many of the arguments used by Gordon and Black, but these arguments did not sway McColl. McColl attacked Gordon and Black's scheme as 'Gordon's gutters'. He argued it was vastly inferior to his plan for gravity-supplied water using channels on high ground. Ominously, he denied there was any need for drainage and attacked Gordon as a 'departmental disability'.[11] McColl was described by his opponents as a man with 'water on the brain'.[12] Throughout the brawl, the tide of public opinion was gradually turning. A new Minister for Water Supply, Alfred Deakin, introduced a Bill in 1883 which allowed for the formation of landholder trusts to undertake water works for irrigation after satisfying all need for stock and domestic use.[13]

In 1884, Gordon and Black produced a second irrigation report.[14] It described a modest irrigation scheme to supply water to the Waranga and Rodney area, to the farmers included in McColl's scheme. This scheme, in keeping with the aims of the 1883 Act, had a capacity for meeting urgent current stock and domestic needs of the district for the current population and had some capacity for irrigation in years of good water supply. Gordon and Black recommended a timber weir, 40 feet high, on the Goulburn River and channels extending as far as the Campaspe River. They argued caution on economic grounds and predicted any economic return on irrigation development would not be forthcoming until at least the next generation. Gordon and Black's new scheme was not popular with the farmers of the area. They had formed the United Echuca and Waranga Waterworks Trust in hope of achieving something better. The trust hired an engineer, Stuart Murray, to design a larger scheme capable of supporting irrigation. Murray's plan was submitted to the government in 1884. He predicted that two-thirds of the Goulburn's water would be available for irrigation under his scheme. The report was rejected by Gordon in his position on the Water Conservancy Board. This was Gordon's last act in the battle. Deakin, his Minister, was now a firm supporter of irrigation. Deakin's response to the report of the United Waranga and Echuca Trust's report was swift. The Water Conservancy Board was dissolved.[15] Deakin established a Royal Commission on irrigation and took extraordinary steps to ensure it produced the right answers. He appointed himself a Royal Commissioner, Stuart Murray as secretary, and promptly resigned from the Ministry.

In 1885, the Royal Commission handed down the expected finding. A glowing picture was painted of the future of irrigation. Murray was commissioned to produce a series of reports on specific irrigation schemes. Deakin presented a new irrigation Bill to parliament in 1886. It allowed major irrigation works to be designated as 'national works' with government paying the infrastructure costs. The cost of other works were to be paid by landholder trusts authorised to take government loans and levy rates on participating landholders. If the farming community could not pay for irrigation, then the state would pay the bill. Both the government and opposition were competing to be seen as stronger irrigationists. The Bill passed, although not without some opposition to the government paying for 'national works'. The River Goulburn Weir Act was approved the same year. Murray and the United Echuca and Waranga Waterworks Trust were to get their weir.

While waiting six years for a weir to be approved, the shires and the trust had constructed supply channels and had begun a stopgap pumping scheme. By 1891 when the Goulburn weir finally delivered water, the United Waranga Waterworks Trust was a financial disaster, with mounting interest bills, channels destroyed by yabbies and deteriorating irrigation structures. Only one rate had been struck, in 1890. This single rate was too small to pay even one year's interest on the outstanding loans. The rate was opposed by farmers who were dissatisfied with service and the delay in the construction of the weir. Many refused to pay. When taken to court, local sympathetic magistrates supported their refusal. Over the following decade, irrigation trusts across the state were placed in a similar position. The United Echuca and Waranga was one of the few trusts to actually supply water. Colin Swinburne Martin summed up the situation in his history of the Shepparton Irrigation District:

> After the passing of the Irrigation Act, 1886, country districts vied with each other for State loans to construct waterworks in the same fashion as they had during the contemporary railway fever. Political lobbying by rural members of parliament led to large sums of State loan money being expended on useless works. Sound review of proposed works by the competent engineers of the Department of Water Supply was often denounced as bureaucratic humbug. Deakin and the irrigationists in press and parliament hailed each new scheme as a victory for enlightened agriculture.[16]

The rush to irrigate rivalled the frenzy of the land speculators profiting from the contemporary development of railways in Melbourne.

In the late 1880s trusts similar to the United Waranga and Echuca Waterworks Trust sprang up along the Loddon River. The Reverend De Garis chaired the Tragowel Plains Trust, which planned to build a small dam on the Loddon and divert water to Tragowel Plains farms. From the beginning it was clear there was not enough water to go around. The Loddon River stopped flowing in dry years when the need for water was greatest. The trust legislation was not a solution and De Garis and others continued their lobbying, this time for a dam on the upper Loddon River to ensure a flow of water through summer. De Garis' politicking helped to ensure that the Loddon River was not overlooked when Deakin and McColl embarked on their irrigation plans and 'national works'. In 1891 the government opened the Laanecoorie Reservoir on the upper Loddon. The reservoir was to release water into the Loddon in summer, with local irrigation trusts diverting and distributing the water.

The increased water supply did little to improve the situation on the Loddon Plains. The dam increased the demand for water. Other trusts sprang up along the Loddon. The existing Tragowel, Twelve Mile, Meering, Leaghur, Swan Hill, Boort East, Boort North and Pyramid Hill trusts were joined by the Mincha, Calivil and Pompapiel trusts. These private irrigation trusts attempted to share a very limited resource in the Loddon River, but descended into bickering and intrigue.[17] Downstream trusts considered there was not enough water to supply all the needs of households and farm animals. Upstream trusts saw water flowing down the river and argued that it was time to make water available for irrigation.

The trusts achieved little in the way of irrigation, but achieved much in their approaches to the banks. Most trusts borrowed heavily. With the financial crash of 1893, they faced insolvency and a legacy of debt. With the severe depression the banks closed and loan funds dried up. Successive governments sought to impose some degree of financial discipline on trusts, buying cooperation with the incentive of relief from some debts. The situation was beyond redemption. In 1899 the Turner Government passed the *Water Supply Advance Relief Act*, which was an admission that the trusts could not be made solvent. The debts of the trusts were gradually taken over by the state. Of £2 395 000 in loans, the trusts repaid only the interest on £364 000. No capital was repaid. Gordon and Black were vindicated. Even Stuart Murray was chastened into the view that irrigation works would take time to become profitable and patience was needed.[18] The resumption of the debt was opposed by urban interests, but there was no real choice. A contemporary chronicler reflected the popular mood:

> . . . the huge expenditure was assumed to enhance the value of irrigated lands by at least one pound per acre . . . it follows, as a result of the financial breakdown, that the property of the irrigated farmer was increased in value by the involuntary contributions of people who had no share in the benefits.[19]

Why did the trusts fail? The blame was partly due to engineering aspects and partly due to the farmers themselves. Works were not well planned or maintained and were often based on over-optimistic estimates of capacity. The financial planning and management of the trusts was unsound. Farmers found the change from dry to wetland farming not always to their liking. They did not realise the level of skill required for irrigation farming. Irrigation had a stigma. Previously the only irrigation had been carried out by the Chinese coolie with a stick across his back and a bucket either side of his shoulders. The farmer aspired to the squatter lifestyle rather than that of the Chinese. He wanted irrigation as an insurance against drought. In good years many saw no reason to use water and no reason to pay for it. The least obvious reason for failure was that irrigation was unprofitable with the existing farming methods. This doomed the trusts to failure, even without their other enormous problems. The irrigation farmers were irrigating native pastures. These produced little benefit under irrigation. Irrigation could not raise the profits of farms sufficiently to justify the investment, even with good management.[20]

Closer settlement schemes

The depression of the 1890s burst the irrigation investment bubble. Just as in the 1840s, foreign capital had flooded into the country in the optimism of the 1880s. When the financial merry-go-round came to a stop, those left holding the debts had little chance of being repaid. The irrigation dreams should have evaporated like a puddle under the hot summer sun. The financial failures only dampened public enthusiasm for closer settlement and irrigation for a brief couple of years. Ironically, the depression that precipitated the collapse of the irrigation trusts was the catalyst for a rebirth of closer settlement enthusiasm. The depression was a disaster for the working class in the large capital cities. Empty houses lined streets where a multitude of homeless slept. Banks collapsed and many small depositors lost their savings. In 1892 and 1893 hungry families demonstrated in the streets, demanding food. For many the only alternative was to leave the cities and tramp elsewhere in search of work.

The hardship and injustice suffered by the working class reinvigorated the dream of the financially independent farm settler among those concerned about social justice and those concerned about social order. The merchant classes worried over the loss of population and also looked to settlement on the land to stabilise the colony. But dreams of further settlement were thwarted by drought. The longest recorded drought occurred during 1895 to 1902. In 1901, the year of federation, the drought was particularly severe and returned the question of rural water supply to the political agenda. Further government action was sought to protect the rural settler. The government abolished the beleaguered trusts in 1905 and, in 1906, established a new body, the State Rivers and Water Supply Commission. With the exception of the Mildura irrigation settlement, all future irrigation and stock and domestic water supply works in Victoria were to be placed under the management of this commission. The political imperative for the new body was to closely settle the countryside by securing the water supply.

The last act of the United Echuca and Waranga Waterworks Trust was the beginning of work on a dam in the Waranga swamp to store water from the Goulburn River and to extend the reach of the irrigation system to the Stanhope area and beyond. The State Rivers and Water Supply Commission continued the work and the new Goulburn River dam was finished in 1909. The lessons of the early trust failures had been partly learnt. The chairman of the new commission, the American Elwood Mead, decided that farmers in the areas serviced by his dams would have no choice but to use the water supply. There would be no option of using water only in a dry year. All properties in irrigation areas were to be rated, and farmers would pay for their water allocation whether or not it was used. This ensured a stronger financial base for development. To justify the favoured treatment for those areas lucky enough to receive water, there had to be some benefit for the rest of the colony. Mead advanced the idea of using irrigated districts to grow lucerne hay and thereby protect the whole of the State from drought:

> More important than all other considerations is the fact that the most useful functions of irrigation canals . . . is to lessen the losses and hazards of dry years, to save money and to relieve the misery of helpless starving dumb animals. This purpose will never be fulfilled as long as land under canals is used as pastoral areas. When dry years come the irrigator is protected, but he is in no position to extend help to the pastoralist on non irrigated land. Irrigating pasture land creates no surplus forage; to do this we must grow and store hay. For this purpose no crop equals lucerne hay.[21]

Hay production was hard work; it was not the life of the grazier. Mead realised that the current settlers would not take easily to the new farming methods. A new type of settler was needed: one prepared to work as an irrigation farmer rather than a grazier. Land was to be compulsorily purchased from the large landowners and subdivided to support a new influx of settlers. Mead promised a profit from the State Rivers and Water Supply Commission by 1911.

Although the government had not learned to be objective about the prospects of irrigation, many others had. The initial enthusiasm of the population for land settlement had waned over the previous decades of failure. There were not enough local aspirants willing to take up land blocks. The head of the State Rivers and Water Supply Commission travelled to Europe to seek new settlers willing to take assisted passage. In six years the land under closer settlement irrigation expanded from 35 000 hectares to 80 000 hectares.[22] The responsible authorities were confident that the majority of settlers would stay on their farms and succeed. This aim was met. However, the administration of the settlement scheme by the Closer Settlement Board masked the economic realities that had doomed the trusts. The State Rivers and Water Supply Commission could not mask the financial failure of their plans and parliament established a series of Royal Commissions.

The Royal Commissions into closer settlement policy in 1913 and 1915 did not measure success in the same way as the State Rivers and Water Supply Commission.[23] They concluded that the expense of irrigation development and closer settlement was not warranted by the results. The blame this time was laid on the schemes themselves. Much of the land, which was exhausted and ill suited to irrigation, had been purchased at too high a price. Often the land was poorly drained. Unsuitable land was often offered to the government for a quick sale. Political opportunism, rather than engineering and agronomic considerations, also determined the purchase of some land for irrigation development.

The schemes themselves were designed to attract farmers with too little capital to develop the land. Mead had dishonestly advertised to unsuspecting overseas settlers. His pamphlets promised plenty of water for a further 30 000 settlers when the supply had already collapsed in the 1914 drought. One of his pamphlets showed an attractive cottage of a Shepparton settler, inferring the new settler could aspire to this comfort. The cottage actually belonged to a Shepparton professional man. No settler lived in such comfort. The failure was summed up in the Legislative Assembly by E. C. Warde:

> We have spent millions of money in endeavoring to settle the land. It is most remarkable that, up to the present we have not succeeded . . . It is evident that our closer settlement operations are one of the most gigantic failures we have ever embarked upon.[24]

No sooner had the findings of the 1915 commission been released, than another closer settlement scheme was on the drawing boards. The dream did not die. This time it was pressed by new priorities associated with Australia's participation in The Great War. Australia relied on voluntary enlistment. With the first soldiers returning from the front in 1915, the government became concerned that the presence on the streets of un-employed or unruly ex-diggers was discouraging others from enlisting. Soldier settlement was quickly conceived as a means of occupying the returning heroes. From its conception as a necessity for recruitment, the soldier settlement idea grew to be seen as a reward for the returned servicemen. At the end of the war nearly 80 000 battle-hardened men were due to return from the horror of trench warfare. The 8000 hectares of unoccupied Water Commission settlement blocks offered a way of reabsorbing the returning soldiers into society. By allocating the land to the soldiers with easy purchase terms, rural life would convert the city subversive into a country conservative. The blocks that, five years pre-viously, had been described as monumental failures were now land fit for heroes. Populating the empty land also helped to ease the then current paranoia about the vast land-hungry population to Australia's north. In Victoria the government settled soldiers on old and new developments: Red Cliffs, Woorinen, Sale, Maffra and Shepparton.

The yeoman dream was reborn and the lessons from the past were ignored. Successful applicants were required to have even less capital than in previous schemes. Irrigation was to be a tool in the new program.[25] Over 2000 soldiers were settled on irrigation farms. Many had been declared unfit for army service, yet somehow they were to be fit for farm life, for stacking lucerne hay.

Yeomen in Stanhope

One of the schemes in the soldier settlement program was the Stanhope subdivision. In 1915 work was started on enlarging the Waranga Basin. The large properties of the Stanhope area were acquired and divided into small allotments of 16 to 24 hectares. The first soldier settlers arrived in

Irrigation in Victoria

SETTLER'S HOME 1910

Progress
in an
Irrigation
District
through
an
Orchard

SAME HOME 1922

Irrigable
Holdings
Available
for
Settlement

IRRIGATING THE ORCHARD

Consult
Local
Officers
at each
Irrigation
Centre

Full Particulars may be obtained from the State Rivers and Water Supply Commission, Melbourne, Australia

Dry Farming Areas

*FREEHOLD
FARMS—Any
person 18 years
of age may take
up Crown Lands
or Closer Settle-
ment Lands.*

PREPARING THE LAND

*DRY FARM-
ING AREAS
on resumed lands
are made avail-
able from time
to time to meet
the demand.*

CROWN LANDS may be taken up as under :—

1st Class Land :
Maximum Area, 200 acres ; Minimum Price, £1 per acre
2nd Class Land :
Maximum Area, 320 acres ; Minimum Price, 15 - per acre
3rd Class Land :
Maximum Area, 640 acres ; Minimum Price, 10 /- per acre
4th Class Land :
Maximum Area, 1280 acres ; Minimum Price, 5 /- per acre
Class 4A Land :
Maximum Area, 2000 acres ; Minimum Price, 4 /- per acre

Closer Settlement of Resumed Areas

Dry Farming Areas up to a capital value of
£2,500 may be taken up under Conditional
Purchase Lease, repayments extending over a
period of 36½ years, with interest at 5% per
annum on the unpaid balance of the
purchase price.

For Plans, etc., and General Information apply to The Secretary for Lands, Melbourne, Victoria, Australia

Advertising 'closer settlement' blocks

Stanhope in 1920. John McEwen, later a long serving leader of the Country Party and Deputy Prime Minister, was one of the early settlers. Included among the settlers were a significant number of British officers. The first task of the settlers was to dig the supply channels. The wages for this task supported the settlers until the first water was delivered in 1921.

The settlers began their new rural life as dairy farmers. The earlier argument that the irrigation areas would be a drought insurance for the rest of the state had been discredited. High prices for butter after the war gave hope that this time the irrigation settlements would prosper. The government now accepted that cows would graze the irrigated lucerne growing in the settlers' paddocks, producing butter for export rather than 'drought-proofing' lucerne hay. Initially all was well, but in 1925 the world butter price dropped. The war-devastated European farmers were again producing their own butter. Europe had little need for the settlers' butter. There was no money to be made in dairying.

Many Stanhope settlers were forced off the land. Farms were too small to make a living. Those who stayed suffered extreme hardship and harassment from government inspectors trying to ensure repayment of debts. It was a similar situation in settlements across the state and eventually the clamour became too great for the government to ignore. Another Royal Commission investigated closer settlement. The Water Commission tried to portray the complaining settlers as lazy grumblers. The Royal Commission declared closer settlement a failure. The commission noted that the lessons of the past had not been learned, but accepted there were mitigating circumstances. It stated that in a mood of initial compassion the previous experience of closer settlement was 'quietly ignored'.[26]

In Stanhope commodity prices were only part of the problem. A major factor was lack of drainage. After severe floods in the late 1920s a number of farmers left their blocks. The lack of drainage led to a more subtle series of problems. Many farms were waterlogged for much of the year. Lucerne would not grow on waterlogged soils. On some lower lying properties salt appeared on the soil surface.[27] In nearby Shepparton there were similar problems in the settlers' orchards. The settlers were learning the harsh reality expounded by the Victorian Engineer for Irrigation, A. S. Kenyon, almost twenty years earlier, but largely ignored in irrigation development: 'drainage and irrigation are inseparable, drainage being the more important'.[28] Without drainage, soil salinity was an inevitable outcome of irrigation.

11

A bitter legacy

The Greek mathematician Archimedes exclaimed *eureka* on his discovery of the use of fluid displacement to solve a problem of counterfeit currency. His thoughts crystallised when he stepped into a public bath and saw it overflow. There is a parallel fluid displacement at work in the Murray Valley riverine plain. While A. S. Kenyon recognised the problem early, the public 'eureka' has come very late. The Murray and Darling rivers flow across the Murray Basin, which is a huge area of subsidence involving south-western New South Wales, north-western Victoria, and the contiguous part of South Australia. The present landscape was developed over 120 000 000 years. Floods on the Murray and Darling came from two directions. The Darling collected Queensland summer monsoonal waters, while the Murray collected the spring melting snow waters of the Australian Alps. Occasionally two floodings occurred simultaneously, with a resulting colossal flood. The rivers were able to disperse these floods and life would eventually return to normal. The natural drainage system of the Murray Basin has never been adequate for the flood of irrigation water that drought-stricken settlers and enthusiastic visionaries let loose on the riverine plains. The legacy of inadequate drainage has been hidden below the ground. In more recent times it has come to the surface.

The fine sediments of the alluvial plains provide the twist to Banjo Paterson's poem:

> They made him a bet in a private bar,
> In a private bar when the talk was high,
> And they bet him some pounds no matter how far
> He could pelt a stone, yet he could not shy
> A stone right over the river so brown,
> The Darling River at Walgett town
>
> He knew the river from bank to bank
> Was fifty yards, and he smiled a smile
> As he trundled down; but his hopes they sank,

For there wasn't a stone within fifty mile;
For a saltbush plain and open down
Produce no quarries in Walgett town.[1]

Paterson's character won his bet, and had his drinks on the locals, by bringing his own stone. There are plenty of stones in the Murray River Basin; where they exist, they are many metres below the surface. The flat riverine plains of the Murray Valley look monotonous and featureless from the surface. The valley is a low lying, saucer-shaped basin. Underneath its surface is a complex jigsaw of sediments. Ancient rivers draining into what is now the riverine plains created deep valleys in the bedrock. Over many ages the valleys were gradually filled with coarse sand. As the valleys filled, the slope of the land lessened and the rivers flowing through them slowed down. The slower rivers dropped finer and finer particles over the coarser sediments. Today the coarser sediments are underground ribbons of sand and gravel several kilometres wide and up to 60 metres thick. They are covered by 100 metres or so of fine sediments. These underground ribbons of coarse sand are called 'deep leads'. The deep leads provide a drainage pathway for water from the highlands towards the interior of Australia. South of the River Murray water from the Avoca, Loddon, Campaspe and Goulburn highlands has filled the deep leads, and slowly flows north towards the ancestral Murray River.

With the old deep valleys filled, the land was flat and the ancient streams were no longer confined to deep riverbeds. They wandered across the landscape, shifting their course after major floods or erosion events. The shifting streams created small deposits of sand and gravel, shoestring sands, on top of the generally finer clays of the area. In time other fine sediments covered the shoestring sands, leaving underground shallow aquifers capable of holding large amounts of groundwater. Today the shallow aquifers criss-cross the landscape at depths varying from a metre or so below the surface to 30 metres or deeper. Some shallow aquifers hold quite fresh water. Others have very salty water. The saltier water is generally much older than the fresher water. These aquifers hold the key to irrigation salinity control.

The salt in the aquifers is common salt-shaker salt; chemically, it is sodium chloride. Most of this salt comes from rain water. Rainfall contributes between 5 and 40 kilograms of salt per hectare per year to inland south-eastern Australia. Coastal areas receive much higher salt loads from rainfall. Although 5 kilograms in a year does not seem a great deal of salt when spread over 1 hectare, over millions of years this led to a large accumulation of salt. The Kerang region in Victoria has up to 7000 tonnes of salt per hectare in its deeper groundwater layers.

The easiest way to understand the problems created by irrigation is to think of the whole Murray–Darling Basin as a bathtub filled with soil. Before irrigation, grass and trees grew in the tub. Rainfall that fell in the tub was disposed in three ways. It poured out of the basin overflow (the traditional runoff through the Murray mouth was less than 2 centimetres per year), evaporated from the top layers of soil, or was sucked out by grass and the woodland trees. Drainage and evaporation only remove water that is close to the surface of the soil. Water that seeps into the soil below the reach of the sun can still be sucked out by grasses or trees. This process, called evapotranspiration, can absorb water from throughout the root profile, possibly as deep as 10 metres in a native forest or grassland. The rainfall that seeps beyond the watertable settles at the bottom of the bath, slowly seeping out of the plug hole — the deep lead aquifers. The top of this pool of settled water is called the watertable. The water cycle in our bath system was in rough equilibrium. The rainfall coming in was matched by the water leaving by evaporation, evapotranspiration and drainage. The watertable fluctuated around its long-term average depth, depending on rainfall variations.

Irrigation settlement dramatically upset this balance. Each irrigation settlement was like a smaller version of the bathtub model. Irrigation more than doubled the amount of water pouring into these baths. The extra water had to go somewhere. More flowed out the overflows. More seeped into the ground. The deep-rooted native grasses and the sparsely spaced trees that were growing in the bathtub have been replaced with shallow-rooted irrigation pastures. More water entering the ground has seeped through the shorter root zone down to the watertable. With more water entering than leaving the watertable, inevitably, the watertables have risen. The rise in the watertables has not been regular. There are dramatic rises in wet years. The watertables recede in dry years, but rarely as much as the increment produced in the wet years. The long-term trend is a steady rise.[2]

When the watertable reaches within a metre or two of the surface, capillary action draws water to the surface, where it evaporates. The extra water leaving by evaporation will bring our bathtub system back into balance. Unfortunately, the watertable and capillary action also bring salt to the surface. Evaporation removes the water near the surface, but not the salt. The result is concentration of salt in the surface layer of soil, the root zone for grasses. The salt, which is toxic to grass, stops grasses drawing water from the soil and they die of water stress, despite the abundance of water in the watertable close by. The high concentrations of chloride ions are also toxic, burning the plant tissue. High watertables do not need to be highly saline to pose a threat of soil salting. Even the freshest of groundwater will contain a component of dissolved salt. When the apparently fresh watertable is close to the surface, the salt is concentrated at the surface. Greater concentrations of salt in the groundwater speed the process of salinisation.

Dragging the feet on drainage

One method of forestalling salinity problems in irrigation areas is to increase the amount of water flowing out of the bathtub by improving the natural drainage. This will decrease accessions to the watertable by quickly clearing winter and spring flooding. When farmers were first settled on irrigation farms in the Murray Valley the enthusiasm for irrigation was not matched by enthusiasm for drainage. In Victoria, the State Rivers and Water Supply Commission had known of the dangers of salting because many warnings had been issued by government scientists.[3] Despite the warnings, the established pattern was delivery channel construction in the periods of development enthusiasm and belated drainage works only when salting or waterlogging threatened an irrigation settlement.

The early part of the 1900s was dominated by the development ethos shared by successive governments. By the 1930s government and public enthusiasm for irrigation development had waned. Australia was having difficulty selling its agricultural produce overseas. Unemployment was again a problem, with the same dislocation as in the 1890 depression. This time, very few believed putting the unemployed on farms was a solution. Instead, the government gave sustenance payments in return for work on government projects. One of the sustenance projects was extending the drainage lines in the Shepparton Valley. Drainage development had lagged behind supply-channel construction. Soldier settlers in some districts complained of waterlogging and salinity. 'Susso' work on drains killed the proverbial two birds. The Second World War ended the sustenance works. Large areas of the Shepparton district remained undrained.

With the war's end a sense of new optimism and development fever pervaded the country. Soldier settlement was revived. The mood was best exemplified in the national commitment to build the Snowy River Scheme. This and other water conservation schemes became something of a sacred creed in the nation's consciousness. Activists such as B. A. Santamaria once again called for closer settlement. Santamaria's argument for closer settlement, a plea for a revival of the old vision of a 'sturdy yeomanry', was a unique mix of green politics, jingoism and redemptive theology. He argued that the higher birth rates in rural Australia justified an increase in the rural population as a national priority:

> The land comes first. Without the extension of rural settlement we cannot hope to establish that reservoir of population which is Australia's first national need . . . The defence of rural life, the stabilisation of our farming and regional institutions so that farm families can live their lives in security, must take precedence over every other policy designed for the preservation of Australia as a nation of European stock.[4]

Santamaria and the National Catholic Rural Movement championed a vision of cooperative communities of partially self-sufficient families on small blocks. Two major obstacles stood in the way of Santamaria's vision: the market and the climate. He chose to ignore the former obstacle by arguing that if the market was allowed to dictate settlement 'there would be no settling new families on the land . . . the aim of policy would be to restrict settlement'.[5] Climatic limitations were not so easily ignored. Strongly seasonal rainfall patterns meant much of Australia's farmland was ill suited for this 'independent farming'. Santamaria argued that irrigation was a priority for national investment:

In those rural districts . . . where there is an insufficient regular rainfall to make Independent Farming practical, the deficiency will need to be overcome wherever possible by national plans of irrigation and water conservation . . . While great difficulties no doubt exist, too much attention to the difficulties and costs involved may magnify what are no more than difficulties into impossibilities, and sap the ardour of those who retain the vision and the vitality to master the technical obstacles in their quest for the achievement of a great national objective.[6]

Half a century earlier such rhetoric could have come straight from the impassioned irrigation advocacy of Alfred Deakin. While Santamaria had little direct influence on irrigation policy, his enthusiasm for water conservation reflected public and political support for irrigation. In Victoria the State Rivers and Water Supply Commission enlarged many of the major supply channels supplying the Shepparton district. The Eildon Reservoir on the upper Goulburn was rebuilt, increasing its capacity eightfold. This almost doubled the amount of water available for irrigators. Despite the massive capital works program, little progress was made in the area of drainage.[7] The lack of government interest in drainage reflected the lack of interest of the irrigation constituency. Pressure for improved drainage tended to be localised in areas where there was persistent flooding or where salinity problems became evident.

By the 1950s it was clear that surface drainage could not be relied on as a complete solution to salinity and waterlogging difficulties. The wet seasons, which caused devastating brown rot outbreaks in the Goulburn Valley, also caused serious waterlogging losses on Mooroopna orchards. In Gippsland watertables rose and threatened irrigated dairy farming land.[8] In both cases the government responded with a sub-surface drainage program, lowering the watertable by pumping groundwater to the surface. The crucial key to this 'sub-surface drainage' was the existence of an aquifer, or permeable sand layer, beneath the farm. A spearpoint well, a line of connected pump inlets, can be buried inside a shallow aquifer. When water is pumped from the spearpoint well, water from the surrounding shallow groundwater table seeps into the emptying aquifer. With regular pumping the groundwater table is lowered over a wide area around the aquifer.[9]

The crucial problem with sub-surfacing drainage is the quality of the water pumped from the aquifer. The groundwater under the Mooroopna orchards was fresh. The fresh groundwater was mixed with the orchards' water supply in summer and poured down the drains in winter. In Gippsland the groundwater was salty but the drainage option was also relatively simple. The saline water was disposed by a pipeline to the nearby sea. For Shepparton the groundwater pumping presented new problems. The Goulburn Valley was much farther from the sea than Gippsland. In the 1960s it was acceptable to dispose of saline drainage into the Murray River. By 1980 interstate agreements precluded such disposal.

Adelaide's deteriorating water supply

The Murray–Darling is Australia's largest river system. Together with its tributaries, this system drains about one-seventh of the land area of Australia. The mouth of the Murray is the only outlet to the sea for the surface and underground drainage water from a vast area of agricultural land. Before European settlement the Murray and Darling flowed strongly after seasonal rain. During drought both rivers stopped flowing, leaving water in deep pools. When freshwater flows were low, the rivers became salty from the seepage of naturally saline groundwater. In his diary the explorer Sturt described the discovery of the lower reaches of the Darling River. His men rushed to the river to quench their thirst, only to find it too salty to drink.

To meet the demands firstly of river navigation, and later for irrigation and urban water supplies, the river system has been progressively regulated. The Murray River no longer flows as low or as salty as in previous droughts, or as often in flood. Today, 80 per cent of the water used for irrigation in Australia is applied to land in the catchment of the Murray.[10] The river also forms the major water supply for many of the urban centres in its lower reaches. It provides about one-third of the water used by the one million inhabitants of the city of Adelaide.[11] It is a water supply that, at times, approaches brackishness. Adelaide's problem is starkly obvious when we look at the changing salinity along the Murray River. In its upper

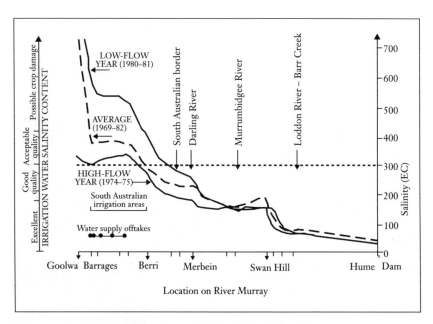

Figure 11.1 The increasing salinity of the River Murray
(*Source*: Department of Resources and Energy, Canberra, 1985)

reaches the tributaries are fed by rainfall in the highlands, and river salinity is low. The River's salinity increases markedly in its middle reaches around Barham and Swan Hill, largely as a result of saline drainage associated with irrigated agriculture as well as salinity in dryland catchments (see Figure 11.1). Between Mildura in Victoria and Morgan in South Australia, the salinity doubles as a result of natural and induced inflows of saline groundwater along this reach of the river. The river salinity at Morgan has been increasing by at least 2 per cent per year since 1970.[12] More ominously, because of time delays in the aquifer responses, even if the saline inflows from the irrigation areas could be held constant, it is expected that saline groundwater inflows from the Mallee lands, cleared of their former vegetation in the past seventy-five years, would continue to increase for the next fifty years.[13]

Increasing concern at the quality of Adelaide's water supply has complicated attempts to control salinity by digging drains. Drainage of saline irrigation areas will increase the salinity of Adelaide's water supply in two different ways. If drains are dug deeply enough to intercept the watertable, they will fill with saline groundwater. Even where drains are shallow enough to avoid direct contact with the watertable, they can still cause downstream problems. The first autumn rains falling on saline irrigated land dissolve salt that has accumulated on the soil surface through summer. Where this water can drain away downstream, it carries with it a slug of salt.

By 1980, concern over the downstream impact of saline drainage had made groundwater pumping more problematic. For farmers wishing to pump saline watertables there were only two alternatives for dealing with

A salt evaporation basin

the saline groundwater effluent. Moderately saline water could be mixed with fresh irrigation water and used for irrigating pastures. Highly saline groundwater could now only be disposed into an evaporation basin. In the 1980s these were the choices that faced the farmers of Stanhope.

Stanhope's reaction to salinity

The small Victorian township of Stanhope sits astride the Midland Highway next to a butter factory. Friesian cows graze on pastures around the town. It is in many ways a typical irrigated dairying district. From a starting point of too many farms with too few acres, Stanhope sorted out many of its early problems; small properties have been amalgamated, maintaining farms at a viable size. In Stanhope, at the eastern end of the Murray Valley irrigation area, the worst drainage problems were alleviated by the construction of the Deakin Main Drain in the 1930s. This provided a drainage path to the Murray River at Echuca. Although not providing protection against major floods, it drained rainwater and irrigation water from properties and reduced the waterlogging problems. With drainage in place, the government felt secure in blaming any remaining salt problems on poor irrigation practices borne of ignorance. The superficial problem of settler ignorance of effective management techniques was tackled by funding agricultural research and a better advisory service. Despite the digging of a major drain, salt has remained in parts of Stanhope. In the wet winters of 1978 and 1981 there was a dramatic rise in watertables throughout the Murray Valley Basin, including the area around the Deakin Main Drain at Stanhope. Further salting developed around the low lying parts of this area.

In 1981, a Gippsland dairy farmer purchased a property near the Deakin Main Drain. He soon found he had a salt problem in the back paddock, right next to the drain! Something had to be done or the farm would be financially unviable. Following advice, the purchaser worked first to improve the drainage from his paddocks, but this had little effect. The salt remained. The next step was to install a spearpoint pump to lower the watertable. A shallow aquifer lay under the salty paddock. Unfortunately the water in the aquifer was very salty. This water was good only for pouring down the drain, an option disallowed by interstate agreement. The spearpoint pump could not be used. There was only one solution, that which had been used many times elsewhere in irrigation areas when salt became a serious problem: to publicise the problem in the hope of stirring government action to control the salt. A minority of farmers formed a community action group to publicise the salt.

The vigorous campaign to prod the government into action placed some Stanhope landholders in a dilemma. They had to decide between their desire to get something done about salt and the possible impact of adverse publicity on the value of their land. Normally, salinity only affects the value of irrigated property if the salinity is obvious to the buyer.[14]

There had been little publicity of salting in the area in the past twenty years. Uncontrolled rumours about salt would lower land values. When the salinity was first publicised, two-thirds of the farmers in the district believed publicity would lower land values.[15] Some of the minority in favour of publicity believed that here was a silent conspiracy to protect land values. They told stories of salinised paddocks being ploughed up and passed off to prospective buyers as ready for sowing. Other farmers said salinity was something that had always been there.

To encourage government action the Stanhope action group had to cultivate support from the rest of the community and convince the politicians that there was strong local pressure for something to be done. The long standing salinity around the low lying Stanhope depression was not a problem for most of the farmers living nearby. The salinised land looked unattractive, but the productive loss was small as the land was unirrigable. The soil salinity along the Deakin Main Drain was more extensive and lay on land that was irrigated. Here, the publicity about the salinity surprised some landholders into sharing the concern of the members of the action group.[16] For some older farmers the salinity was not something to be concerned about because the rising salt always came after wet years and retreated in the dry years. Local farmers had to weigh up the risk of lowered land values against the benefits from government assistance as a result of publicising the problem. By 1982 most farmers agreed that the problem should be publicised.[17]

Much of the Stanhope area is underlaid by fresh reusable groundwater in shallow aquifers and one solution to the salinity problem was groundwater pumping. After another wet winter in 1982 and a significant watertable rise, there was a wave of installations of spearpoint pumps to tap this fresh water. The enthusiasm was not based on a concern about salinity, but the security of irrigation water supplies. The 1983 season turned out to be one of the driest in the history of Victoria. Irrigation farmers were forced to start irrigating their pastures much earlier than normal. Farmers were concerned that the drought would extend for more than one season and threaten the following year's channel water supply. The groundwater pumps were installed to allow continued irrigation if water restrictions were introduced in the following year.

The following season saw a return to normal rainfall. Most farmers no longer pumped groundwater for irrigation supplies because it cost less to use channel-delivered water than to run a pump. Although many farmers believed that the government should do something about salt, few farmers with pumps were sufficiently concerned about the future threat of salinity to incur the cost of pumping water.[18]

Part of the explanation of irrigators' reticence to undertake personal action to control salinity was the difficulty in detecting early signs of salting. Recognition of salt damage is not always easy. The endpoint of soil salting is inescapable: barren land. This is how salinity is presented in the media: dramatic photographs of barren salt pans broken only by the skeletal tree trunks.

The first stages of the process that leads to this final degradation are much more subtle, much less perceptible, changes in the vegetation and soil. The first effect of salting is on the most productive grasses and particularly on salt-sensitive white clover, which then grows less vigorously. As salt levels increase, the salt-sensitive clovers are gradually replaced by more-salt-tolerant grasses, by which time farm production has been significantly affected.

Salt was not seen as a personal threat by farmers who were not suffering obvious losses at the time. Farmers had to see obvious signs of salt on the productive parts of their land or their neighbours' properties before they would take action to control it. Farmers needed to know the early signs of salt to recognise salt on their land. In 1982 most Stanhope farmers were able to recognise the middle and later stages of salting.[19] But only half of the landowners on farms where the early stages of salting had been measured recognised the early signs of salt damage. Several farmers believed bare ground was an early sign of salting. Even knowing the early signs did not guarantee recognition of salt effects. Some who knew what to look for did not see existing low-level salt on their properties. Recognising salt was more than just interpreting grass cover on a paddock. Recognition entailed emotional acceptance as well as physical recognition because acceptance of the existence of salinity entailed acknowledging a risk of losing the farm.

A third reason for the lack of interest in salinity control was that farmers beyond the immediate locality of salinised farms did not realise the extent of the salting in the district, and so discounted its seriousness. The community as a whole was not aware of the degree of salt damage that its members were admitting privately. Almost a quarter of farmers said their farms were significantly affected by salt, yet two-thirds of Stanhope farmers believed that less than 10 per cent of the local farms were so affected.[20] Until the formation of the action group, farmers with salt on their farms did not often talk publicly of their problem, whether from fear of a stigma of poor farming or poor purchasing judgment, or a desire to protect the land market. The lack of discussion limited local appreciation of the extent of salting.

The Stanhope and Girgarre Salinity Action Group lobbied for government-sponsored groundwater pumps and a pipeline to the sea to dispose of the saline groundwater. The government responded by hiring a consultant to produce a plan to control salinity in the district. The consultant proposed three government-run groundwater pumps. Two would pump fresh water into a drain to be reused by farmers downstream. A third pump was to remove salt water. Without removal of the salt water, the other fresh-water aquifers would eventually become salty as the first two pumps drew in water from the nearby unpumped salty aquifer. The salty water was to be disposed into an evaporation basin on the farm over the salty aquifer, with the state and federal governments agreeing to share the construction cost. It remained to be determined who would pay the operating costs.

For a decade, a vision of 700 government-owned groundwater pumps protecting the Goulburn Valley had been optimistically nurtured by the leaders of the irrigation industry. The Stanhope district became the experimental test for the dream. The Rural Water Commission argued that the government should go ahead with the proposal and bear most of the cost because of the value of the information it would produce. Local irrigators in favour of the basin argued that the government should pay the operating costs because of the benefits to the state. This was a view similar to the argument that irrigation development was in the national interest, used by Stuart Murray and Alfred Deakin almost 100 years earlier.

Although the government agreed to bear the cost of basin construction, there was unease in the local community. Some worried that the basin would leak salty water into fresh aquifers and surrounding land, or that the basin would exhaust the underground water, which was seen as drought insurance. Farmers who pumped water from the Deakin Main Drain downstream of Stanhope were concerned their water supply would deteriorate.[21] Despite the opposition, the evaporation basin was installed, protecting the seventeen immediately surrounding properties by controlling the depth of the saline watertable. The basin has been a technical success and its value has been capitalised into the value of the protected properties. From a wider economic point of view, the basin was not a worthwhile investment. Further, the farmers who have benefited have not contributed fully to its cost. As Henry Giles Turner remarked ninety years earlier, after the collapse of the irrigation trusts, 'the property of the irrigated farmer was increased in value by the involuntary contributions of people who had no share in the benefits'.[22]

New rules for government intervention

The Stanhope evaporation basin was perhaps the last significant Victorian irrigation structure to be built according to the old rules originally developed by Alfred Deakin: the government providing the capital and the irrigator the operation and maintenance cost. The old arguments were no longer persuasive. Economists such as Bruce Davidson had argued that irrigation was an uneconomic investment from the national point of view.[23] In the early 1980s a new Labor government came to power in Victoria. A Labor government had not controlled the Treasury benches in Victoria for over a quarter of a century. The new government abolished the State Rivers and Water Supply Commission and formed two new bodies. One, the Rural Water Commission, was responsible for the delivery of rural water. The other, the Department of Water Resources, had responsibility for water policy. The new government broke a long established tradition in appointing an economist instead of an engineer as head of water policy. The economist, John Paterson, had introduced a 'user pays' philosophy to the Hunter River Water Board, cutting water demand and delaying the need to start building a controversial new storage. His charter was clear: Victorian irrigators would have to learn to live with irrigation charges based on the user pays principle. The government announced that water charges would rise by 2 per cent in real terms every year until the full cost of providing the service was met. The full cost of the service was deemed to include an annual depreciation charge based on the modern cost of replacing the current water infrastructure rather than the historically incurred cost. This policy of current cost accounting leads to significantly greater depreciation costs than historical cost accounting. Finally, the government imposed a requirement that all government assets earn a *real* 4 per cent rate of return for the government. These policies meant that the cost of irrigation water would increase considerably.

By the 1980s most of the major Victorian rivers had at least one dam built on them. The only remaining unexploited river was the Mitchell in Gippsland. The new rules for funding irrigation development made it clear that the proposed Mitchell dam would not be built. North of the Great Dividing Range, the irrigation lobby had long since abandoned the quest for further water storages. The interest was now focused on salinity control and better drainage. Salinity control required a means of disposing of salty groundwater. Victoria wanted to dispose of groundwater into the Murray River but South Australia would not support this. Paterson and his Department of Water Resources was able to find a solution. Using an economic model, all parties were convinced that everyone could benefit by a scheme that allowed the upstream states to dispose of salt into the Murray River in return for paying for schemes to stop salt entering the Murray downstream.[24] The major source of salt in the Murray is groundwater flowing into the downstream reaches of the river. The cheapest method for attempting to reduce the salinity of the river is to remove this groundwater before it reached the river and dispose of it in evaporation basins.

The interstate agreement on salt disposal gave the Victorian government an opportunity to do something about drainage and salinity control.[25] The government believed the local community should bear a greater share of the cost of any future drainage works. Because of local controversy over establishing evaporation basins in Stanhope and other areas, the government asked local community groups to develop draft plans to control salinity in their respective areas. The groups were to consult widely with their community and produce plans that had community support. Plans without community support would not be funded by government. The government set broad guidelines within which the plans were developed. A crucial government guideline was that beneficiaries of any salinity control works should pay the cost of those works. The new rules for irrigation development also applied to salinity control.

Dilemmas of community management

In the Shepparton region the Salinity Pilot Program Advisory Council was responsible for preparing a salinity management plan. The council of local representatives worked with the help of consultants and government scientists. The council predicted that, if management of the irrigation scheme were not modified, the area of irrigated land with high watertables would increase from 31 per cent to 55 per cent of the total irrigation area by the year 2020.[26] The consequences of no action were seen not only in terms of land degradation, but also in terms of the effect on the local community. Use of groundwater and economic models speculatively supported the case that, without salinity control, by the year 2025 average farm earnings might fall to 58 per cent of average weekly earnings.[27]

The community group, convinced of the future threat of salinity, faced several political difficulties. Its members had demonstrated to themselves that something needed to be done about watertables. In this concern they were not representative of local landholders who were less concerned about the long-term threat of salinity. Further, the government-imposed guidelines for developing proposals for the management of salinity did not conform with local farm community expectation that farmers should only be expected to pay for operation and maintenance costs. These guidelines for community cost sharing were likely to result in a plan in which local farmers were asked to pay significant sums to control a threat about which they were only mildly concerned. It was unlikely such a plan would be accepted by the local community. Yet the government had indicated that it would only support a salinity plan if it had wide community acceptance.

The Shepparton Advisory Council adopted four strategies to overcome this potential conflict. It mounted a large publicity campaign to convince the community that salinity control was important for the district. The campaign focused on irrigation farmers and local urban opinion leaders. Bank managers, service clubs, churches, estate agents and local government received particular attention. The campaign aimed to lessen local resistance to the cost of salinity control and keep political pressure on the government.[28] The second strategy was to spread the cost of salinity control as widely as possible. The council proposed that farmers, local government, state and federal governments should share the cost equally. Third, the council developed a plan to protect all the irrigation region from salinity: if all farmers were protected from salinity the cost could be spread equally across the region, avoiding the divisive arguments about who benefited and who paid, which had bedevilled the Stanhope evaporation basin scheme. Finally, the members of the council used their salinity plan to mount arguments against the government's financial guidelines for the economic analysis of the impact of the management plan.

The first Shepparton draft plan proved too expensive for the government to accept. A plan protecting all local farmers against high watertables required extensive capital investment. In 1989 one-third of the Shepparton region was served by high-standard drains built by the government. In the proposed plan these drains were extended to a further 14 per cent of farms, with the remainder of the region serviced by lower standard community drains built and maintained by the community with a government construction subsidy and a capital cost of $223 million. Groundwater pumping was to protect large areas of the region. Four hundred existing pumps were to be used to control groundwater, with 426 new public groundwater pumps and 365 private pumps to be installed. Fifty evaporation basins were to be built. In comparison, the Stanhope scheme consisted of three public pumps and one basin. The estimated capital cost of the proposed sub-surface drainage was $83 million. Areas without pumpable aquifers were to be protected by tile drainage in the future when cost-effective methods had been devised.[29]

The State Government disagreed with the economic analysis and the cost-sharing strategy. It requested a redrafted plan. The government wanted a plan that identified the most economically attractive areas for salinity control. This conflicted with the Advisory Council's aim of spreading the cost across all irrigation farmers, because such a plan would place the greatest cost on the largest beneficiaries. A new plan with fewer govern-ment-built drains, fewer government groundwater pumps and new rules for sharing costs was developed. The new plan brought the contentious issues of cost sharing and equity into the open. The new plan also placed a much greater emphasis on the environmental costs of salinity; several wetlands that would be destroyed by rising watertables were identified within the region.

In 1990 the State Government responded positively to much of the revised plan, and the Federal Government also agreed to provide some funding. The major criticism of the plan came from the Australian Conser-vation Foundation, which argued that too few trees would be planted. Members of the Goulburn Valley Tree Group, who were members of the Shepparton Advisory Council, had ensured that the salinity management plan incorporated a proposal to plant trees on 5 per cent of the area of the Goulburn Valley Irrigation District. The aim for 5 per cent derived from the practical recognition that irrigation farms are small and intensively farmed. Tree planting by farmers has been concentrated on road reserves, channel and drain easements, around dams and along paddock boundaries for shelter. The Conservation Foundation recommended that the aim of 5 per cent tree cover should be increased to 30 per cent. It was not clear how the local farmers on small irrigation properties could farm profitably with 30 per cent tree cover or how this would lower watertables on the rest of the farm.[30]

Economic analysis of the Shepparton plan showed that the benefit to the state was marginal, but there were substantial regional benefits. If the plan was not implemented, large areas of land and environmentally signifi-cant wetlands would be harmed.[31] The government determined that its support would be contingent on community cost sharing and required the local community to pay 53 per cent of the total cost of the program.[32] This cost-sharing insistence changed community preference for surface drain-age from expensive government-built drains to cheaper community drains. The government rejected the proposal to cover 5 per cent of the region with trees as shelter belts or as agroforestry, proposing that further re-search work was needed before such an approach could be justified. The government also proposed that the groundwater pumping strategy should be carried out in several priority districts of size similar to that in the

Stanhope project. Unlike the Stanhope project, the new cost-sharing rules were to be applied. The beneficiaries of the pumping were expected to pay a greater share of the capital cost, including the cost of building down-stream evaporation basins as compensation to South Australia for the disposal of saline water into the Murray River. Farmers who pump from saline aquifers, thus protecting their neighbours' fresh aquifers, will need to be subsidised by their neighbours who are pumping from fresh aquifers. The difficulty of determining the beneficiaries of pumping is still unre-solved.

The minority of farmers actively concerned about salinity in irrigation areas, when faced with the problems of achieving consensus, must feel pessimistic at the prospect of farmers reaching agreement on the installa-tion of a groundwater pumping scheme. In 1988 most irrigation farmers were unwilling to use existing groundwater pumps for watertable control:

> Although most landowners are aware that groundwater pumping does provide some degree of salinity control, this is not reflected in their water use behaviour or farm practices. There is no evidence that landowners pump groundwater specifically to control rising watertables. Two possible scenarios emerge . . . Firstly that landowners do not fully understand and appreciate the benefits of pumping groundwater in terms of salinity control. Secondly, for many land-owners their view of the threat of salinity does not warrant spending the time and money involved . . . While surface water remains cheaper and more conven-ient to use, it is unlikely groundwater use will increase.[33]

There have been positive developments in some localities. A group of Tongala farmers formed a cooperative pumping scheme to reclaim and protect their land from salt.[34] Attitudes to salinity have changed as a result of community education programs. Over the first two years of the Shepparton salinity management program, the number of irrigation farm-ers who thought their farms might be affected by salinity in the future increased from 39 to 61 per cent.[35] While most irrigators may concede they face a future salt problem, such recognition is unlikely to reconcile them to meeting the new costs of the irrigation drainage legacy.

After 100 years of irrigation, irrigators face some sobering realities. In one Victorian irrigation district the likely cost of salinity control could be $10 per megalitre of irrigation water used by the beneficiaries — almost doubling the current cost of irrigation water. In the past, the 'benefit' of the low cost of irrigation water has been capitalised into high values for irrigation land. Higher water prices will encourage more efficient water use; and this will lower watertables. Efficient irrigation will require new investment and a higher water price is likely to result in lower land values. Politically, it is difficult to double the price of irrigation water. Such a large change in the structure of farm costs will change the number and nature of irrigation farms. Saving the land will not save all the farms. The new rules have raised the stakes for salinity control, and with raised stakes there will inevitably be some bitter disagreement. Nowhere has this become more apparent than in the Victorian Campaspe West irrigation settlement.

Sharing the cost

The Campaspe West settlement lies on the western extremity of the Shepparton region, next to the town of Rochester. Because of its more westerly location, the groundwater is saltier. The saltier groundwater has brought the cost-sharing issues into sharp focus. The settlement was the last hurrah of Victorian closer settlement enthusiasm. The last properties were released in 1975 and 1976. Within a few years high watertables and soil salting developed on several farms in the north of the settlement. Two farmers abandoned their blocks. The salinity on these farms was an indication of a greater problem developing on the settlement as watertables gradually rose.[36]

By 1982 the extent of the salt problem was well explored, both physically and in terms of its economic impact on farm incomes. Low-level salinity affected clover production on properties in both the north and south of the settlement. The areas of serious salting were nearly all in the northern half of the settlement, which was lower land with a higher watertable. The dairy farms on most of the salt-affected land were profitable, but production was maintained by using more inputs.[37] Despite the evident effects of salinity, community perceptions of the problem differed. Farmers across the settlement disagreed about the cause of the salt, whether it would spread, whether it was serious and even whether many farmers were suffering because of the salt. Many farmers in the north believed salting was serious and likely to spread to the rest of the settlement. Most farmers in the south believed the salting was a scattered problem, possibly serious on some farms, but unlikely to spread.[38]

By 1985 some of the farmers affected by salt had installed pumps and were pumping saline groundwater. Some of the groundwater was recycled on the farms and on a temporary basis some was disposed into drains. Sometimes the amount of channel water needed to dilute the salty groundwater coming from a single pump was greater than the water right on an individual farm. The water needed to be diluted over several farms during the irrigation season and disposed into a drain during the winter months. This required a district plan.

Faced with the gradual destruction of some of the farms in the northern settlement a minority of northern landholders lobbied to be included in the first wave of the Victorian Government 'Salt Action' plans. They hoped this would provide a means for resolving their groundwater disposal problem. Members elected to the community working group represented differing interest groups. Those from the north of the settlement, whose constituency was affected by salt, expected salt to spread to the rest of the settlement. They hoped the cost of containing encroaching salinity would be borne as widely as possible. Those from the south of the district believed salt was a problem confined to the north of the settlement. Representatives of two other groups who diverted irrigation water from the Bamawm drain or the Campaspe River were concerned they would be adversely affected by the disposal of saline water.

The nature of the problem meant that one of the four interest groups must lose. The northern representatives were committed to saving the northern farms by pumping groundwater. The pumped salty groundwater had to be disposed somewhere and agreement reached about sharing the costs. The summer disposal of salt water into the Campaspe and Bamawm systems was unacceptable to the diverters of river and drainage water. Disposing saline drainage into the Campaspe River in winter was not environmentally acceptable. There were too few farms to effectively recycle the saline water in the northern settlement. The working group proposed a system of groundwater pumps to protect all the Campaspe Irrigation Settlement with high watertables. The saline groundwater was to be pumped to the top of the settlement and mixed with the channel supply to levels that would not significantly lower production. Thus all farms would share the burden of increased salt load at a level that would have little effect on individual farms.

The working group considered most of the irrigators on the settlement were responsible for the rise in watertables so the whole irrigation district should share the salty water. This fulfilled the directions of the government's cost-sharing guidelines by identifying the whole of the settlement as polluter and justified the whole of the settlement paying for the cost of salinity control through an increase in the price of irrigation water.[39] Recognising the potential difficulties this strategy could entail, the working group first sought community support for the principles that underlay their plan. They conducted a census of the opinions of all the settlement farmers. The census appeared to show a dramatic change in community perceptions of salinity. By 1989 nearly all the seventy settlement farmers agreed salt was a serious district problem and acknowledged that all settlement irrigators contributed to this problem. Most accepted that an increase in the salinity of the channel water supply would not adversely affect production.[40]

The path from agreement in principle to acceptance in practice did not follow the expected course. As the details of the working group's proposals for recycling salty water in the delivery channel and for equal sharing of costs became better understood, irrigators not on the working group became apprehensive, and then alarmed. Many who had agreed with the basic principles had not anticipated they would be asked to bear the cost of those principles. The limitations of community consultation became clear when perceived self-interest was threatened. Opponents of the plan demanded a ballot and a public counting of votes, where the plan was defeated overwhelmingly. Community opinion was still similarly split as in 1982, with the differences now more public.

The government responded by proposing to support only the elements for which there was community agreement. It agreed to the establishment of groundwater pumping in the north of the settlement over five years, with pumps to be installed in the south at some unspecified later date. The government proposed that half the contentious salty groundwater be disposed of down the Bamawm drain and half recycled in the northern settlement water supply, excluding the southern settlement from the plan.

The conflicting interests aroused by salt disposal in Campaspe West should not have been unexpected. Problems arising from perceived self-interest are widespread in matters of irrigation.[41] The core of the salinity problem is that often the cause of the salinity is on one property and the result on another. This is also true for some erosion problems. To solve these problems requires cooperation and an agreement about who will bear the costs. In the past the usual method has been for the government to pay the costs. Ideally, the polluter should pay where the polluter can be identified. Where the polluter cannot be identified, the beneficiary should pay. The experience of Campaspe West shows the kinds of problems that can be expected when attempts are made to identify and charge either the polluter or the beneficiaries, particularly when exacerbated by the intractable problem of salt disposal.

Who were the polluters of the Campaspe West settlement? To some, they were all irrigators in the settlement. To others it was the government who established the irrigation scheme, or the upstream dryland farmers whose salted properties drain salt into the irrigation supply dams, or the downstream farmers using the Waranga Mallee channel, which potentially contributed to the watertable. To those in the southern settlement the polluters were the irrigators on salt-affected properties who wanted to share their salt. There was similar uncertainty and ignorance about identifying the beneficiaries and, more specifically, identifying the level of benefits accruing to different groups. Very few were willing to identify themselves as polluters, and those who perceived no obvious benefits for themselves strongly believed in the beneficiary pays principle. Salt-affected farmers strongly believed in the polluter pays principle. Such problems cannot be resolved to everyone's satisfaction. The technical information to resolve the arguments generally does not exist. Thus, the reactions of those involved depend on what individuals see in it for themselves. Some community leaders have realised that salinity prevention will not be achieved solely by consensus.[42]

The pessimistic prognosis for tackling diffuse land degradation problems, such as irrigation salinity, is that the problem has to develop to a stage where nearly everyone in a land catchment is suffering some disadvantage before concerted community action can be effective. At such a point everyone is pushed across the threshold of believing that they are personally invulnerable. Most irrigators in the irrigation areas we have considered are still able to go about their business without being seriously inconvenienced by encroaching salinity. The legacies of earlier inadequate attention to drainage and the inappropriateness of earlier irrigation development are yet to have their full effect. Elsewhere in the Murray–Darling Basin there is land where the pessimistic salinity prognosis has already happened.

12

Historic misjudgments

In 1836, Major Thomas Mitchell stood atop a small pyramid-shaped granite hill at a place close to the centre of the Murray River Basin. He surveyed a remarkably flat plain, which extended in all directions (see Plate VIII). Like other inland explorers, Mitchell dreamed of discovering a rich, productive heart of Australia. Maybe this was that heartland? The mostly treeless plain with its lush grass cover greatly impressed Mitchell. He envisaged how the land might be transformed by men, animals and irrigation:

> I ascended a rocky pyramidic hill . . . Its apex consisted of a single block of granite and the view was exceedingly beautiful over the surrounding plains, shining fresh and green in the light of a fine morning. The scene was different from anything I had ever before witnessed, either in New South Wales or elsewhere. A land so inviting, and still without inhabitants! As I stood, the first European intruder on the sublime solitude of these verdant plains, as yet untouched by flocks or herds; I felt conscious of being the harbinger of mighty changes; and that our steps would soon be followed by the men and the animals for which it seemed to have been prepared . . . This seemed to me a country where canals could answer, the better distribution of water over the fertile plains.[1]

Today the Victorian town of Pyramid Hill sits at the base of its name-sake hill and the sign at the town's southern entrance proclaims 'Pyramid Hill: The Major's Vision'. Yet if you drive through the town and continue north or west on a wet winter's day the land smells of the sea, although the sea is more than 200 kilometres distant. Another underground sea is only a metre or two away. The Tragowel Plains watertable is very salty and in mid-winter after heavy rain may rise to within centimetres of the soil's surface. The surrounding land bears the salty scars of the watertable. How did Mitchell's vision turn to salt?

Mitchell did not know he had surveyed the plain at its best. He was travelling, during the winter, in a good season when rain was sufficient to grow impressive stands of native grass. The squatters who followed Mitchell in the 1840s soon learned what was not obvious to a visitor passing in one good year. The drover Hawdon, one of the first Europeans after Mitchell to cross the plains, recorded in February 1838 that he could scarcely find sufficient grass for stock, having to rely on reeds in depressions.[2] It was clear to Hawdon that the average summer in these plains was long and dry. Water was short and grass for stock soon dried off. George Hepburn was one of the first overlanding squatters to travel from the north, following Mitchell's dray tracks. Mitchell's drays had travelled over moist ground, leaving deep ruts. Hepburn's dray left no mark in the dry ground beside Mitchell's ruts:

> We reached the Murray River in about fourteen days. When the Major crossed the country it had been very wet, but many places where we encamped had been destitute of water when we passed, although the lapse of time was short; the tracks of the drays were deeply cut, and ours, which was moderately loaded, did not make a mark.[3]

Taddy Thompson followed the overlanding trail behind Hepburn. He was even less impressed with Mitchell's discovery. In February 1841 he observed that a series of dry seasons had altered the country over which Mitchell had travelled and 'the fertile region which had presented itself to his [Mitchell's] delighted view had been converted into an arid waste, destitute of either grass or water'.[4] Thompson, Hepburn and the other early overlanders hurried through the central Victorian plains and continued southward to the moister country of the Western District. When the better runs to the south were taken up, the later squatters filled the gaps around Pyramid Hill.

The squatters on the plains had drawn the short straw. In dry seasons the grass disappeared and the waterholes receded. In 1851 the only place stock could drink was in waterholes in the Loddon or at Lake Boga, many miles to the north-west.[5] Land beyond the fringes of the river was useless. The wet years brought grass, but also brought floods. Water flowing north from the Great Dividing Range slowed as it reached the plains and spread out over the runs.

Presaging future problems, the Tragowel Plains and the land on the Loddon Valley Plains had a high level of natural salinity. To the east of Pyramid Hill, on the Terrick Plains, the natural vegetation until about 1860 was salt-tolerant pigface (*Dysphyma* spp.). Edward Curr described a trip he made from Tongala to Mount Hope Station (north of Pyramid Hill):

> . . . the pigs' faces were covered with ripe fruit, so, naturally, as we had been living on mutton and damper for months, we indulged in them more than we should have done . . . The plain, for the thirty miles we followed it from the Campaspe to Mount Hope, was one bed of ripe fruit.[6]

With the introduction of the Selection Acts the northern squatters were again less fortunate than the squatters in the Western District, who had been able to safeguard their land using loopholes in the early Acts. By the time the northern lands were released, most of the loopholes had been removed. Despite attempts at subterfuge and trickery, the squatters could not stem the invasion of the selectors. Perhaps, instead of trickery, the squatters should have relied on the truth: the land was harsh and would break many a selector's heart.

In the 1870s small settlements sprang up across the plains: Macorna, Mologa, Tragowel, Mincha, Terrick and Durham Ox. The settlers around these towns tried to make a living growing wheat. The seasons seemed to conspire against them. The droughts and floods that bedevilled the squatters bedevilled the selectors as well. In drought the wheat withered and settlers had to make the long trek to the Loddon River to water their horses. By 1880 the district was in a sorry state. The land between Bendigo and Echuca was 'grain sick' — depleted of nitrogen by constant cropping. The settlers who survived the decade were mostly destitute. A few 'boss cockies' had accumulated large properties. The lot of the remaining settlers was hardship and poverty.

Like the squatters before, the selectors had found Mitchell's vision of a grass-covered plain to be misleading. On one point there was strong agreement with Mitchell. What the land needed was irrigation and a better distribution of water! It was water shortage that made selections unprofitable and life a misery. Local minister of religion, E. C. De Garis of Durham Ox, had taken the role of social activist in defence of his parishioners. Inspired by Benjamin Dods' proposal to irrigate the northern plains of Victoria, he organised a campaign to bring stock and irrigation water to the Tragowel plains, hoping irrigation would alleviate the hardship his parishioners suffered. When legislation was passed allowing groups of farmers to form waterworks trusts for commandeering and distributing water, De Garis chaired the Tragowel Plains Trust.

After the failure of the waterworks trusts the newly formed State Rivers and Water Supply Commission took over the responsibilities and financial commitments of the trusts. State investment in irrigation infrastructure gradually increased the water available to Tragowel Plains farmers. In 1912 a channel from the Waranga Basin reached the Loddon River. The Loddon farmers no longer had to rely on the limited water from the Loddon River. Water from the more prolific Goulburn River was now channelled across the plains and diverted into the Loddon River. The channel placed increased demands on the storages on the Goulburn River. Within two years a drought exposed the overcommitment of the available water. In 1927 the State Rivers and Water Supply Commission built a dam at Eildon on the upper Goulburn River. It also built a new channel supply from the Waranga Mallee Channel into the middle of the plains. This provided a gravity supply across much of the plains. Irrigators no longer had to lift water out of the Loddon, merely open a gate on the channel and the water flowed across their land. 'Gordon's gutter' was now superseded.

The Tragowel Plains were different from irrigation districts to the east of the Riverine Plain. In the Goulburn Valley the Water Commission was creating new communities to use the new irrigation water. It was buying land, subdividing, supplying water and finding new settlers to farm the land. Each block was allocated a legal right to sufficient water to irrigate the whole of the block. This was generally in excess of 1 acre foot of water per 1 acre (equivalent to 30 centimetres of rainfall per year). On the Tragowel Plains the commission supplied water to the existing settlers. This difference had profound implications for the way farming developed on the Tragowel Plains.

The Water Commission could not supply enough water to the Tragowel Plains to water all the land. Rather than supplying a few farms with all the water they needed, as in the Goulburn Valley, the commission shared the available water between all the farms with access to its channels. The commission granted water rights at the rate of 1 acre foot for every 5 acres. This was insufficient water to irrigate the whole of a farm. The channel system distributing the water to the larger Tragowel Plains farms was longer and more complex than channel systems in the Goulburn Valley. It was harder to manage and the water delivered to the Tragowel farms arrived more irregularly than in the Goulburn Valley. The Tragowel Plains were flat, so farmers could start irrigating with very little preparatory work; a supply channel and branches were all that were needed. When the farmer dug a series of holes in the branch channel, water flowed out over the plain. This practice came to be known as 'wild flood' irrigation.[7] Once the water left the branch channel, it was uncontrolled. Although the land was generally flat it was riddled with 'crabholes', small circular depressions between 5 and 50 metres across. With wild flood irrigation, water collected in the crabholes leaving the higher areas of the paddock dry.

The State Rivers and Water Supply Commission encouraged irrigation farmers to grow lucerne on their irrigated paddocks. In the mid-1920s the Department of Agriculture established a demonstration lucerne plot in nearby Cohuna. But farmers who tried to grow irrigated lucerne on the plains were often unsuccessful. Lucerne did not grow well under wild irrigation methods and it did not grow well in the wet crabholes. The roots succumbed to waterlogging and the leaves were scalded as the ponded water warmed in the summer sun. Irrigation crusted the surface of the heavy Tragowel clay soils and the delicate lucerne seedlings often could not break through the crusted surface.[8]

The Tragowel farmers developed their own style of farming. In spring they used their irrigation water to irrigate wheat. In summer and autumn the wheat did not need water. If water arrived in the channel, to produce a 'green pick' of native pasture, it was directed into crabhole depressions and dry swamps, where it often remained until it percolated away or evaporated. The more adventurous farmers used their summer water to irrigate sorghum, ambercane or millet, crops that could cope with waterlogging. Farmers applied as much as possible of the irregularly delivered water to ensure the crop was tided over until the next watering. Water ponded in the sorghum paddocks for days. Lucerne did not thrive with this style of irrigation.[9]

When government advisers changed their emphasis from lucerne to perennial white clover pasture in the 1930s, the Tragowel Plains farmers did not change their traditional practices. White clover was better able to stand waterlogging, but it needed to be watered regularly through summer otherwise it would not survive. White clover pasture was unreliable because of the irregular Tragowel Plains water supply. Consequently, Tragowel farmers followed the example of the dryland farmers farther south who were planting Wimmera ryegrass and subterranean clover. This *annual* pasture grew in autumn, winter and spring, dying off in summer each year and re-establishing itself in autumn when the rains fell. By irrigating in autumn, the Tragowel Plains farmers could grow subterranean clover by creating an artificial break to the seasons. Within fifteen years Tragowel Plains farmers had planted 6000 hectares of subterranean clover pasture, converting land from barley, lucerne and native pasture.[10] At last there was a crop that could replenish the soil nitrogen mined by previous generations of farmers. The land described as grain sick in the 1880s was finally being replenished.

Misjudging salinity

Soil nitrogen was only one of the problems on the plains. The local activists who conceived the scheme for irrigating the Tragowel Plains made the mistake of Thomas Mitchell, seeing what they wanted to see in the landscape, not what was there. The landscape held warnings of the likely impact of irrigation schemes. When the drover Hawdon crossed the plain in 1838, he found little grass, but much 'salsuginous pigsface'. He crossed level plains of saltbush (*Atriplex* spp.) and pigface (*Dysphyma* spp.) between the Campaspe and Loddon rivers.[11] These were signs of salt quite close to the soil surface. The squatter Edward Curr satisfied his wanderlust by taking long rides from his station near Echuca. In his later written recollections he told of seeing 'saltbushes ten feet high' near Mount Hope, on the north-eastern edge of the plain, and lower saltbushes over wide areas to the west.[12] At that time oldman saltbush and cottonbush (*Kochia aphylla*), with sparse grass, extended from the middle reaches of the Loddon River to beyond the Murray River.[13] With hindsight we can surmise that irrigating this country was bound to cause salinity problems. The first settlers learnt about the land through experience, but they had no experience of salinity.

The Tragowel Plains are a natural flood plain. In places the slope of the land is less than 1 metre in every 2 kilometres. The plains are drained by many ill defined small creeks and anabranches of both the Loddon River and Bullock Creek. Originally, when the heavy rains came, every five years or so, the floodwaters were very slow to clear. Settlement brought roads, levee banks and, eventually, irrigation channels. All interrupted the tenuous natural drainage lines, exacerbating the winter flooding of the plains. Irrigation further worsened the drainage. Summer irrigation water flowed into the normally intermittent creeks. This radically changed the ecosystem of the creeks. Dense cumbungi weed does not survive in intermittent streams, but its dense thickets quickly dominated the now perennially wet shallow creeks. They slowed the drainage, spreading the area of permanent inundation, spreading the cumbungi habitat, and further exacerbating the winter flooding. Despite the summer shortage of water, the winter flooding was unwelcome, waterlogging the soils and hindering access to paddocks for long periods. Crop and pasture production was reduced and prolonged flooding increased the amount of water reaching the watertable.

Irrigation itself exacerbated the watertable problems. Irrigation ensured large areas of the plain had wet sub-soil before the natural autumn break. When the winter rains came there was no water storage capacity in the soil and winter rain penetrated the soil directly to the watertable. Farmers irrigated natural pasture by wild flooding of undrained depressions. Water ponded in crabholes, causing water to percolate into the watertable.

In the 1880s the watertable was 8 to 10 metres below the surface. In 1890 the watertable depth was 4 metres; by the turn of the century it was only 2 or 3 metres. By the 1920s the first visible salt appeared on either side of the Macorna channel in the north of the plains, where the drainage was almost totally cut off. The salt proceeded to spread north and south from the channel. The plains were flat and the salt spread widely, unhindered by local topography. There was little elevated land above the influence of the watertable.

Salinity reclamation

By the early 1930s salinity was a serious problem not just on the Tragowel Plains, but on much of the irrigation country around Kerang. The Water Commission started a belated program of drainage for the Kerang East area, progressively extending the drains upstream and promising that the drains would reach the Tragowel Plains. The State Department of Agriculture placed an agricultural scientist, Alan Morgan, in Kerang with the task of discovering how farmers could reclaim their saline land.

Morgan hoped to reclaim salt-affected land by regular watering throughout the irrigation season. The regular watering would leach salt from the soil surface down below the root zone of pasture grasses. To do this he had to grow white clover pasture, which needed to be watered throughout summer. This approach required a dramatic improvement in the inefficient local irrigation methods. Wild flooding, which oversaturated some ground and left higher ground dry, was not compatible with salt reclamation. The high unwatered ground was not leached and so stayed salty despite irrigation. The low ground was waterlogged and did not grow healthy white clover.

Better water control was to be achieved with border check layout, which divided the farm into rectangular watering bays. The supply channel was at the top of the bay and the drainage channel at the bottom. The sides of each bay were raised mounds of earth called checkbanks. The checkbanks stopped water flowing sideways out of the bay, forcing it to flow down the

Levelling irrigation land with a buckboard and horses

bay to the drain. The surface of each bay was levelled using horse-drawn implements like the buckboard and the scraper. This levelling could fill in small crabhole depressions and level small rises. Removing cross slopes on bays was beyond the capacity of the buckboard. This meant bays had to be built to fit with the slopes of the farm. The government advisers recommended bays no wider than 15 metres to ensure that water covered the whole of the land in the bay.

Irrigators had in the past been unconvinced of the need to layout their farm with border check. The narrow bays and limitations of farm topography meant a farm was covered with many small bays, often pointing in differing directions. Laying out the farm was time consuming and expensive and irrigating was time consuming. Farmers had to water each bay in turn by digging and then filling a hole in the supply channel. The State Rivers and Water Supply Commission first demonstrated the system before the First World War on two demonstration farms in the Goulburn Valley. The farms were closed within a few years because irrigation farmers did not believe they could achieve the same results with private resources.[14] Farmers chose instead to persist with wild irrigation, or to layout their farms with much longer and wider bays. Farmers overcame the difficulties of cross slopes on the wider bays by raising small banks across the bay at intervals of about 20 metres. These transformed the bays into a series of shallow basins. At watering, each filled in turn, watering all the way across the bay, before overflowing into the next small basin. The

Department of Agriculture advised against this style of watering with the observation 'under this system there can be no pretence of uniform watering, since there can be a foot of water near the bank and an inch at the further reaches'.[15]

Farmers in the Tragowel Plains were particularly reluctant to layout their farms according to the recommended border check style.[16] They used most of their irrigation water in a couple of waterings in the autumn over a much larger area of pasture than irrigation farms elsewhere in the state. Where they did layout to border check, their bays were wide and long, sometimes over a kilometre long, taking over a full day to water.[17] The large farms and subterranean clover pastures made small border check bays specially unattractive.

Morgan set up demonstration plots on farms to show the importance of improved layout if a farmer wished to reclaim salty land. Pasture sown on a saline property at Kerang East was soon established, as regular watering washed the salt down below the root zone. After three years paddocks of lucerne and pasture were growing where before there was nothing but saltbush and salt, though the productivity was below the level achieved elsewhere.[18] Later attempts to grow crops and subterranean clover on neighbouring paddocks were less than successful. Regular watering of the first paddock had raised the watertable in neighbouring paddocks. The only crops that could be grown next to the successfully reclaimed pasture were those that the farmer watered throughout the summer.[19] Subterranean clover, which died off in summer, could not survive next to the reclaimed land. While the subterranean clover was dead and unwatered in summer, evaporation drew salt to the soil surface.

At another demonstration paddock at Tragowel, whenever the new pasture germinated it was quickly eaten by rabbits, which had a natural habitat in the Dillon bush and lignum on the adjacent salty land. In summer the perennial pasture was a magnet to the rabbits. After one poisoning around the demonstration block Morgan and the block owner found 3000 rabbit carcases. Many local farmers regarded the rabbits as a bigger problem than the salt.[20]

The war brought Morgan's official, but not unofficial, experimental work to an end. He managed to continue to supervise the reclamation until 1943 by working in his own time at weekends. In 1947, Morgan summed up the results of the experiment:

> The Department has advocated for many years, from the standpoint of efficient utilisation of irrigation water, the fullest application of the principles of:- (a) effective grading; (b) short, narrow bays; (c) sowing of permanent crops and pastures; (d) maintenance of fertility by adequate topdressing with superphosphate; (e) scientific grazing management. These principles apply even more forcibly to areas affected by salt. The reason is obvious — the aim of reclamation by irrigation is to leach the harmful mineral salts to a depth at which they no longer affect the plants to be grown, and this is most satisfactorily accomplished where the irrigation water is evenly applied and the amount not sufficient to cause any appreciable rise in the watertable of any adjoining unirrigated area.[21]

But the message was not as clear as it seemed. Agricultural advisers lamented the low use of superphosphate in the Kerang area. For most farmers there was no point in fertilising the land until the salt had been washed away.[21] The experiments had shown the importance of maintaining cover on land with high watertables to slow the rate of salting, and yet the first step in re-layout was to remove all vegetative cover to grade the land. The experiments had not demonstrated that careful watering would eliminate the risk to neighbouring paddocks. On the experimental plots watering one paddock brought up salt on a neighbouring paddock. It was impossible to predict where this seepage would occur, so the farmer had to water the whole farm at once.[22] But Tragowel farmers did not have enough water to irrigate all their land. Better layout was not the answer to this dilemma. Measurements on the plots showed layout did not lower watertables.[23] The only contribution of layout was to improve leaching. To leach salt to the lower soil depths it was necessary to irrigate consistently with sufficient water to wash the salt through the soil. The water had to go somewhere. If it did not raise the watertable then it increased the drainage flow of salty water. The greatest contradiction was the observation that 'irrigation is required to undo the damage done by irrigation', when it was clear that irrigation was the cause of the problem. Morgan noted that where farmers were reclaiming their salted land 'the point has been reached where more irrigation water is required than can be guaranteed by the existing storages'.[23] It was difficult to see how farmers on the Tragowel Plains could reclaim more than a quarter of their land.[24] Yet increased water supplies raised watertables and caused more salting.

The gospel of efficient irrigation

With all these contradictions, Tragowel farmers were slow to respond with effective land layout. Advisers complained that it was the exception to find a farm with good layout. Tragowel farmers had developed a reputation for conservatism, for 'sticking to poor layout', 'eschewing perennial pasture for summer crops' and being 'sparing with superphosphate'.[25]

The prescription for efficient irrigation and perennial pasture did offer Tragowel farmers the opportunity to decrease waterlogging on their irrigated pastures and eliminate unproductive weeds like rush and sedges. Improved irrigation efficiency also offered the chance to reduce the amount of fresh water flowing into the local creeks and nourishing the cumbungi menace, thus easing the State Rivers and Water Supply Commission's difficulties in distributing water across the plains and, in turn, increasing the area of pasture that farmers could sow.[26]

In 1940 the Pyramid Hill farm competition was supplemented by a water efficiency competition in which the prize was free water from the Water Commission. Harold Hanslow judged a number of the competitions. He advised:

Continued watering in excess will spoil the land, and areas subjected to it will soon become artificial swamps. Rushes and yellow weed will replace the better vegetation. Excess water implies waste of water and aggravation of drainage troubles. All Pyramid farmers should fall into line and adopt the border system of layout for irrigation.[27]

The response was less than overwhelming. During the war there were about ten entries a year in the farm competition.[28] Superphosphate was rationed and labour was short. Tragowel farmers had a reputation for sparing use of superphosphate before the war and rationing during the war was based on pre-war use. After the war the number of competition entries dropped to three.[29]

The reluctance to accept the prescription lay in the unavoidable risks for Tragowel farmers. If a saline paddock was reclaimed by re-layout there was a strong risk that the surrounding land would become more saline. On the experimental blocks Morgan had been forced to water the whole of the block to prevent differential seepage. On the plains where there was not enough water to water all of each of the farms, the closest alternative was to continue with annual pasture irrigation over the widest possible area. We have already seen that re-layout of annual pasture was less financially attractive than re-layout of perennial pasture. Harold Hanslow, an irrigation farmer himself, diagnosed the cause of the Tragowel Plains conservatism as due to the necessity of partial irrigation in the Pyramid Hill district. Inadequate water rights and the extended water distribution meant that supply at frequent intervals was impracticable.

The conservatism of the Tragowel farmers was in part an outcome of the decision of the early irrigation planners to share water widely in the interests of 'social justice'. If the plains had been transformed by closer settlement into more, smaller, concentrated settlements, it would have been easier to establish border check irrigation; but the cost would have been the social dislocation of remaining settlers without water, who would have been forced off their 'uneconomic' farms. With the benefit of hindsight, encroaching salinity has produced an inexorable attrition of farmers from the plains. Given the geology, topography and natural salinity of the plains the recommended irrigation practices were unlikely to have produced a result much better than now exists.

Looking to re-layout irrigation land on farms to reduce accessions to the watertable overlooked the huge problem of the leaking water-delivery systems. During the Second World War concrete and iron were almost unobtainable. The Water Commission could not effectively maintain its already leaky supply structures. In the Torrumbarry system to the north only one-quarter of the water that left the Torrumbarry Weir arrived on farms. The other three-quarters either evaporated, leaked or seeped into the watertable before it reached the farms. The long supply system in the Tragowel area probably leaked at a similar rate. It is no coincidence that the worst salting developed first along the Macorna channel in the north of the district. The channel probably contributed its share of this leakage.

In 1950 Alan Morgan was still lamenting the prevalence of wild irrigation methods, wide and long bays, inadequate fertiliser applications and inefficient water use in the Kerang region.[30] Only 20 per cent of irrigable land was laid out to the recommended standards of the day. There was optimism that this would change.[31] Tragowel farmers were abandoning the old summer crops. Rising wool prices meant there would be more spare cash for farm investment. The rubber-tyred tractor promised to revolutionise the business of paddock re-layout from the rough and ready methods of horse drawn implements.

While the focus of agricultural advisers was on salt reclamation and re-layout, the interest of the irrigation community was on increasing the water supply. Tragowel Plains farmers played their part in lobbying the government for both the enlargement of the Eildon Reservoir and the enlargement of the Waranga Mallee channel, which brought the water to their district. The government acceded on the first request, but not the second. The enlargement of Eildon Reservoir dramatically increased the amount of water available to Tragowel farmers, though they still had insufficient water to irrigate all their land.

The prospect of this extra water on the plains raised concern in a few quarters.[31] In 1952 one scientist estimated that 20 per cent of the land around Tragowel and Macorna was so salinised it grew no vegetation.[32] A further 10 per cent of land grew only poor medic and ryegrass. Perhaps up to half of the district was suffering salt damage. In the mid-1950s soil surveyors laboriously tramped over the plains taking soil and sub-soil samples. They found large areas of the plains soils lay over extremely salty sub-soils.[33] They predicted any rise in the watertables would mobilise vast quantities of salt, bringing extensive salting to the surface.

The mobilisation of this salt was hurried by a cruel coincidence. In the 1950s there were very wet years, which produced bumper crops in the Mallee and brown rot devastation in the Goulburn Valley. On the Tragowel Plains the rains flooded the plains season after season. As watertables rose, the salt spread. The extra Eildon water arrived in time to further stress the watertable balance. By the early 1960s the salt problems in the Kerang area prompted a new round of investigations. The Water Commission appointed an engineer, Alan Coad, to assess the area of salted land and find economic solutions to the problem. He mapped the area of salt and warned that large areas were at risk of imminent salting. However, the economic solution proved more elusive.

A little to the north of the Tragowel Plains two agricultural scientists took Alan Morgan's reclamation work a step further. They adapted a technique used by the Municipality of Kerang, which had been forced to pump groundwater to control watertables on construction sites.[34] The researchers irrigated the soil from above while pumping out the salty groundwater from below. The salty groundwater poured down the drain. The pumping was successful. Pastures grew on land that had not borne crops for 30 years.[35] The scheme was successfully repeated on the Kerang Irrigation Research Farm. Today it is hard to believe that the lush pastures on the community farm were once saline, windswept land. But the pumping solution was no solution at all. What was physically possible was neither responsible nor politically acceptable. The groundwater of the Tragowel Plains is extremely salty, sometimes saltier than sea water. The real problem was not the pumping but the disposal of this saline effluent. The salty groundwater from just these two experimental bores had little effect downstream. If every farmer with a salty farm was to follow this example the repercussions would be felt all the way to Adelaide. The Murray River would become a saline drain. The pumping was stopped.

The extension of the Kerang East drainage system on the northern borders of the Tragowel Plains was also stopped because of concern at the downstream impact of salty groundwater. The deep drains, which had snaked through the Kerang East district, flowed salty throughout the summer. Extending the drains into the Tragowel Plains would have in-creased the salt flow. The Tragowel Plains had an improved water supply, but was left with rising watertables, poor drainage and little hope for a solution. The few tree crops grown in the area vanished from the land. The last peach trees around Pyramid Hill disappeared about 1960.

Learning to live with soil salinity

There is not a lot that farmers in the north of the Tragowel Plains can do to reduce the salting on their properties. They have had to learn to live with salinity. Tragowel farmers developed a strategy borne out of their experience of salinity. The first part of the strategy was to be very con-servative when investing in improving the farm. Old fences were repaired rather than being replaced with new ones; water reticulation was rarely upgraded. Fertilisers were used sparingly; and new equipment was pur-chased rarely.[36] There have been good reasons for following this conserva-tive investment strategy. Salt-affected land does not produce the profits of unaffected land and cannot support as many overhead costs. Farm debt levels need to be lower to forestall the cyclic downturns in commodity prices or the more recent experience of high interest rates. Investing in farm improvements on salt-affected land has a greater than average risk of investment failure.[37]

The second part of the strategy was to spread irrigation water over all the farm. Because there is insufficient water to irrigate all the land through summer, the majority of Tragowel Plains farmers irrigate most of their properties in the spring and in the autumn only. More income could be generated by concentrating water on only part of the farm, thus maintain-ing a continuous perennial pasture.[38] But concentrating water on perennial pasture requires investment in fertilisers and re-layout, breaking the secu-rity strategy of not investing in development. There is also the risk that large parts of the farm will become more saline once that land does not have irrigation water flushing salt from the surface several times each season.

The third part of the strategy has been to invest in extra land, particularly when commodity prices are low. Improved profitability has been more likely to come from buying extra land rather than *improving* existing land. Not everyone has been able to follow this strategy. A minority able to save cash for the right time has built up very large undeveloped properties irrigating extensive annual pasture.

The results of these strategies are easily visible. The Tragowel Plains do not look a prosperous farming area. The deterioration of farm equipment and low investment levels are obvious. Fences, sheds and houses are often dilapidated. Farm productivity is obviously low. The farmers of the Tragowel Plains have regularly been exhorted to pull themselves up by their boot straps. In one study of cropping management in the Pyramid Hill area agricultural extension officers concluded that 'disappointing yields are a result of a number of management inadequacies'. In other studies extension officers have observed that low investment is a reaction to poor responses to previous investment.[39]

The salting of the Tragowel area has influenced more than the pattern of farm investment decisions. It has in some ways shaped the nature of the community as a whole. Because of the reputation of land in this area, potential outside buyers are more often repelled than attracted to the area. As a result there have been comparatively few outsiders buying into the area.[40] Innovative farmers have often either left willingly or paid the ultimate price of innovation in a saline area — they have forfeited their farming investment.

Sharman Stone, a sociologist with family roots in the area, was in a unique position to study the impact of salt on the Tragowel Plains community. She concluded that the local community has not maintained the skills and innovativeness of other local communities. It has not been able to protect, maintain or develop resources in competition with other towns.[41] Boort, in the same shire as the town of Pyramid Hill, provided a contrast to the Pyramid Hill and Tragowel Plains area. The farming land around Boort town is responsive to innovative farming practices. The soil is better, there is little salt. Farmers in Boort have produced Australian record sunflower crops. They have a history of quick adoption of new ideas and techniques on their farms.[42] This innovativeness spills over into community activity. The people of Boort have done well in attracting facilities to their end of the shire.

The continuing encroachment of salinity means that even the historically successful strategies are becoming unsustainable. Today the northern half of the Tragowel Plains Irrigation Area has some of the worst irrigation salting in the Murray–Darling Basin. By 1988 only 12 per cent of land in the northern third of the irrigation district around Macorna was capable of growing white clover.[43] Farther south this figure rose to 34 per cent around Mincha and 53 per cent south of Pyramid Hill. Such salt levels would be calamitous in most other irrigation areas. In 1989 a third of land had salt levels capable of supporting only the most salt-tolerant plants such as saltbush and salicornia. The impact on farm production and profitability has been great. The production per megalitre of water on the worst affected farms is half the production of farms with limited salting.[44] Watertables in the north are at an equilibrium. Most water flowing to the watertable simply evaporates back again, so the watertable fluctuates with seasonal conditions. The future for this area seems to be one of gradually increasing salting.[45] Although salting in the southern Tragowel Plains is not as bad as the north, watertables are rising and there will be increased soil salting in the future.

Contemporary visionaries

By the 1970s much of the Tragowel Plains was either not layed out or remained under very poor layout, in which farms were divided into very many small irrigation bays. Drainage was often unreliable or even non-existent. The large number of bays made irrigation a time consuming task. As property sizes increased with farm amalgamations, irrigation workload increased. The uneven slope of irrigation bays reduced production, and inadequate drainage often meant the lower ends of paddocks were con-stantly under water.

The introduction of more powerful tractors in the 1960s meant larger volumes of soil could be shifted. Where properties were regraded the irrigation bays were made much larger and there were fewer of them on a farm. Watering became a less arduous task. With fewer, longer bays it was easier to achieve an effective drainage system.

There still remained the problem of getting an even slope in the bay. The results of manual grading were not always reliable and farmers were reluctant to embark on a significant alteration to the existing topography. This difficulty was solved with the introduction of laser-beam-controlled landforming equipment in 1977. Grading could be simply controlled to produce a very even slope. It provided the key to radical re-layout of irrigation farms. A farm of many small bays could be transformed into a farm with a small number of wide and long bays with no cross slope.

Laser grading is expensive, costing over $1000 a hectare. Land that is laser graded is taken out of production for at least a year. Despite this, the innovation was rapidly adopted. In the ten years after the first laser grading in the area in 1977, nearly 60 per cent of farmers on irrigation properties around Kerang laser graded some of their property.[46] A laser graded re-layout of pastures greatly eased the burden of watering for irrigation farmers. Farmers also seemed to gain increased production from re-layed paddocks because of improved drainage. Dry high spots and waterlogged low spots in the paddock were eliminated. Irrigation was easier, so irriga-tion was often more timely.

In the 1940s Morgan had promoted grading and re-layout to ensure there was efficient leaching over the paddock. After the arrival of laser graders the belief grew that grading would help control salting by limiting accessions to the watertable. The belief prevailed that poor layout was the greatest factor in causing high watertables and subsequent salting in the

A laser grader levelling irrigation land

region. With a more level paddock, it seemed natural to assume that watertables could be better controlled. This idea spread through the farming community. By 1982 probably 60 per cent of farmers in the irrigation areas of northern Victoria believed laser grading would prevent the encroachment of salinity. A third of farmers believed existing salt damage could be controlled by laser landforming and re-layout.[47] It is surprising to look back on such a pervasive belief and realise there was little scientific evidence to support it.[48] The results of laser grading on the nearby Cohuna farms had been impressive. The Victorian Government offered 'salinity loans', which were used by some Tragowel Plains farmers to finance investment in land re-layout. This support for laser grading dovetailed neatly with another recommendation to Tragowel Plains farmers to grow more permanent pasture.

In the 1960s additional irrigation water storages allowed the government to more than quadruple the amount of water allocated to the Tragowel Plains. The capacity of the Waranga Mallee supply channel was not increased. This caused a problem for the Tragowel Plains farmers who watered mainly annual pastures, using most of their water allocation in the autumn. Each autumn the majority of Tragowel Plains farmers require large proportions of their water allocation within a short period when the delivery system cannot cope with the demand. Water is rationed by allowing each farm a fixed proportion of the farm's water right within a certain number of days. In dry autumns there is a long period between waterings of annual pasture, which often begins to die between autumn waterings, resulting in lost production. The minority of farmers with large areas of perennial pasture do not suffer from this rationing. They use their water allocation gradually throughout the summer irrigation season. They grow a much smaller area of annually irrigated pasture, and when autumn comes they need a smaller proportion of their water right to maintain the pasture.

The drought of 1982–83 led to particularly severe water rationing on the Tragowel Plains. Despite the hopes of some that upgrading of channels could solve the rationing, the only economically justifiable solution lay in changing farming practices. The local agriculture and water supply departmental authorities recommended that irrigators increase the area of perennial pasture and decrease the area of annual pasture, thus lessening the impact of rationing.

In the following years the area of annual pasture actually increased slightly. Not everyone believed that switching to perennial pasture would solve rationing problems. There was a fear that, if rationing still occurred, perennial pastures would be much more vulnerable. Changing to perennial pasture required major changes in farm enterprises and the use of laser grading. Perennial pasture required slopes steeper than the Tragowel Plains natural gradient. On many properties concentrating water on perennial pasture required radical laser landforming. But by the mid-1980s some Tragowel Plains farmers were discovering that laser grading was not necessarily a good investment.

In the first few years after its introduction, farmers on the Tragowel Plains were as enthusiastic about laser grading as farmers in neighbouring districts.[49] The initial enthusiasm was soon replaced by caution. The results of re-layout were often disappointing. Pasture regrowth after laser landforming was much slower than expected. Sometimes pasture did not re-establish at all. On salt-affected land the salt-line on paddocks did not retreat down the paddock as expected; sometimes it advanced up the paddock once vegetative cover had been removed by the grader. These difficulties were compounded by historically high interest rates. Laser grading seemed a safe investment with a 12 per cent interest rate and an assumed single-year loss of production after landforming. The investment turned sour as interest rates rose to 22 per cent and the unproductive period after laser landforming over high watertables stretched to two or three years.[50] A financial counsellor mediating between banks and financially pressed farmers passed a damning judgment on laser grading on the Tragowel Plains:

> Reclamation of land within 1.2 metres of the watertable has become a rich man's hobby and a poor man's demise. The cost of various initiatives to alleviate the problem is usually considered too high.[51]

The most vocal proponents of re-layout noticed that many farmers re-laying out land were not putting in the necessary inputs of gypsum, nitrogen and superphosphate. How could they expect the paddocks to recover as quickly as elsewhere? The disappointed investors pointed out that it made no sense continuing to apply superphosphate when there was no response. In a contemporary survey the Tragowel Plains farmers were found to have had very mixed success with laser landforming. Landforming had not been a burden for a minority who had few debts, but the majority

---

Laser grading and Tragowel Plains farmers

Successful developers (5 per cent of Tragowel Plains farmers): These farmers had laser landformed a large proportion of their farm. They had a high profit from their farm. They had a long experience of farming on the plains and a low farm debt. They have landformed without going into heavy debt. Most planned further farm development.

Developers at risk (20 per cent of farmers): These farmers had re-layed out as much land as the successful developers. They had high gross farm incomes, yet had the second lowest profit of any group. They made an average loss of $5000. Only a quarter of these farmers made a profit, because of their high debt level and slow response from landforming.

Land purchasers (5 per cent of farmers): These farmers had large properties, made little farm development investment and had few debts. Their farms were profitable; they followed the strategy of investing in land when the opportunity arose.

Low input farmers (12 per cent of farmers): These were the oldest farmers. They had few debts and were sceptical of investment in laser landforming. They had the lowest profitability of any group, yet had the lowest stress levels. Their strategy for survival was to save money when times were good and 'pull in the belt' when times were bad. While they were making little money at the time of the survey, they had few commitments to banks and knew they could survive until conditions improved.

Intending developers (25 per cent of farmers): These farmers saw their farm layout and drainage as problems. They intended to layout land in the future, though few had whole-farm plans. Most were comparatively new to the district. They had a high household income, often from off-farm work.

Discouraged developers (33 per cent of farmers): Most members of this group believed there was too big a financial risk with laser landforming. They believed they did not have the financial capacity to bear the risk, though they were not financially different from the intending developers. Many had whole-farm plans, so knew what was involved. They also had an average ten years greater farming experience on the plains than the previous group.

who had undertaken laser grading were suffering acute financial stress. (See box: *Laser Grading and Tragowel Plains Farmers*).[52]

Laser grading came to be seen as a risky proposition, a risk many with salt-affected land were not able to afford. Without laser grading, there were difficulties in concentrating irrigation water on perennial pasture. Farming cautiously and living with salinity seemed to be the only option for farmers in the most salt-affected areas. A small group of Tragowel Plains farmers met to discuss how they could solve their salinity problems. From this meeting evolved a community working group similar to the one that formed in the Campaspe West Irrigation District. The group saw itself as planning for the long-term future of the Tragowel Plains Irrigation District. Rising water prices and continued economic pressure meant the old strategy of conservative low-input farming offered no long-term future for the Tragowel Plains. Farmers following this strategy had the lowest incomes on the plains. There was also a group of farmers planning to landform in the future; without safeguards there was every risk many would share the financial circumstances of those who had already lost money from re-layout. The risks needed to be taken out of development.

The Tragowel Plains community working group was supported in its work by a group of public service scientists working under the State Government's Salinity Program. The prevailing expectation of the supporting research and extension officers was a re-statement of the earlier departmental recommendation to concentrate water on perennial pasture. New technology allowed a refinement of this recommendation, which previously had been received with some deserved ambivalence. More precise local measurement of soil salinity provided a new way to look at farms on the Tragowel Plains. With the new EM meter (which was something like a metal detector) an electromagnetic flux of the soil was used to measure salt levels. It enabled cheap mapping of salt levels on properties. The EM meter was used to make a comparison between the amount of salt on farms, farmers' perception of salting damage and the actual farm productivity.

The extent of salt damage was greater than many had anticipated. For many farms, the limiting factor to improved production was not shortage of water, as had previously been assumed, but a shortage of unsalted land. In the north there were many farms with little unsalted land. Many farms on the Tragowel Plains had insufficient low salinity land to offer any prospect of long-term viability under current patterns of agriculture. Farmers were watering large areas of land from which they could hope to make little income.

The use of the EM meter demonstrated the inability of both farmers and advisers to perceive accurately the extent of salting in a paddock.[53] Land that had not been fertilised in many years looked remarkably similar to land with moderate salinity, both being covered by rushes and unproductive grasses and growing no clover. This explained why some people claimed success at reclaiming land and others claimed reclamation was financially untenable. Laser grading and fertilising the unsalted and unfertilised land produced quick results. The same treatment of ground of similar appearance but with moderate salinity levels was less successful. By the time the salted land responded, the debt and interest to be paid made the investment problematic. There were rewards for concentrating water, but only if it was concentrated onto the better unsalted land. But the identification of this land was not always obvious. The EM meter provided a cheap method of identifying appropriate land.

In their most recent attempts to come to terms with salinity Tragowel Plains farmers are now encouraged to survey their farms for salt with the EM meter. For most farmers this identifies which land is safe to develop. For those farmers with insufficient unsalted land to allow any development the only long-term options are to accept a declining standard of living or sell their land. For most farmers, the EM survey will indicate whether there is enough unsalted land to provide a chance of financial viability in the medium term. In the longer term, with the inexorable rise of the water table the Tragowel Plains will become more saline; and the low profitability of farms will mean that this land may return to being nearly as empty as when Mitchell saw it.

As well as recommending more precisely monitored concentration of irrigation water and the use of shallow farm drains, members of the community working group were concerned that the land that was no longer irrigated would quickly degenerate into salty wastes. The government was asked to extend the existing subsidies for tree planting to include saltbush planting. Much of the northern Tragowel Plains was originally treeless and covered with saltbush. Planting saltbush would be a return to the original native vegetation and the advice of earlier advisers.[54]

The vision inherent in the community salinity plan is a well drained plain composed of patchwork of permanent green pasture and saltbush. It is a continuation of the strategy of learning to live with salt. At Bears Lagoon, at the southern and less saline end of the Tragowel Plains, there is an alternative way of farming the Tragowel Plains without irrigation water. Bill Twigg planted large areas of his farm to lucerne, an exotic replacement for the perennial native vegetation destroyed over a century ago. The lucerne has been planted sparsely to encourage plants to put down deep roots. The lucerne is rotationally grazed and wheat crops are sown into the lucerne paddocks using minimal-till techniques. In this way, lucerne has taken over the role of subterranean clover or medic in the normal wheat rotation. The aim has been to develop a totally 'organic' farming system. The barrier to others imitating the Twigg system may be the skill required to manage it.

The Twigg system could be called a conservationist vision of the plains — organic dryland farming. If organic dryland farming were adopted across the plains (the northern plains may now be too saline for such a system), the farms would be far larger, and there would be fewer of them. The local community would be smaller — a state of affairs the Tragowel Plains community has attempted to avert. This helps to explain the conservation movement's frustration with community involvement in salinity planning. The priorities of the farming community are differently ordered from the conservation movement's priorities. Maintaining the local community is seen as a vital component of sustainable agriculture.

The most recent strategies to attempt to live with salinity on the Tragowel Plains represent a radical revision of Thomas Mitchell's vision in 1836. The current judgments may be no more enduring than Mitchell's impulsive pronouncements. The current strategies provide no guarantee of future prosperity. Any reasonable prognosis must face the unspoken inevitable — people will continue to leave the saline northern plains and the population will decline. This is a continuation of an existing trend.[55] Many small settlements of the selection years have long since vanished; but concern about encouraging the decline of the rural community is strong. At best, local landholders can hope their strategies will slow the inevitable.

There is another message we can draw from the history of the Tragowel Plains. In the early 1980s the accepted wisdom was that re-layout was a general solution to salting.[56] Enthusiasts promoted laser landforming with an almost evangelical fervour. With hindsight, it is clear a reliance solely on laser graded re-layout has been a doubtful solution for the Tragowel Plains. In more recent times the language of enthusiasm has no longer been heard when laser landforming has been discussed, but elsewhere in Australia other advocates have promoted other solutions to land degradation with equal fervour. The real lesson is that many of these currently proposed solutions for land degradation control are untried, untested or unprofitable. Those who promote them with evangelical fervour should be aware of the consequences for farmers as well as for the environment when things do not work out as planned.

13

A tale of two settlements

Straddling the middle reaches of the Murray is one of the most diverse farming districts in Australia. The river flows through the heart of the Mallee country and its dry Mallee wheat and sheep farms. Close to the river and its anabranches and tributaries are irrigated farms. In the east around Kerang are grazing properties, where irrigators produce medium wool and fat lambs. In the same area there are farmers who grow irrigated crops: wheat, sunflowers, maize or sorghum. On the Tyntynder flats west of Swan Hill are irrigated dairy farms. On cultivated and cleared Mallee ridges, which overlook the river and lakes of the district, are five horticultural settlements producing grapes, stone fruit, citrus, vegetables and melons.

The people who farm the land are as diverse as the crops they produce. There have been six distinct waves of migration into the mid-Murray district. The Aboriginal people were first. They farmed the land with fire. The first European settlers were the opportunistic squatters who fought and dispossessed the Aborigines.[1] The squatters, in their turn, were displaced by the selectors and later the soldier settlers. After the Second World War came the Sicilians and Calabrians. The past decade has seen the arrival in some settlements of a new wave of migrants: the hobby farmers. When visiting a farm in the Swan Hill region today you may meet a descendant of a soldier settler or selector, a Sicilian settler who left a life of peasant poverty to build a future for his family in Australia, or an urban professional who works in Swan Hill and farms at the weekend.[2]

The diversity of produce and farmers is matched by the diversity of settlement histories. There were the usual closer and the soldier settlement schemes: Woorinen and Tyntynder. In this area government had no monopoly on property settlement. There were also privately developed schemes motivated by dreams of profit for the developer as well as the settler. Such were the origins of the horticultural settlements of Koraleigh in New South Wales and Tresco in Victoria. The settlement of Goodnight in New South Wales had its origins in well meaning nepotism. It was subdivided to settle a growing extended family. Nyah has an even stranger genesis. Theocracy and socialism are not often linked with Australian farming, but the establishment of the settlement of Nyah combined elements of both in a dream to recreate a village lifestyle for the urban unemployed.

This chapter concerns two settlements with fundamentally different beginnings: Tresco, born of speculative opportunism; and Nyah, born of idealistic communalism. It is a tale of capitalism and communalism. It is also a story of the unintended consequences of settlement, which these communities shared: the salinisation of land that followed irrigation and the environmental consequences that now threaten major wetlands and wildlife habitat.

The cooperative village

Nyah grew out of the boom and collapse of the Australian economy in the 1890s. The 1880s were heady years for anyone in Melbourne with capital to invest; property speculation was rife. It was an exaggerated version of the 1980s. Like all financial bubbles, the 1880s boom eventually burst and the paper tycoons lost their fortunes. It was not only the tycoons who lost. Many ordinary working people who had shared little of the boom became ordinary non-working people. Unemployment increased dramatically and homelessness became commonplace. Row on row of tenant houses in working class suburbs stood empty because the unemployed could pay no rent, while the innumerable homeless families slept where they could. Not surprisingly, public confidence in capitalism waned. New social movements appeared, dedicated to organising society in ways that would avoid the hardship, injustice and stupidity of the depression. The Labor Movement was one of these. Another was the Tucker Village Settlement Association. The Tucker Association did not have the longevity of the Labor Movement, but in the 1890s it had a far greater influence on government policy. The Tucker Association was formed by a group of Melbourne clergymen and churchgoers. The association proposed the creation of village settlements across the countryside as an alternative to the mass unemployment and poverty of the cities. The dream was of cooperative communities of God-fearing part-time farmers and labourers. To create these communities, land would be subdivided into blocks, too small to allow settlers to develop into full-time farmers, but large enough to provide protection against depression unemployment. The settlements would govern and develop themselves as cooperatives. The Tucker Association lobbied for legislation to allow the creation of their village settlements. In 1893 a Land Act was passed that allowed for subdivision of settlements and a government subsidy to help settlers to establish themselves on their blocks. That same year a village was created on the banks of the Murray about 33 kilometres north-west of Swan Hill. The founders expected the village to develop using the nearby river for irrigation. The largest farm was 20 hectares — quite appropriate with a little irrigation water but, without water, it was inadequate and small compared with the mile square (259 hectares) selection blocks in the nearby Mallee wheat country.

Soon after the first settlers arrived, a nursery was growing apple and vine stock for settlers to plant their orchards. The settlement lay on sand dunes 20 metres above the river. The cooperative had sufficient funds for a pump to lift water from the river, but not to pipe the water to the settlement. The settlement was named after the politician Taverner; it was hoped Taverner would help with the task of gaining government assistance for laying the pipes. The rhetoric of cooperative management was discredited in the capital city. The government had lost taxpayers' money in the collapsing waterworks trusts. Government policy now favoured individual farmers over cooperating groups. The pump quietly rusted. By the federation drought of 1901, villagers were carting water from the river in drays.

With such an unsatisfactory water supply, dreams of orchards evaporated like puddles in the hot Mallee sun. The villagers grew lucerne hay on their small plots and changed the name of their settlement from Taverner to Nyah, recognising the failure of their flattery. As the founders had envisaged, the settlers could not support themselves on the proceeds of the small blocks alone. The villagers needed off-farm work to provide for their families. The Tucker Association idealists had overlooked a major difficulty. There were no local employers. Villagers were forced to travel great distances for work and stay away from their farms for long periods. Many properties were abandoned to the rabbits. The village dream foundered.

The decline of Nyah coincided with the regrowth of government interest in closer settlement. By 1910 the newly created State Rivers and Water Supply Commission was looking for sites to develop closer settlement schemes. In 1911 the infrastructure of pipes and channels, of which the early villagers had dreamed, appeared as part of the transformation of Nyah into a commission-sponsored closer settlement scheme. Optimism and untempered enthusiasm were eagerly expressed in the local Swan Hill newspaper:

> Nyah is considered by many people best able to judge to be a second Mildura, and rightly so on account of the potentialities of its rich soil . . . As an irrigation settlement Nyah has natural advantages which would be hard to surpass . . . owing to its wealth of soil resource and its excellent geographical position for despatching its produce by river, soon by rail . . . The vines and fruit trees in well laid out and beautiful orchards present a pleasing spectacle, predicting the ultimate prosperity which the landowners will enjoy.[3]

Time eventually unmasked most of the naive reporter's predictions. Such propaganda did not convince many outside Nyah. By 1914 new settlers had purchased only three of the forty Nyah closer settlement blocks. The few Nyah farmers watched the slow progress of their settlement with apprehension. The unoccupied blocks proved a financial burden on both the government and the settlers. The State Rivers and Water Supply Commission saw no reason for restraint. It developed the nearby settlement of Woorinen in 1914. Nyah found itself competing with

Woorinen for the very limited pool of settlers. Woorinen was closer to the town of Swan Hill, and the terms of purchase were more attractive than at Nyah. The commission enticed experienced settlers from downstream Mildura to Woorinen and used their presence to advertise the success of the area.[4] This was of little help to Nyah.

The First World War helped to fill the Nyah blocks. By 1916 the government wanted to settle soldiers on the land. Nyah was an already developed settlement with too few settlers. Nyah was expanded and the government filled the blocks with returned soldiers. The raw soldier settlers experienced the same problems as their comrades on irrigation blocks across the country: low prices and salinity. Because there was no market, peach and apple trees were replaced by vines. Dried grapes were transportable and there was initially a market of sorts for them. As European agriculture recovered and other nations expanded their own plantings of grapes, world sultana prices collapsed.

Salinity appeared on some Nyah blocks. A few higher blocks were drained with tile drains, but for most soldier settlers there was little done about salinity. Some farms were on lower clay soil, which was most prone to salting and could not be drained. Elsewhere tile drainage was quite possible, but settlers could not agree on whose property the collector drains would be placed.[5] It took thirty years for Nyah, the settlement founded on cooperative ideals, to agree on a drainage line.

By the mid-1920s it was clear that soldier settlement was a financial disaster. In Nyah it was also a land degradation disaster. During the Royal Commission into failed soldier settlement in 1925, both farmers and the State Rivers and Water Supply Commission were keen to allocate the blame for the salinity. In Nyah and Woorinen a dozen soldiers had walked off their farms, many of them because of salt. In the majority report the failures in Nyah and Woorinen were blamed on the settlers' lack of capital and lack of skill. In a minority report, the soldier settlers' representative pointed out that the 1916 inquiry into failed closer settlement policy had warned about the dangers of irrigating certain soil types, but that this had been ignored.[6] Blocks of land identified as unsuitable for closer settlers in 1914 later had been sold to unsuspecting returned soldiers. The soldiers agitated for compensation for all settlers sold unsuitable land. This set the tone of relations between farmers and the State Rivers and Water Supply Commission for the next fifty years.

The private enterprise settlement

Like Nyah, the seeds of the Tresco development were born in the boom and bust days of the 1890s. Unlike Nyah, which was a reaction to the worst excesses of Victorian era capitalism and depression, Tresco was part of the excess. Lake Boga is a shallow lake on the edge of the Victorian Mallee, a few miles south of the Murray River. It was fed by intermittent flood flows of the Avoca River. In the early 1890s Mallee leaseholders with frontage to

Lake Boga observed the financial success of the Melbourne property speculators and saw how their leasehold frontage to Lake Boga could be transformed into speculative profit by subdividing it as an irrigation settlement.
The only limitation to their scheme was the necessary government approval for the pumping of irrigation water from Lake Boga. In the corrupt
years of the 1890s this was not an insurmountable problem. The share-
holders register of the Lake Boga Development Company included several
prominent government politicians who could argue in the company's favour when Cabinet approval was sought for pumping from the lake.

The leaseholders launched their development before the supply of
water was resolved. Taking up their pre-emptive rights, they bought the
freeholds of their waterfront leases and subdivided them into town blocks.
Other leasehold land controlled by the company to the south and east of
Lake Boga was subdivided and offered to the prospective settlers on the
promise of imminent pumping approval and continued long-term leasehold, although the company could guarantee neither. Prospective buyers
arriving on the railway were taken around the planned settlement in the
company of politicians with shares in the project, giving credence to the
claims of imminent government approval. Blocks with no water and no
security were sold to naive settlers. The speculators and politicians had not
counted on the commitment of Premier Deakin to the Mildura settlement
farther downstream. To protect Mildura from the competition of settlements closer to Melbourne, he refused approval for pumping from Lake
Boga. The Lake Boga Company collapsed. Speculators and purchasers
were left with worthless leasehold blocks.

After the collapse the remaining leasehold land passed into new private
hands. The State Rivers and Water Supply Commission was interested in
the land as a possible site for closer settlement, but the Closer Settlement
Board refused to become involved or support the development of the land.
The board believed the land was a salinity risk. There was saltbush in the
area, a sure sign of salt danger; and nearby wheat land was naturally salty
in some places.

The Closer Settlement Board's dismissal of the land did not deter the
new owners from dreams of irrigation development. They formed a new
company with the patriotic name 'Australian Farms'. The investors included financiers, leaseholders, politicians, journalists and even a couple of
members of the Closer Settlement Board. They hoped to ensure government approval for water rights and good publicity to attract settlers. The
group declared that they were not in the business of trafficking farms, but
were offering the settler with capital both good land and good water. The
company members knowingly overlooked the decision of the Closer Settlement Board. The possibility of profiting from the government money
supporting assisted migration blinded the investors to the capability of the
land. The scheme was publicly supported by the State Rivers and Water
Supply Commission who described it as an opportunity to compare the
merits of private and public development of land. The granting of water
right from the lake caused disquiet among nearby Mallee settlers whose
stock and domestic water supply was drawn from Lake Boga. The Mallee

farmers were concerned that there would not be enough water in a dry year for both themselves and the irrigators.

The new settlement of Tresco covered two Mallee ridges and the clay saltbush flats between the ridges. The location was chosen because it was close to Lake Boga. The company had warnings that irrigating saltbush flats was a risky investment, but eliminating the flats from the settlement would have meant more pumps and another ridge to develop. Water was drawn from the lake in three stages of pumping and distributed by wide, cheaply constructed, horse-scooped channels.[7] Surface drainage was installed but the drainage water was returned to Lake Boga, 300 metres from the inlet of the irrigation supply pump, ignoring the potential danger of salinising the lake. For the settlers, the terms of settlement were not nearly as generous as state-sponsored schemes. There was no interest subsidy and the repayment period was shorter. Despite these drawbacks, there was little difficulty in selling the blocks. By 1914, while most of Nyah remained unsold, half the Tresco blocks had been sold.[8]

Calamity respects no ideology. Tresco suffered the same hardships as Nyah. In the severe 1914 drought the Murray River and its tributaries dried up and the Mallee stock and domestic water supply could not cope with the demand. Angry Mallee selectors attacked the government and the State Rivers and Water Supply Commission for having agreed to provide a water allocation for Tresco that could not be guaranteed. The drought also increased the salinity of Lake Boga, compounding the bad decisions on land development, drainage and channel construction.

Many of the Tresco farms were established as citrus orchards. A viable citrus farm was smaller than a viable grape or lucerne farm. Other farms were planted with doradillo grape vines. The price of oranges plummeted after the war. Doradillo grape growers found they could not transport their grapes to the distilleries before rot set in, so they received low prices for their poor quality fruit. The predicted salinity problems emerged with the end of the First World War. The settlers discovered that citrus trees were extremely sensitive to salt. It would have been hard to choose a less suitable crop for an area likely to develop salinity. The experience of Mr A. Sloane's orchard was typical. It was planted in 1915 and purchased in 1919 when its citrus groves looked to be some of the best on the settlement. The trees deteriorated rapidly from 1921 and within three years two-thirds of the trees in the orchard were dead.[9] Land on the flats was abandoned as it became obvious that watertables were within 30 and 60 centimetres of the surface and nothing could be done to save the trees.[10] By 1925 half the citrus trees in the settlement were dead. Salt appeared above the flats on the dunes. To farmers with no experience of salinity, the destruction of their orchards was hard to understand. The cause of their losses was clear. Salinisation in Tresco was inevitable without outfalls for tile drainage.[11]

One settler wrote to the settlers' newspaper in Melbourne, *The Australasian*, inquiring what he should do about salinity. The reply must have distressed this correspondent reading between the lines that he had been duped by Australian Farms:

To H. D. A., Lake Boga: Salt occurs nearly always when land has been irrigated for some time. The irrigation water dissolves the salt in the soil, brings it to the surface by capillary action, and as the water is evaporated by the heat of the sun, salt is left on the surface. When there is good natural drainage, instead of being brought to the surface the salt solution is carried off in the drainage. Where salt shows itself, drainage is the proper corrective . . . This is very expensive, but several orchards in your district have already been drained in this way, a machine having been used to excavate the trenches . . . Some of the lakes and lagoons in your district hold water which contains a fair amount of salt, and when this is applied to the land with no natural drainage the effect is soon disastrous. The old maxim that getting the water onto the land is only half the water in irrigation, is well exemplified in this instance. The other half is getting the water off again.[12]

In 1919 Australian Farms bought more land to the west of Tresco, funded by the same group of wealthy speculators, politicians and public servants.[13] The government's policies for assisted immigration offered another chance for astute businessmen to make a profit. Government policy was driven by a xenophobic desire to redistribute the white races to protect the Anglo-Saxon heritage in Australia. The government was willing to pay British settlers, especially army officers, advances of £300 to migrate to Australia to farm the empty open spaces. A director of Australian Farms travelled to India to attract retiring British officers as settlers. Learning from some past mistakes, the new settlement was established on the dune ridges, with the intervening dune flats left to act as evaporating basins for the farms on the dunes. The company attracted forty-five British officers to Tresco West with the promise of three years sustenance of £75, three years free water, expert supervision and a three-year moratorium on loans for planting, piping and fencing farms. Tresco West was dubbed 'Blighty' by the residents of Tresco.

The new scheme was doomed from the start. A more unsuitable bunch of farmers would be hard to find. Most of the officers had been invalided out of the army, yet somehow they were expected to have the stamina to establish and run a farm. The officers brought British upper and middle class expectations of farming to the hot dry Mallee sand dunes. They dressed in riding jodhpurs in the hot Mallee sun. They employed labour, rather than doing the work themselves. Recreation consisted of expensive weekend trips to Melbourne, an unheard of luxury for most Mallee settlers, or paper chase fox hunts. The blocks could not provide for their expectations or style of life. The local opinion was that the settlers brought colour to the district but were totally unsuited to the life on the blocks:

Looking back they were square pegs in round holes — I don't think they'd ever cleaned their boots before. They'd gone straight from public schools to commissions in the army, servants galore in India, and their clothes! They were delightful people — dressed for dinner, had paperchases at the weekend as we didn't go for foxhounds. They got up all sorts of theatricals . . . in those days fancy dress balls were very popular and these people were outstanding in the costumes they devised. They gradually lost all the money they had invested and started to trickle away.[14]

By 1925, only five of the forty British homes were occupied. Much of the settlement was deserted. The original Tresco settlement was in worse shape than Tresco West. Settlers abandoned blocks that, with large debts and nothing growing on them, were worthless. The remaining growers on higher ground at Tresco West had pulled out their citrus trees and planted sultana grapes, which were less sensitive to salt. In 1923 the world market price for dried grapes collapsed. The Tresco growers were left with no profitable crop for their depleted farms and they sought redress. The block owners felt deceived and took the Australian Farms Company to the courts. In 1925 the company collapsed and there was no money for compensation. The State Government took over responsibility for running what was left of the settlement. The State Rivers and Water Supply Commission had supported the scheme and it fell to the government to pick up the pieces.

Picking up the pieces

On taking over responsibility for the fiasco at Tresco, the government formed a scientific committee to investigate the cause of the salting and recommend solutions. The local settlers distrusted both the Closer Settlement Board and the State Rivers and Water Supply Commission. The commission had supported the Tresco development and members of the Closer Settlement Board had held significant shares in Australian Farms. The settlers expected investigations by either of these bodies would lay the

Furrow irrigation about 1900 using narrow, deep furrows

blame on the landholders, rather than the company. Scientists from the Department of Agriculture and the newly formed Council for Scientific and Industrial Research were brought in to do the work. The scientists conducted a soil survey, measured watertables and conducted tests on irrigation methods. Their conclusions partially vindicated the settlers. The salinity on the flats was extensive and caused by a regional watertable raised by heavy irrigation, leaking channels and a saline water supply. Less than 40 per cent of the settlement was salt free.[15] There was nothing that could be done about this land.

Vindication was sweet, but of little financial value. More important for the growers, the investigation showed that their own irrigation methods were causing localised salting on the dune slopes above the flats.[16] Growers were irrigating by running water down V-shaped furrows ploughed between the rows of vines. The furrows were long and narrow. These were methods that had been advocated in the *Journal of Agriculture* and were based on irrigation practices developed in Arizona. The Arizona practices were designed to slow water travel in furrows to better saturate the soil and to prevent degradation of soil structure and wastage of water flowing out the end of furrows.[17]

The Arizona techniques may have been appropriate for the clay soils of the Arizona Research Station. They were entirely wrong for sandy Mallee soils, where irrigation water took a long time to reach the end of the furrows because it seeped quickly into the sand. As a result, during irrigation the top ends of the furrows were a quagmire, from waterlogging, and the trees at the bottom of the furrows were watered for only a couple of hours. The narrow furrows also leached salt towards the trees.[18] The scientific committee recommended Tresco farmers flood irrigate their groves using short runs to leach the salt from all the soil and shorten irrigation times.[19]

Improved irrigation practices gave hope for the properties on the sandy dune ridges. For the flats the outlook was depressing. Despite the abandonment of many of these low-lying properties, irrigation still was pushing the salt front up the lower slopes of the dunes. Even while the scientific committee was publishing its results, orchard trees were dying.

In the years before the Second World War both Tresco and Nyah struggled with their salinity problems. Tresco farmers prevaricated on the advice given by the scientific committee. In Nyah the salt problem was less serious. Some areas of low-lying land had been abandoned because of salinity, but salinity was not as widespread as in Tresco.[20] There was a consensus among the Nyah farmers and scientists that tile drainage would be economic and effective. Unfortunately there was no consensus on the manner of implementation and Nyah farmers argued over where the drain would be located. During the war years the price of dried fruit rose because of decreased production in Europe, as it had in the First World War. Farm incomes also rose and many 'blockies' along the Murray Valley had a rare rest from worrying about making ends meet. This must have given them time to think about other problems such as saline soil.[21] More money may have helped the Nyah irrigators finally come to some agreement about drainage; but drainage works had to wait until the finish of the war when men and materials were available to do the work.

While the end of the war promised some relief from Nyah's salt, weather and developments in the dried fruit market soon threatened the financially fragile settlers with ruin. There was a series of wet and cold summers and autumns. The fruits didn't dry, but rotted on the vines. When the warm summers returned, the price of dried fruit collapsed. The dried fruit settlements around Swan Hill could not compete with the warmer Sunraysia down the river, let alone the rest of the world. In 1950, the Australian Dried Fruits Association asked the Federal Government for help to restructure the settlements around Swan Hill. Four years later a committee began an inquiry and produced an unflattering report. The supposed natural advantages of Nyah for fruit production did not exist as it was too cold and too wet by world standards.[22]

The slightly cooler weather at Swan Hill meant the grapes ripened three weeks later than in the Sunraysia, significantly increasing the chance of rain falling on a mature crop. Rain has little effect on immature grapes, but splits ripe grapes, which then quickly rot. While noting the poor state of many of the blocks ravaged by salinity, the committee considered there was enough good land remaining to justify saving the settlements. This could only be achieved by diversification to new enterprises and a massive restructuring of the settlements. There were too many small farms.[23] For the two settlements the recommended solution was drastic; nearly one-third of the farmers needed to be convinced to quit their farms for compensation of £200 an acre. The farmers remaining were to be encouraged to diversify their production.

New crops and new settlers

The Mid Murray Dried Vine Fruits Area Inquiry had made it clear the Tresco area could not compete with the Sunraysia District's dried sultanas. Changes in road transport provided new opportunities. The 1960s was a period of unprecedented investment in road building across the country. New bitumen brought the Tresco area within four or five hours drive of the Melbourne wholesale fruit market. Growers could pick during the day and load their fruit onto a truck in time to reach Melbourne for the market opening at 5.00 a.m. The improved roads also opened up the Sydney market. This transformed Tresco's southern and easterly position from a competitive disadvantage to an advantage over the more distant Sunraysia area when marketing fresh picked table grapes in the two largest cities in Australia. Tresco farmers planted fresh grape varieties. Their Nyah neighbours stayed predominantly with dried vine fruits.[24] The block owners found they could sell their grapes to the newly emerging home wine making market — a market, created by the post-war influx of European migrants. Post-war migrants transformed the Tresco district in an even more direct manner — they became the new wave of settlers.

Before the Second World War a small number of Calabrian and Sicilian settlers bought blocks on the Woorinen and Tresco settlements. From this vanguard the word spread back to Italy about the opportunity to own a farm in Australia.[25] The farms were in many ways similar to the farms found in southern Italy. The climate was hot and dry in summer, and the main crop was grapes. After the war, more Sicilians and Calabrian families made their way to Swan Hill. These were hard working, ambitious people. They aimed for a better life for their families. They worked up the farming ladder: first working on the farms as labourers, then as sharefarmers, until eventually they were able to buy a farm, without using credit. The prices they paid were often considered high by local standards; it seemed that whenever a property was placed on the market there was an Italian to buy it.[26] By 1990, almost half the family names in the settlement register of properties were of Italian or Greek origin.

For Tresco West this was a total contrast to the original middle-class English officer farmers. The new settlers had little education and little command of the English language. Their world view was of a world of limited good; one person's gain is someone else's loss. They mistrusted credit and were frugal with money. These new settlers had no pretensions or misunderstandings about the nature of these blocks. They knew the only way to succeed on the blocks was through hard work.[27] Local Australian opinion was that the new migrants mercilessly farmed their blocks and families. Often the newcomers purchased blocks with salt or drainage problems; but by buying and combining blocks, the migrants were doing some of the work of restructuring recommended in government reports.

Beating the salinity

In Tresco few had taken up the Department of Agriculture's 1930s suggestion of flood irrigation. It may have been a good idea for citrus groves, but it was impracticable with narrowly spaced vines. One local block owner, Mr T. Malcolm, came up with the idea of irrigating with broad furrows. Instead of running water down several narrow V-furrows, it was directed down wide flat furrows separated by narrow V-shaped hills on either side of the furrow. Water was too hard to control in a single broad furrow, so Malcolm invented a broad wooden sweep, which formed two narrower broad furrows under each tree. This gave good control of irrigation water and leached salt from most of the surface area of the orchard.[28]

Other farmers developed Malcolm's ideas, evolving a system of three broad furrows between rows of vines. This leached nearly all the soil profile and controlled the salinity. In 1956 a local garage owner started manufacturing an implement to form the furrow system. By 1961 the Department of Agriculture was recommending the system. The concurrence of local experts and government scientists, and the existence of a locally made implement, increased the credibility of the system. By 1970, the broad-based furrow technique was used on 90 per cent of the farms.[29]

The government officers took the method a step further by recommending that farmers irrigate with a heavy flow first to get water down to the end of the furrow, then reduce the flow so that it seeped in equally along the length of the furrow.[30] This avoided the overwatering problems at the top of the rows. The wheel of government advice had turned full circle from the earlier Arizona-derived recommendations. Irrigators quickly recognised the value of dual flow irrigation. It was a simple change and was soon commonly used across horticultural districts.[31]

Salinity in the water supply remained a contentious problem. Tresco farmers were aggrieved that Tresco's drains still poured back into Lake Boga, near the intake for their water supply. Various scientists assured the farmers that their water supply was quite adequate, but their assurances were not believed. In the wet year of 1956, with extensive local flooding, Lake Boga broke its banks and flooded the abandoned saline flats of the Tresco settlement and flushed back into Lake Boga. Growers claimed they were being supplied with salty water. The Water Commission, while disputing this claim, was reluctant to test the water. Eventually a boat was sent out onto the lake to survey the salt concentrations. The results vindicated the commission rather than the farmers.[32] In 1960 a new drainage system for the Tresco district was completed. Drainage water was diverted from Lake Boga into nearby Lakes Round and Long, transforming them from freshwater lakes into evaporation basins.

Four years of salt damage after the 1956 flood had farmers in the mood for quick action to beat the salinity. Local experience had proven wrong the 1930 scientific committee's advice that tile drainage was unprofitable. Most farmers knew of tile drainage on local blocks that seemed to have worked. The key was to minimise the cost of construction. Here local experience proved indispensable. Within eight years, two cooperative drainage groups had been formed and nearly all the farms on the Tresco settlements were drained with tile drains. At the same time, some farmers replaced their irrigation channels with pipes.[33.]

The saline water supply remained a problem; the channels in which it was delivered leaked; and the scheduling was inflexible. Half the water supplied to the district seeped away before it was delivered to the farms. In 1966 the government announced it intended to build a new pumping point upstream of Lake Boga and planned to reconsider the way it scheduled water. Promises to build new concrete supply channels and automatically pumped drainage sumps in low lying areas soon followed.

In the space of a decade the growers of Tresco had seen a revolution in their fight against salinity. The water supply was better protected against salt. The supply channels didn't leak. Many farmers had replaced farm channels with pipes. The new furrow irrigation methods assured a more even watering over the whole farm, leaching salt into new underground drains, which carried the drainage water off into a new district drainage system, and eventually into two new evaporation basins. The changes on the farms had come about from local experimentation and improvisation, and had been achieved without the use of credit, but with subsidised government assistance. Some farmers even began to talk of reclaiming

salinised land. A disparate group of farmers had overcome cultural, financial and knowledge difficulties to achieve a dramatic turnaround in their
fight against salt.[34]

What to grow and who to grow it?

The restructuring recommendations of the 1950s were not popular with
the Nyah settlers. While farmers in the region were unhappy with their
financial situation, few were willing to sell their farms and leave for an
uncertain future as labourers in the city. Some were willing to sell their
land, but only when the price of their properties rose. In the short term
settlers remained on their properties.

The wartime consensus on drainage had cleared the way for the State
Rivers and Water Supply Commission to build a drainage system into
which farmers could connect their own farm drains. By 1956 many farmers, lacking the income to finance investment in drainage, had still not
constructed drains on their own properties, despite obvious salt and erosion problems.[35]

Low incomes also meant that Nyah settlers were unable to diversify
their enterprises. Nyah maintained its dependence on the sultana. Today,
Nyah shares its land between wine, sultana and table grapes, whereas
Tresco and Woorinen have almost abandoned dried sultana production.
By the 1960s, nearly three-quarters of Australia's sultana production was
sold on the export market, but Australia's grapes made up less than 10 per
cent of the total world trade in sultanas. Australian production had little
influence on the world price. If there was a glut of production from Turkey
or Greece, the Australian price would fall dramatically. The Sunraysia,
Riverland and Nyah districts suffered similar hardships from a frequent
cycle of low prices. Poverty on the farm was the inevitable result. Every
time there was a price slump, there was another government inquiry.[36]

In 1976 the Industries Assistance Commission concluded that 30 to 40
per cent of Sunraysia growers had little hope of becoming viable, and
commented that the situation was worse in the mid-Murray. It recommended quota protection and rural adjustment assistance, including
concessional loans and small grants to those leaving the industry. The
Industries Assistance Commission criticised furrow irrigation because of
the limited area of land one farmer could water. It favoured a pressurised
pipe delivery system and conversion to the new drip irrigation, supplying
water to the trees through drippers in a plastic supply pipe. With spray or
drip irrigation one person could manage 40 hectares, rather than 12 hectares. The commission argued that furrow irrigation increased accessions
to the watertable and led to more salting.

Water could be left in the pipes for prolonged periods between irrigations
without it leaking away. Individual growers would then be able to grow
vegetable crops that needed watering outside the regular irrigation season.
This would release Nyah from the strictures of a channel irrigation system

and growers would be able to choose from a wider choice of crops instead of receiving low returns from dried fruit. A local committee lobbied successfully for this major investment. The pipeline proposal was to be the means of saving Nyah from depression.[37]

By the early 1980s a large pump on the river delivered pressurised water through pipes to every Nyah farm. But growers at Nyah were still unwilling to diversify their enterprises. Low prices for farm properties attracted buyers seduced by the dream of rural utopia, and the cycle continued. After each slump the same properties remained unviable, but different people owned them. During the mid-1980s the Industries Assistance Commission, which had a decade earlier pinned its faith on pipelines, decided more drastic action was needed. It proposed a vine pull scheme so that the farms would then be planted to some other crop. Growers were offered money to pull out their vines in return for signing a contract not to replant vines in the same paddock.

To some Nyah growers the vine pull scheme offered an opportunity to be paid to pull out old vines that should have been replaced years ago. To others it was an opportunity to get quick cash. On some properties the vines were pulled and the paddocks left bare. Sometimes the properties were put on the market. Owners were left with a water bill and no income.[38] Other growers replaced their vines with peaches and nectarines; as production increased the price of peaches and nectarines fell. The problem of inadequate farm income moved from one horticultural commodity to another.

Nyah today still struggles with the main problem that dogged the settlement in its first years: the need for high value, profitable crops. A small core of larger, economically viable properties are farmed by long-term residents. Another group of small farmlets are owned by long-term residents who are either retired or work in Swan Hill during the week. None depend on their farm for their livelihood. Most properties are too big to be hobby farms and too small to be viable farms. The urban illusions that created the Nyah village are still pivotal in shaping the destiny of these properties. The ownership of many of these properties changes frequently. A new resident moves in, often with little experience of farming, hoping to succeed where others have failed, or attracted by housing that is cheaper than in nearby Swan Hill. The arduous life of a part-time job in Swan Hill, long hours on the block and a poor reward for the effort soon lead to disillusionment, and the property is again offered for sale.[39] In such circumstances there is an inability to make improvements to the property. There are few experienced, progressive farmers in the area who can provide an example for others to follow.

At Tresco there is a different challenge: not what to grow, but who to grow it. The post-war European migration wave washed over Tresco but missed Nyah. The first generation of Mediterranean migrants are now ageing and are ready to retire. The post-war migrants who found their way to Tresco were ambitious for their families. They appreciated the value of education in building a better life for their children. Their children are often now working in professional jobs. Few children of either Italian,

Greek or Anglo-Saxon settlers wish to follow in the footsteps of their parents working the arduous life of a blockie.[40] Neither are there Italian sharefarmers waiting for their chance to become land owners. The latest set of migrants, the urban-employed, part-time farmers and retirees, have passed by Tresco. This is not an area likely to attract urban dreamers seeking an idyllic rural lifestyle. Though the vines and citrus trees on the hilltops are green and healthy, the abandoned flats at the bottom of each farm are bare, salty, windswept and unattractive. Prospective buyers from Swan Hill or Melbourne quickly find their way to the much more attractive Woorinen, or the significantly cheaper Swan Hill Flats.

Retiring Tresco settlers typically like to sell the farm and move into a town near their children. For intending retirees the price of the average Tresco block is too low to allow retirement near the children if they have taken jobs in Melbourne. The next best hope is Swan Hill, and even there some cannot afford the most modest house. The future for Tresco may be a small number of large and viable farms and an increasing number of trapped retirees without the service support found in Nyah.

Sanctioned anarchy

In both Nyah and Tresco the historical battle against soil salting on the farm has been won. The key was the installation of underground tile drainage. Historically, irrigation developers have shown little concern for drainage water once it has left the farm. This has not always led to wise decisions. At Mildura, wetlands destroyed by drainage now leak large amounts of salt into the Murray River. At the more recently developed neighbouring settlement of Nangiloc drainage is slowly destroying riverine wetlands. In Nyah the water drained into the Nyah State Forest. Nyah has been fortunate. There seems to have been little effect on the Murray River from this drainage. Tresco drainage water ended up in two small lakes, Lake Long and Lake Round, sacrificed as evaporation basins. Planners will eventually need to think about the limited life of evaporation basins. They will eventually fill with salt and must be either abandoned or, in favourable circumstances, harvested. Tresco's basins are not likely sources of harvested salt. A more pressing matter is the question of whether the evaporation lakes should have been sacrificed in the first place. In the Murray Valley horticulture areas there are community pressures on irrigators to reduce their production of saline drainage by increasing the efficiency of water use.

One way to reduce the demand for drainage is to change from furrow irrigation to more efficient irrigation methods. More recently, in all the Murray irrigation settlements, horticulturalists have been encouraged to install modern, pressurised micro-irrigation systems: over-tree sprinklers, under-tree sprinklers or drip irrigation. The hope is that this will enable farmers to use less water — decreasing the amount of water reaching watertables, flowing into evaporating basins or into the Murray River. Up and down the Murray, from the South Australian Riverland to upstream Tresco, many farmers have taken little notice of such recommendations. They have retained their furrow irrigation systems.[41]

Most growers have not installed pressurised irrigation systems because of the cost of the investment. Because many mid-Murray fruit growers have low incomes they are neither willing nor able to undertake such an investment. In the Tresco settlement in particular the generation of 1960s settlers was steeped in the tradition of avoiding debt and paying with cash. That way of doing business was part of the Calabrian and Sicilian culture.[42] In the 1960s the settlement solved its own salinity problem with government support but without debt. These farmers have been unwilling to incur debt to solve downstream salinity. There does not appear to be a financial advantage in micro-irrigation on small to medium sized farms. Part of the reason for this lies in the manner irrigation water is sold. The price of water has risen substantially in recent years as the government has sought to recoup the cost of delivering the water. Despite this, most Tresco and Nyah irrigators have nothing to gain financially from using water more efficiently. All the farms have a water right and the owners are obliged to pay for their water right each year, whether or not they use the water. Few farmers in either district use more than their basic water right, so inefficient water use is not penalised.[43]

For Tresco irrigators the environmental impact of its drainage is insignificant compared with the more pressing environmental problem with the Tresco supply system. For the growers, the salinity of the water can still be a problem. To upstream farmers the demands of Tresco horticulturalists cause flooding or limit the ability to control soil salting. To environmentalists the supply system is a major intrusion on important wetlands. The objectives of reducing water salinity, controlling soil salinity, reducing flooding and protecting the wetland environment create a complex challenge.

The Torrumbarry Irrigation Scheme (Figure 13.1), supplying the Tresco settlement, is an unintended tribute to Gordon, the bureaucrat who lost the debate over the development of irrigation in Victoria in the 1890s (see Chapter 10). The winners of the debate advocated the distribution of water through channels along the high ground. Gordon advocated the cheaper alternative: distribution through existing watercourses. His detractors labelled his scheme 'Gordon's gutters'. The Torrumbarry system looks suspiciously like Gordon's gutters. It developed as a topsy turvy scheme, integrating the demands of private developments like Tresco, soldier settlement requirements and existing precedents claimed by farmers.

The irrigation water supplied to Tresco is stored in the Torrumbarry Weir on the Murray River between Echuca and Kerang. Water is diverted from the weir along the National Channel to what was once Kow Swamp. From Kow Swamp the water travels along the Pyramid and Box creeks through Johnson and Hird's swamps to the Kerang Weir on the Loddon River. Here Torrumbarry water mixes with Loddon River water. There is an outlet on the side of the weir, which releases water into what was once a flood overflow of the Loddon River. The overflow used to flow through

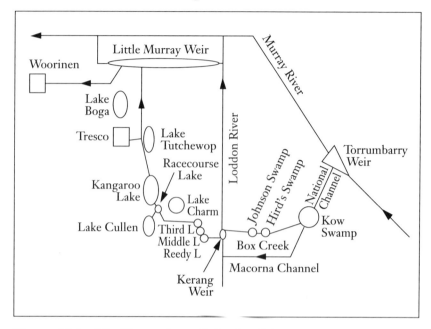

Figure 13.1   The Torrumbarry Irrigation Scheme

a series of lakes: Reedy, Middle, Third, Racecourse and Kangaroo, Tutchewop and Lake Boga before finding its way into the Murray River via one of its anabranches, the Little Murray River. In high flows the water used to spill into the adjacent Lakes Cullen and Charm. Today the overflow system is heavily regulated. Irrigation water flows through Reedy, Middle, Third and Racecourse lakes to Kangaroo Lake. From Kangaroo Lake it flows through a channel into the Little Murray River. Tresco pumps its water from this channel. The Little Murray River is dammed at either end to create a weir to store the Torrumbarry water. Water from this Little Murray Weir flows down another channel to supply the Woorinen horticultural district.

This complex supply arrangement has been at the heart of Tresco's water salinity problems. Torrumbarry water is initially fresh, but as it travels west it becomes progressively saltier as it mixes with saltier water flowing north in the rivers and creeks it crosses. Farmers using the fresh water upstream grow pasture. Tresco (and Woorinen), at the end of the Torrumbarry system, is supplied with the saltiest water to irrigate stone fruit and citrus trees, which are much more sensitive to salt than pasture.[44] A major priority of the management of the Torrumbarry system is to minimise the salinity of water reaching Tresco and Woorinen. This is achieved at the cost of increased flood risk for farmers along the lower Loddon River. The centre of this conflict is the Kerang Weir, on the Loddon River, which is used to push water across the Loddon River into the overflow to Reedy Lake and downstream to Tresco. The gates on the weir are also used to keep Loddon floodwaters out of the overflow. Farmers on the Loddon flood plain feel they are being asked to bear an unfair burden of flood risk. They believe that farmers along the overflow are getting an unfair advantage.[45]

The nub of the wetlands problem is the disruption to the normal flow of the river and lake system. Some lakes never dry out, killing trees and habitat that had adapted to seasonal wet and dry conditions. Some lakes are now rarely filled, becoming dry and progressively saltier. Others are kept full, but are never flushed, thus becoming more saline. The result is a series of problems along the length of the supply channel, presenting difficult decisions and complex tradeoffs. Duck Lake is a small wetland that has been cut off from the natural floods that used to intermittently flood and flush it. It is now extremely salty, acting as a discharge point for the surrounding farm land. Reconnecting the lake with the natural drainage system would eventually repair the wetland habitat, but large volumes of salt would be flushed into the Murray River. The increased pressure on the groundwater underneath the lake would threaten the surrounding farmland with salting.[46]

Too much, rather than too little, water is a problem in the Third Marsh. Levees built below the marsh protect farmland from flooding, but the

A Kerang wetland habitat. In this case the habitat value has been enhanced by irrigation development.

levees prolong the period the marsh is inundated each season. The levees were augmented by a sill built during a drought in the 1970s to maintain water and duck numbers for shooting. Changes in the upper catchment have increased the frequency of flooding in the Avoca River, further compounding the problem. Dying trees around the marsh indicate dramatic changes in the catchment.[47]

Irrigation is not always environmentally destructive. It may enhance some habitats. Third, Middle and Reedy lakes, once intermittently filled, have been transformed by the regular summer flow of fresh water. Trees around their edges have suffered, but irrigation has enhanced the wildlife of these lakes. The three lakes are breeding refuges. The permanent shallow water has provided a home for a prolific number of waterbirds. Together with nearby Lake Cullen and the Avoca Marshes, these lakes provide a major breeding ground for migratory waterbirds, which Australia has agreed to protect under international treaties.[48]

The Little Murray Weir is another example of conflicts between land and water quality and fortuitous environmental benefits from irrigation. The weir has raised the level of the Little Murray River during summer. This has helped raise watertables and increase soil salinity in the surrounding land. If the Little Murray Weir were removed, the salty local watertables would drain into the Little Murray River and into the Murray, raising salinity downstream. Removing the weir would destroy a valuable fish habitat. It seems that the unnatural state of the river is more desirable for Murray cod than the natural state.

Lake Charm, at one side of the main flow of Torrumbarry water, formerly only filled during high floods. It has no outlet; water flowed into the lake as floods rose and flowed out again as floods receded. The remaining water slowly evaporated and became saltier, until the next flood, when the lake would be flushed. When irrigation came, Lake Charm was cut off from the rest of the lake system by a regulator. The regulator allowed irrigation water into the lake to supply the irrigators around its edges, but prevented floods from flushing salty water from Lake Charm into the water supply system. The lake has become progressively saltier. The increasing salt levels threaten the amenity and recreation value of the lake as well as being of even greater concern to those irrigators who draw water from Lake Charm.[49]

Lake Tutchewop was once a flooding buffer and a high-value wetland.[50] The lake has been taken out of the lake system and is used as an evaporation basin. Cohuna was saved from salting early in the century by digging drains and deepening Barr Creek. Barr Creek now drains water from the salty watertable under Cohuna. It is the largest single point source of salt entering the Murray River. In the 1970s the Water Commission built a channel to carry salty water from Barr Creek to Lake Tutchewop. Some of the salty water that once flowed into the Murray River is now evaporated from Lake Tutchewop. The lake no longer fills in floods, so the flood peaks on surrounding land are higher. Local farmers are concerned the lake may leak salt onto their land.[51] Environmentalists see its destruction as vandalism.

Following the heated conflict over the development of further evaporation basins the Victorian Government changed its policy and asked a representative group of local community interests to help find an acceptable balance between the competing demands of water quality, soil salinity, flooding and habitat concerns.[52] Our discussion of the issues has only touched on the complexity of the tradeoffs that are involved.

The settlements of Nyah and Tresco, established 100 years ago with the best and the worst intentions, have both continued to face social problems and environmental problems, which are inextricably bound together. At Nyah and Tresco the long battle with salinity on the farms appears to be at a truce. Salinity has been controlled by installing underground drainage, with government assistance to remove the drainage effluent and improve water delivery. Unlike irrigation of the Tragowel Plains, drainage was possible because the topography allowed it and the crops grown were, at least sometimes, of higher value. The disposal of saline effluent has caused other land to be sacrificed. These settlements have been unable to isolate themselves from larger systems of influence. They exist in a milieu of low world prices, particularly for Nyah's produce, and the increasingly saline water supply to Tresco, which cuts a swathe through wetlands in the lower Loddon Valley. The unintended consequences of settlement that these communities shared have indeed been complex.

The long history of the problems faced by these two settlements suggests that they are not easy problems to solve. The physical, ecological and social relationships involved in such land-use problems are a complex and intricate web. As with other intractable examples of land degradation, economic, social and environmental problems aggravate each other. Today's problems often have complex antecedent causes. Political solutions to such problems, other than solutions involving very large government subsidies given by the courtesy of taxpayers, are very often difficult to achieve.

Conclusion

In Australia we regularly hear calls for more sustainable land use. These calls suggest that there should be a better way of doing things. The problem is to know what this 'new order' might be and what needs to be sustained. The first step is to identify what is unsustainable in the present order.

Agricultural industries face threats to their sustainable use of land from many directions. There are threats from depleted soil fertility or structure, unbalanced water cycles and rising watertables, uncontrolled diseases and pests, limited genetic diversity, problems of low income and social dysfunction. The lesson of history is that attempts to deal with one of these problems in isolation will often introduce or exacerbate other problems. There are forced trade-offs at every turn. Increasing soil fertility may be in conflict with the need to control soil acidity. Increasing fertility may increase the population of soil-borne pests and diseases such as potato wireworm, pasture cockchafers or brown rot in peaches. It may enhance eucalypt tree dieback and encourage chemical use. In conservation cropping there may be a trade-off between improving soil structure and increasing plant water use. It is also necessary to find a balance between chemical dependence and soil protection. In orchards, contour planting and contour ploughing to prevent erosion encourage fungal diseases fatal to the trees.

The need for sustainable farm incomes can be in conflict with other conservation objectives. There is continuing pressure on farmers to become more efficient. To protect their resource they may need to make investments that produce a monetary return only in the long term, and yet in the short term there is a need for sufficient income to finance these investments. The struggle to maintain financial viability has meant that since the 1920s farms have been growing larger. The number of farmers has been decreasing and this has been accompanied by the decline of smaller rural communities. With improvements in transport, smaller country towns have decreased as larger country towns have maintained or increased their size.

How an observer places priority on the differing and conflicting goals will be a reflection of the observer's values. The challenge posed by the debate over sustainable resource use is to find consensus on the relative importance of current income, future income from our productive resources, natural ecosystems, and the social and economic viability of local communities. Often we can only maintain one at the expense of another. In the passage of time we have considered in this book, dominant community values have been in constant evolution — from social survival to economic growth to environmental concern.

Changing values and perceptions of the land

The first Aboriginal inhabitants saw themselves as a part of the land. Their firestick farming created open forests and grasslands, increasing the marsupial stocking rate. Their culture and systems of land management developed over tens of thousands of years. A mere 200 years ago new settlers from Europe stepped ashore at Port Jackson. The new arrivals did not see themselves as part of the land. For most of the early European settlers in Australia, the garrisoned convicts and the later emigrants from the British Isles, Australia was harsh and inhospitable. The settlement of Australia was a solution to a growing prison population and the burgeoning and volatile ranks of the poor and unemployed. The British were motivated by a desire to pre-empt colonial rivals expanding in the southern hemisphere rather than pulled by any perceived attractiveness of the land. To the early British view, this peripheral colonial world and its inhabitants were made for exploitation. It was a world where constraints of British law and Christian morality, scarcely able to control the brutality of everyday life at home, were easily ignored.

Joseph Banks may have found Botany Bay to be a mine of scientific novelty but his advice to the British Government that a fledgling convict colony could support itself there proved to be ill informed. In the first years the settlers had to import virtually all their food. The early convicts were reluctant to become farmers and the free settlers were ignorant about appropriate farming practices in a land where the seasons were upside down and the rainfall was uncertain. The history of European agriculture in Australia was to be dominated by this uncertainty over rainfall.

The first major agricultural development was the anarchic pastoral expansion of the squatters. These opportunists spread across the countryside. Much of their early pastoralism was exploitative, with the understandable aim of making a fortune and returning to the home country. The impact was devastating. They dispossessed the native population. The grasslands were overgrazed and debilitated. Few squatters held tenure long enough to notice or record the changes on their runs. The obvious exploitation by the squatters was eventually ameliorated by more socially constrained self-interest as social structures developed. Successful squatters, with security of tenure, indulged their genteel aspirations by building permanent housing, even mansions. The unstated assumption of these self-appointed gentry was that their social position and the resources on which it depended were founded on a permanent system of farming. It was social, not ecological, pressures that stressed the squatters' farming system beyond the limits of its own resilience.

The gold rush brought a new influx of settlers, seeking wealth by panning and digging the yellow metal. As they exhausted the alluvial resource, the diggers turned to the land, believing ownership of the soil offered freedom from poverty and unemployment, and an independent stake in the country. The great Australian dream of making land more accessible to its citizens was born and the first seeds of future closer settlement were sown. Despite many ill-informed and sometimes well-intentioned attempts, Australia never conformed to the popular vision of small farms and pioneer families. Harsh lessons taught to hapless pioneers by an unkind climate and the rigours of competitive international trade did not produce a society of sturdy yeomen husbanding the land. The systems used for the maintenance or improvement of soil fertility in the United Kingdom had no application in Australia. The selectors had no choice but to participate in a new form of exploitation, transitory cropping. Community concern on the failure of selection focused on the inequitable social outcomes of this grand experiment, rather than the depletion of the soil's fertility.

Conscious exploitation of the land by the selectors was followed by a period of uninformed exploitation due to ignorance. Renewed closer settlement policies were initially driven by a vision of the land as a fragile, unexploited social resource, threatened by a much larger Asian population to the north. This view was later replaced by a need to both reward and pacify returning survivors of the European war. As Australia approached Federation in 1901 average wheat yields in the preceding decade had fallen to half the levels achieved in the 1860s, partly through exhaustion of the soil and partly through the expansion of wheat farming into marginal lands.[1] Extensive pastoralism had suffered from the destruction of fodder bushes. Land used for intensive grazing suffered from the depletion of phosphorus and nitrogen.[2] The innovative use of superphosphate, new varieties and dry fallow improved yields and offered initial security to the small wheat farmer. The security and durability of the system was illusory, culminating in the erosion decades of the 1930s and 1940s.

After the Second World War came a new period of agricultural consolidation and expansion. Wide community concern over soil erosion helped to initiate government policies to protect this threatened resource. Perhaps an equally strong motivation was the need to protect investment in irrigation headworks. There was a widespread and strong community belief that water conservation was in the best interests of the country, increasing exports and settling the undefended northern shores to forestall invasion. This belief justified a massive government investment in irrigation infrastructure, sometimes developed in the face of strong evidence of the risks of soil salinity and likelihood of poor returns. Somewhat serendipitously, a period of high wool prices encouraged the adoption of new legumes and rotational cropping practices, which did much to overcome the worst of the soil erosion and fertility mining of previous decades. The use of ley cropping returned wheat yields to their initial levels by 1950; by 1990 average yields were nearly double the initial yields at white settlement.

More recently, our communal perspective of the land has been courted by advocates with two opposing systems of cultural values: environmentalism and economic rationalism. Environmentalism and ecologicalism came to dominate the media debate in the late 1980s. The dramatic turnaround in community perceptions of the tree reflects the influence of mainstream environmentalism on community values. Less than forty years ago many people saw the tree as little more than a native weed. Today tree planting is advocated as being synonymous with sustainable agriculture. Likewise, cultural visions of Australia's red centre have changed. What forty years ago we saw as unproductive wilderness is now described as 'green desert'. The ecological movement has reshaped communal perceptions of our resources. At one extreme of the environmental movement there is acceptance of a definition of land degradation that denies legitimacy to agricultural production itself because it represents a simplification of the natural ecological systems. Through the mainstream of the movement this ideology underpins a strong preference for the protection of native remnant vegetation on private land and a preference for 'sustainable rural land use' rather than 'sustainable agriculture'.

Those who subscribe to the economic approach are concerned with sustaining the land as a productive resource to ensure that community well-being is maximised. The land is seen as a limited resource to be rationed between present and future users by the price mechanism. From this viewpoint there is validity in the idea of an optimal rate of land degradation determined by society's preference for current and future income. The economic approach has some difficulty accounting for other social preferences for which markets are not established, such as the value of financially unproductive native vegetation.

Many farmers hold strong views about the value of their land and the need to protect the resource. This tradition provided the basis for the 1989 alliance between the National Farmers Federation and the Australian Conservation Foundation. At the grassroots this alliance is under tension because of fundamental disagreements on the importance of remnant native vegetation and the question of regulatory control to achieve this end. In the early 1990s the practical meaning of an economically sustainable agriculture became starkly obvious to many farmers. Those farmers without a sustainable financial basis faced an uncertain future. Farmers' priorities will be focused on the problems that most immediately threaten financial viability. The sustaining of local communities can also be a pressing problem for rural communities, as we have seen in our exploration of irrigation districts in the Murray Valley. In these communities economic argument in favour of 'restructuring' and 'social adjustment' is generally viewed with disdain if not outright antipathy.

The two hundred year experiment

The history we have recounted can be interpreted as a 200-year experiment to develop a sustainable agriculture in this land. It is unlikely that sustainable systems were of paramount concern in farmers' short-term considerations over this period. However, over longer periods response and change occurred to maintain and develop the system of agriculture that now exists. This does not imply the present system is without problems. Throughout history the goal of sustainable land use has been a moving target. Each generation has faced its own set of challenges. Each generation has defined its own challenges according to existing cultural norms. The progression from ignorance about 'how to survive' to knowledge about how to farm an alien environment has taken some time. There have been incremental advances, midcourse corrections, and feedback about how we are faring in relation to the contemporary goals of society. Our current community stock of knowledge of how to farm is the result of a long process of evolution by trial and error and learning from mistakes. Innovation and development of new knowledge has been an evolutionary process.[3] The experiment will never be completed. Many of the current prescriptions for problems of Australia's land use may not stand the test of time and future new knowledge.

The fate of fallow in Australian wheat cropping is an excellent example of how we have learnt to farm. The practice of bare fallow, introduced when the nutrient status of cropping soils was as its lowest ebb, improved yields. It was thought to conserve soil moisture thus increasing yields. Later, it was discovered bare fallow increased yields by releasing nitrogen from soil organic matter. But bare fallow led to such a serious loss of soil and damage to soil structure that altogether new problems arose. For thirty years, from 1925, soil erosion posed a real threat to the stability of cropping and grazing land. Improved pasture species, changes in management practices and the control of rabbits overcame some of these problems. The continued practice of bare fallow has produced other problems of soil structure only more recently noticed. Excessive cultivation has reduced soil organic matter and created a soil structure less amenable to sustainable cropping. Again, new practices have been developed to overcome this problem. One of the solutions to the earlier problem of fallow and the release of soil nitrogen has now itself become a problem: soil acidity brought about by a build-up of soil nitrogen in a form that acidifies the soil. The lesson is that we are never likely to have a system of cropping that will endure forever.

The history of fallow is consistent with the observation that mistakes are implicit in trial and error learning. We have documented some of the agricultural outcomes that are today perceived as mistakes. In developing our agriculture we have modified the native flora and fauna. The rabbit and other feral pests and weeds were introduced with drastic consequences. At various times in the history of European settlement the critical destruction of deep rooted vegetation and forests, farming beyond adequate rainfall limits or overstocking of rangelands have damaged soils by various forms of erosion and, in some places, encouraged salinisation of soils. The desire to overcome drought and intensively settle a naturally brown land has left an intractable legacy. Since the 1950s the huge expansion of irrigation water applied to the Murray Valley Plains has accentuated long standing salinity problems. By the late 1970s the watertable was within 2 metres of the surface in an area of more than 300 000 hectares in the irrigation districts of north-western Victoria and adjacent New South Wales.

We have solved some problems and have adapted ourselves to live with others. Some problems, particularly those where human settlement is locked into a given form of land use, have proven more intractable. Some human settlements may eventually fail. Despite these mistakes, we should not condemn Australia's agricultural land use experiment as a failure. There have been major gains. Perhaps the most obvious is that seventeen million non-native people now live in the country. Nearly all of the flora and fauna of agricultural production, on which modern settlement and development of the country depended, and still depends, were introduced species: grain varieties, Spanish merino sheep, all the other domestic sheep, cattle and horses, pasture species, cane sugar, rice, potatoes, tobacco and vines. While some of the current research evidence, including some of our own, may engender pessimism about farmers' stewardship of the rural environment, we are optimistic rather than pessimistic about the longer term health of the rural environment. With the notable exceptions of the irrigated agriculture in the Murray–Darling Basin, Western Australia's dryland salinity and the, yet to be solved, problem of acid soils, the prognosis for rural land is better than for some of the earlier periods in Australian agriculture. Much of the current, conditioned reaction to problems of land degradation does not take adequate account of the sum of human experience in these matters. The store of science and technology has continued to grow since Europeans have come to Australia. There are many cases where land use that at one time was unsustainable has subsequently become more sustainable because of advances in science and technology. This has been one of the lessons of the preceding chapters.

Complexity and simple truths

In the debate over the seriousness and importance of environmental issues, there are strong temptations to promote stark assessments of the problems we face and advocate simple solutions, weaving a spell of misplaced confidence. The attractiveness of simple solutions to complex problems is an inevitable outcome of the natural human desire to eschew uncertainty. The trend is further encouraged by the nature of the electronic media. It has led to misguided estimates of the extent and cost of land degradation in Australia becoming oft-repeated and commonly accepted fallacies.[4] Consider one example. A report prepared by the Federal Standing Committee on Conservation estimated that 52 per cent of farmland was in need of treatment for land degradation.[5] The estimate was achieved by sampling areas of 2-kilometre radius across Australia. If any area within a sample was found to be degraded, the whole area was nominated as in need of treatment. This is an understandable simplification. This analysis has been transformed by the Commission for the Future into a public statement that 52 per cent of land is degraded.[6] From this statement it is but a small step to a popular vision of half of Australia's farm land being either erosion gullies or white salt pans. Such a depressing vision is misleading.

The tactics of exaggeration provide something of a conundrum. There are some obvious environmental problems; and there are elements of other possible problems. Community awareness may be a necessary precursor to action to solve the most pressing problems. Despite this, we are concerned that this style of debate polarises the community and eventually runs the risk of being discredited as the wild predictions are disproved by the inexorable march of time. In short, it does not contribute to sustained public opinion.

We have similar reservations over the use of simple prescriptions. Two deserve particular comment. We are uncomfortable with the generally held belief that there has been a dramatic deforestation of the land, that this is a prime cause of land degradation and replanting trees is the obvious solution to many of our problems.[7] The truth is far more complex. Australia was not covered with trees at white settlement and their subsequent removal has not been the root cause of today's degradation. The landscape of Australia is considerably different from the way it was at European settlement. Some places in Australia have a higher density of trees than at white settlement, for example managed forests and burgeoning woody species in the Australian rangelands. Some places in Australia do not have the tree loss since 1788 attributed by some advocates. Firestick farming by the Aborigines, and the vegetative explosion that followed its cessation, have not been fully acknowledged in some of the less sophisticated sections of the conservation movement. The poet Les Murray used some telling prose to describe the management system the Aborigines had maintained for untold centuries: 'the wilderness we now value and try to protect came with us, the invading Europeans. It came in our heads, and it gradually rose out of the ground to meet us.' Europeans created a literal wilderness with guns and smallpox and then, as a consequence of this displacement, an illusory wilderness of thickening forests.[8]

Perhaps our greatest concern is with a widespread belief that the most important task to achieve a more sustainable agriculture is the raising of community awareness and changing of farmers' attitudes to their land. Our retelling of agricultural history helps to provide an understanding of the barriers to the adoption of sustainable land use practices. A few practices have been widely adopted and have been of major importance in sustaining agricultural land use in Australia. The clearest feature of these technologies is that they offered realisable advantages to the landholder. The benefits of superphosphate drilled with wheat were clearly obvious and testable. Farmers saw the results in one year. Improved pasture offered the prospect of dramatically improved production, though because it entailed greater changes in farm management it took longer to be fully accepted. Ley rotations restored 'grain-sick' farm land and improved yields. The implications of these observations are simple yet profound. What is required are profitable and practical conservation farming techniques and management strategies. Where these are not available the best assistance is research directed at producing and promoting practical and profitable solutions, rather than a reliance on evangelical calls to better farming and changing attitudes. General community awareness is needed to maintain support for the funding of this work.

The dangers of simple prescriptions are that they will not encourage the sustained commitment of the social resources required to continue the unending search for sustainable rural land uses. Where resources are allocated, they may be allocated in the wrong direction, as may occur if we place too much emphasis on tree planting instead of establishing perennial pasture to stabilise rising watertables. Another risk of the simple prescription is that it can fuel sometimes inappropriate populist cries for 'works on the ground'. Many of the links between causes and effects of inappropriate land use in Australia are poorly understood. Theories and models are constantly modified by unexpected factors. In Australia the knowledge we have about such matters is relatively small; the resources devoted to developing new knowledge and better understanding of physically sustainable land use are also inordinately small. Often time is required to get to the truth of the matter. Where a problem is characterised by lack of knowledge and uncertainty as to the outcome of possible solutions, sometimes it is less costly to do nothing (until more is known) than to act with uncertainty.

The matters we have canvassed require honest and dispassionate thought. We all need to develop an appreciation of the points of view of other advocates and groups. Building disputes builds barriers to both resolution of existing problems and anticipation of new challenges. Building false dichotomies between the 'ignorant' or 'pillagers' and the 'far sighted' will achieve little in the long run. We should never have the arrogance to assume we know the answers. We have seen how other generations have believed they had the solution and been proved wrong. Why should we be any different? Our understanding today is just as likely to change as past understandings; and to wed ourselves strongly to today's perspective will be to render more difficult the acceptance of new understandings. Believing in our present answers blinds us to the need to adapt. The future is for the adaptable. Evolution is about the survival of the 'fitting', not the 'fittest', those who adapt to ever changing biological niches.[9]

In dealing with these complex environmental problems the scientific discipline will have much to offer. As in the past, many of today's problems will be solved if we maintain the social resolve to do so. If science is to be criticised in relation to matters of environmental management it is its lack of appreciation of time and cultural perspective in determining what is a *problem*. In this field, the scientific experts are, in their own way, amateurs. They bring essential skills and information. But their contributions are to a dialogue of exploration and consensus rather than to a rigid demonstration that conclusively proves something.[10] Attempting to understand most environmental problems requires knowledge to be integrated from more than one scientific discipline. The assessment of such information and testing it against changing value systems is a task for more than scientists. Environmental groups needs to appreciate the complexity of their task.

They need to adhere to scientific standards of argument while debating in a political forum. There is always a tension between political effectiveness and scholarly integrity. Excessive doomsday talk, based on emotionalism rather than the knowledge that is available, will result in the environmental emperors being seen to have no clothes.

We end with a plea for measured debate. Emotional overstatement should not be an accepted tool in environmental debate. Hindsight shows us agriculture has often exploited the land beyond its capacity. This hindsight can be used cheaply. Images of soil exploitation are regularly presented in the media; dramatic gullies, devastated salt patches or dust laden clouds assault us in two second 'grabs' — black and white arguments in full colour. But the story behind these images is complex. This complexity cannot be conveyed in two seconds. It cannot portray the dilemmas of farming. The quick story sacrifices context for impact and difficult dilemmas for simplified, emotional choices. Emotional commitment and fervour may be necessary to bring about change in the face of apathy, lack of concern or conflicting self-interest. Environmental religious fervour, passionately expressed, is no substitute for facts; it merely allows an eloquent exponent to use the facts better. We should not fall for the doomsday debating trick of reducing a continuum of possible policies to two extremes, and then showing the one extreme to be so unappealing that the alternative is seized on with grateful relief.

The Italian writer and chemist Primo Levi, a survivor of the Auschwitz labour camps who wrote dispassionately about human behaviour at its lowest ebb, exhorted suspicion of charismatic leaders and of those who seek to convince us with means other than reason.[11] His advice is pertinent to some of the more emotional assessments of impending environmental catastrophe. Since it is difficult to distinguish true prophets from false, it is well to regard all prophets with suspicion. Levi contended that, rather than embracing *revealed truths*, it is better to be content with more modest and less exciting truths. These one acquires painfully, little by little and without shortcuts, with study, discussion and reasoning. We hope to have contributed to this process. We need to develop the wisdom to ensure that Australia the brown land, or Australia the green land, is an enduring land.

Endnotes

Introduction

1   See Richard Eckersley, *Regreening Australia: The Environmental, Economic and Social Benefits of Reforestation*, Occasional Paper No. 3, CSIRO, Canberra, 1989; and K. Wells, N. Wood and P. Walker, 'Forest and woodland severely modified since settlement: an example of computer mapping of relevance to Greening Australia', *Trees and Victoria's Resources*, vol. 25, No. 1, 1983. Elsewhere we have discussed the early tree cover in Australia, see J. W. Cary and N. F. Barr, 'The semantics of forest cover: how green was Australia?' in G. Lawrence, F. Vanclay and B. Furze (eds), *Agriculture Environment and Society*, Macmillan, Melbourne, 1992.

2   Colin Chartres, 'Australia's land resources at risk' in A. Chisholm and R. Dumsday (eds), *Land Degradation: Problems and Policies*, Cambridge University Press, Melbourne, 1987, p. 8.

3   Bruce Davidson, *Land Degradation: Problems and Policies*, p. 360.

Chapter 1: The early grasslands

1   W. H. Hovell and Hamilton Hume, *Journey of Discovery to Port Phillip New South Wales by Messrs W. H. Hovell and Hamilton Hume: in 1824 and 1825*, A. Hill, Sydney, 1831, pp. 19, 28, 32, 34, 37, 49, 58, 62, 64, 67 & 76.

2   T. L. Mitchell, *Three Expeditions into the Interior of Eastern Australia*, vol. 2, Boone, London, 1839, Chps 7–13; and R. M. Moore, 'Southeast temperate woodlands and grasslands' in R. M. Moore (ed.) *Australian Grasslands*, ANU Press, Canberra, 1970, pp. 169–180.

3   Joseph Hawdon, *The Journal of a Journey from New South Wales to Adelaide, Performed in 1838*, Georgian House, Melbourne, 1952, pp. 19 & 21.

4   T. F. Bride, *Letters from the Victorian Pioneers*, Trustees of the Public Library, Victoria, 1898, pp. 254–9.

5   William Howitt, *Land, Labour and Gold or Two Years in Victoria with Visits to Sydney and Van Diemen's Land*, Longmans, London, 1855 (Letter XX).

6   See F. W. L. Leichardt, *Journal of an Overland Expedition*, Boone, London, 1847, p. 354; J. Oxley, *Journals of Two Expeditions into the Interior of New South Wales*, Murray, London, 1820, pp. 5 &174; J. L. Stokes, *Discoveries in Australia*, Boone, London, 1846, p. 400; also Henry Reynolds, *With the White People: The Crucial Role of the Aborigines in the Exploration and Development of Australia*, Penguin, Ringwood, 1990, pp. 8–9.

7   Edward Curr, *Recollections of Squatting in Victoria* (first published 1883), 2nd ed., Melbourne University Press, 1965, p. 171.

8   A. E. Sharp, *The Voyages of Abel Janzoon Tasman*, Clare, Oxford, 1968, pp. 44 & 111.

9   James Cook, *An Account of a Voyage Round the World with a Full Account of the Voyage of the Endeavour in the year MDCCLXX along the East Coast of Australia*, W. R. Smith and Paterson Pty Ltd, Brisbane, 1969, pp. 638–9.

10   Hovell and Hume, *Journey of Discovery to Port Phillip New South Wales*, pp. 17, 35, 42, 44, 48, 52, 54, 57, 63 & 64.

11   T. L. Mitchell, *Journal of an Expedition into the Interior of Tropical Australia*, Longman, Brown, Green & Longmans, London, 1848, pp. 412–13.

12   R. Jones, 'Geographical background to the arrival of man in Australia', *Archaeology and Physical Anthropology in Oceania*, vol. 3, 1968, pp. 186–215.

13   A team of scientists from the Australian National University Biogeography Department has carried out an extensive study of the sediments of Lake George, near Canberra, Lynch's Crater in north Queensland and Lashmar's Lagoon on Kangaroo Island. See G. Singh, A. P. Kershaw and Robin Clark, 'Quaternary vegetation and fire history in Australia' in A. M. Gill, R. H. Groves and I. R. Noble (eds), *Fire and the Australian Biota*, Australian Academy of Science, 1981.

14.   C. Anderson, 'Aborigines and conservationism: the Daintree-Bloomfield Road', *Australian Journal of Social Issues*, vol. 24, 1989, pp. 214–27. Anderson questions the popular thinking among conservationists that Aborigines lived in harmony with nature. He argues that this thinking is: 'muddled, based on misinformation, and racist. It is muddled because it confuses attributes [of the Aboriginal economy] with attitudes, and assumes that individual action stems only from the latter. The assumption is of a primordial human nature which we have regrettably moved away from through gaining the trappings of civilisation. Individual aboriginal people may or may not have been conservation minded. In their system it did not really matter. Our culture is perhaps the only one which needs to invent and articulate such a concept, because ours is the only one that has the capacity to destroy the environment on a scale that we now have . . . *The harmony with nature* notion is potentially racist in that it prepresents a remnant of the nineteenth century view that Australia's plants, animals and people represented a sort of museum. In this view Aborigines were seen as not only living in nature, but part of it . . . it assumes Aboriginal

behaviour is governed more by instinct than is the behaviour of others.'

15 P. J. Hughes and M. E. Sullivan, 'Aboriginal burning and late holocene geomorphic events in eastern New South Wales', *Search*, vol. 12, pp. 277–8.

16 D. R. Horton, 'Extinction of the Australian megafauna', *Australian Institute of Aboriginal Studies Newsletter*, No. 9, 1986, pp. 72–5; and D. Merrilees, 'Man the destroyer: late quaternary changes in Australian marsupial fauna', *Journal of the Royal Society of Western Australia*, vol. 51, 1968, pp. 1–24; T. Flannery, 'Who killed Kirlipili?', *Australian Natural History*, vol. 23, 1989, pp. 235–41.

17 Anderson, *Australian Journal of Social Issues*, vol. 24, p. 220.

18 *Journey of Discovery to Port Phillip New South Wales by Messrs W. H. Hovell and Hamilton Hume: in 1824 and 1825*, pp. 20, 45 & 61.

19 B. R. Davidson, *European Farming in Australia: An Economic History of Australian Farming*, Elsevier, Amsterdam, 1981, p. 77. Davidson drew the quotation from unpublished transcripts of evidence of State Agriculture and Trade, 1823, in the Mitchell Library.

20 Curr, *Recollections of Squatting in Victoria*, p. 86; Bride, *Letters from the Victorian Pioneers*, pp. 167, 182 & 204; W. Brodribb, *Recollections of an Australian Squatter*, John Ferguson, Sydney, 1978 (first published 1883), pp. 38 & 40; and the transcription of Batman's Journal in Bonwick's *Discovery and Settlement of Port Phillip*, Melbourne, 1856, pp. 180–3, cited in J. M. Powell, *The Public Lands of Australia Felix*, Oxford, Melbourne, 1970, p. 19.

21 The *Argus*, Melbourne, 31 January, 1885.

22 Curr, *Recollections of Squatting in Victoria*, p. 88.

23 Ibid. pp. 27 & 160.

24 J. C. Hamilton, *Pioneering Days in Western Victoria*, Macmillan, Melbourne, 1923, p. 35.

25 J. A. Palmer, *William Moodie: A Pioneer of Western Victoria*, Hedges and Bell, Maryborough, 1973, p. 44.

26 Howitt, *Land, Labour and Gold* (Letter IX).

27 Curr, *Recollections of Squatting in Victoria*, p. 86; Hamilton, *Pioneering Days in Western Victoria*, p. 35; Bride, *Letters from Victorian Pioneers*, p. 34; Alfred Joyce, *A Homestead History*, Oxford, Melbourne, 3rd ed., 1969, p. 65; and P. M. Cunningham, *Two Years in New South Wales*, Colburn, London, 1827 (reprinted by the South Australian Libraries Board, 1966) pp. 212–13.

28 B. Muir, *Pastures and Fodder Crops for the Ballarat District*, Department of Agriculture Victoria, Melbourne, 1985, p. 28.

29 A. E. V. Richardson, 'Development of grasslands', *Journal of Agriculture Victoria*, vol. 22, 1924, p. 196.

30 S. H. Roberts, *The Squatting Age in Australia: 1835–1847*, Melbourne University Press, Carlton, 1964, p. 187.

31 Bride, *Letters From Victorian Pioneers*, p. 25.

32 *The Colonist*, vol. 3, No. 21, April 20, 1837, p. 127.

33 Cunningham, *Two Years in New South Wales*, pp. 211–12.

34 Curr, *Recollections of Squatting in Victoria*, p. 87.
35 Bride, *Letters from the Victorian Pioneers*, p. 34.
36 The utilisation of rainfall water can be expressed as an equation of *water balance*: Rainfall = drainage + evaporation + water used by plants + drainage to the watertable + changes in soil moisture.
37 Salting is more complex where there are clay layers in the soil. Clay slows the rate of infiltration of water to the soil. Clay is normally deepest in gullies and flat areas at the bottom of slopes. Both soil and clay are thinnest at the tops of hills where the rock is often very fractured. It is here, from 'areas of preferable recharge', that water is most likely to get into the watertable. Once in the sub-soil, the water will infiltrate downhill constrained between the solid bedrock below and any constraining upper clay layer in the soil of the lower slopes and flats. The water is stored under pressure in the sub-soil. It will gradually seep out, causing saline flows in the local stream and salting of paddocks. This type of salting most commonly occurs at the break of slope at the bottom of hills on the inland side of the Great Dividing Range. Infiltration rings dramatically demonstrate preferential areas.
38 The worst excesses described by Robertson were most likely concentrated around the numerous local creeks that are tributaries of the Wando River where rates of stocking were excessively high. The land in the area settled by Robertson is extremely prone to mass movement, and hence gully erosion and slippage when the soil, with underlying deep clay layers that overlie impermeable bands of ironstone, is saturated. Over grazing reduces vegetation and causes the soils to be saturated over a longer period each year.
39 Bride, *Letters from Victorian Pioneers*, pp. 34–5. This salinity, possibly salt scalding associated with soil erosion, was on very heavily stocked land near creeks. The back country was more lightly grazed because of the absence of surface water and was probably much less changed.
40 Joyce, *A Homestead History*, pp. 137–9.
41 Ibid., p. 143.
42 Hamilton, *Pioneering Days in Western Victoria*, p. 36.
43 Palmer, *William Moodie: A Pioneer of Western Victoria*, pp. 62–3.
44 This theme is explored in greater detail in Chapter 4.
45 Hamilton, *Pioneering Days in Western Victoria*, p. 38.
46 Howitt, *Land, Labour and Gold*, p. 32.
47 Ibid. Letter XLII.
48 Margaret Kiddle, *Men of Yesterday: a Social History of the Western District of Victoria 1834–1890*, Melbourne University Press, Carlton, 1961, pp. 285 & 291.
49 C. Ramsay, *With the Pioneers*, Latrobe Group of the National Trust of Australia, 2nd edn, Launceston, 1980, pp. 29–30.
50 C. French, 'St John's wort', *Journal of Agriculture Victoria*, vol. 3, 1905, pp. 624–5.
51 Ibid., p. 624.
52 Ibid., p. 630.

53 'Eradication of St John's wort by means of introduced insects', *Journal of Agriculture Victoria*, vol. 19, 1921, pp. 187–8.

54 W. T. Parsons, 'St John's wort: history, distribution, control in Victoria', *Journal of Agriculture Victoria*, vol. 55, 1957, pp. 784–8.

55 Joyce, *A Homestead History*, p. 54.

56 In fact, the rabbits made their home in the stone fences of the western plains, precipitating the destruction of these fences in the name of rabbit control.

57 Kiddle, *Men of Yesterday: a Social History of the Western District of Victoria*, pp. 320–2.

58 Davidson, *European Farming in Australia*, p. 177.

59 Joseph Jenkins, *Dairy of a Welsh Swagman* (ed. W. Evans), Macmillan, Melbourne, 1975, p. 129.

60 *Report of the Committee Appointed to Investigate Erosion in Victoria*, Government Printer, 1938, p. 2.

61 For example, *Journal of Agriculture Victoria*, vol. 41, 1943, pp. 6, 63 & 232.

Chapter 2: Recreating grasslands

1 W. H. Hovell and Hamilton Hume, *Journey of Discovery to Port Phillip New South Wales by Messrs W. H. Hovell and Hamilton Hume: in 1824 and 1825*, A. Hill, Sydney, 1831, p. 46.

2 T. F. Bride, *Letters from Victorian Pioneers*, Trustees of the Public Library, Victoria, 1898, p. 169.

3 *Land of the Lyre Bird*, Shire of Korumburra, 1966 p. 315; Margaret Kiddle, *Men of Yesterday: A Social History of Western Victoria 1834–1890*, Melbourne University Press, Carlton, 1961, p. 291.

4 Eric Rolls, *A Million Wild Acres: 200 Years of Man and an Australian Forest*, Thomas Nelson, Sydney, 1981, p. 114.

5 H. A. Mullett, 'Agriculture — past, present and future', *Journal of Agriculture Victoria*, vol. 53, 1955, p. 291.

6 Mullet, 'Pasture management', *Journal of Agriculture Victoria*, vol. 29, 1931, p. 162.

7 Mullet, 'Pasture top-dressing', *Journal of Agriculture Victoria*, vol. 24, 1926, pp. 385–90; and 'What does a ton of cheese cost the soil?', *Agricultural Gazette of New South Wales*, vol. 2, 1891, p. 152.

8 J. A. Palmer, *William Moodie: A Pioneer of Western Victoria*, Hedges and Bell, Maryborough, 1973, p. 44. Dry is a commonly used term meaning 'not lactating'.

9 Mullett, 'Scientific agriculture in Victoria', *Journal of Agriculture Victoria*, vol. 53, 1955, p. 341; A. E. V. Richardson, 'Science and production', *Journal of Agriculture Victoria*, vol. 15, 1917, p. 102.

10 'Pasture top dressing', *Journal of Agriculture Victoria*, vol. 19, 1921, p. 239.

11 Richardson, 'Development of grasslands', *Journal of Agriculture Victoria*, vol. 22, 1924, p. 197. These estimates should be regarded with some caution.

12 Mullett, *Journal of Agriculture Victoria*, vol. 53, 1955, p. 341.

13 Richardson, *Journal of Agriculture Victoria*, vol. 22, 1924, p. 193. See also the many articles on native grasses in Volumes 1 and 2 of the *Agricultural Gazette of New South Wales*, 1890–91.

14 F. Turner, 'Grasses of New South Wales', *Agricultural Gazette of New South Wales*, vol. 1, 1890, pp. 309–13.

15 'Answers to Correspondents', *Journal of Agriculture Victoria*, vol. 8, 1910, p. 125.

16 Richardson, 'Top dressing of native pastures', *Journal of Agriculture Victoria*, vol. 12, 1914, p. 232.

17 Richardson, 'Rutherglen State Farm', *Journal of Agriculture Victoria*, vol. 17, 1919, p. 376.

18 Richardson, *Journal of Agriculture Victoria*, vol. 22, 1924, pp. 198–9.

19 Richardson, 'Some results in top dressing pastures', *Journal of Agriculture Victoria*, vol. 19, 1921, p. 349.

20 Mullet, *Journal of Agriculture Victoria*, vol. 29, 1931, p. 162.

21 Mullet, 'Subterranean clover', *Journal of Agriculture Victoria*, vol. 23, 1925, p. 704.

22 'Rutherglen experimental farm', *Journal of Agriculture Victoria*, vol. 28, 1930, p. 637.

23 G. D. Duncan, 'Save more fodder', *Journal of Agriculture Victoria*, vol. 53, 1955, pp. 481–2.

24 H. J. Sims, 'The Mallee Championship', *Journal of Agriculture Victoria*, vol. 52, 1950, p. 154; J. M. McCann, 'The Mallee Championship', *Journal of Agriculture Victoria*, vol. 53, 1951, p. 170.

25 B. Elliot, 'The soil fertility story in the Wimmera and Mallee', *Journal of Agriculture Victoria*, vol. 76, 1978, pp. 10–11.

26 E. J. Pemberton, 'Controlling the rabbit menace', *Journal of Agriculture Victoria*, vol. 49, 1951, pp. 127–35.

27 C. W. Douglas, 'Myxomatosis in Victoria', *Journal of Agriculture Victoria*, vol. 56, 1958, pp. 779–90.

28 G. D. Kohn, 'Compound 1080: its use as a rabbit poison in Victoria', *Journal of Agriculture Victoria*, vol. 54, 1956, p. 279.

29 The bounty was $6 per ton 'to stimulate increased use of superphosphate as a means of improving still further the productivity of farm lands and pastures'. Rt Hon. H. Holt, Budget Speech, 1963–64.

30 D. B. Williams, *Agricultural Extension: Farm Extension Services in Australia, Britain and the United States*, Melbourne University Press, Carlton, 1968, p. 140.

31 J. Sharples and N. Milham, 'Longrun competitiveness of Australian agriculture', USDA Foreign Agricultural Economic Report, No. 243, 1990, Table 1.

32 P. A. Yeomans, *Water for Every Farm Using The Keyline Plan*, Second Back Row Press, Katoomba, 1978, p. 25.

33 F. H. Gruen, 'Economic development and agriculture since 1945' in D. B. Williams (ed.), *Agriculture in the Australian Economy*, 3rd edn, Sydney University Press, Sydney, 1990, p. 23.

34 Ibid., p. 25; and *Quarterly Review of Rural Economy*, vol. 10, 1988, p. 428.

35 Gruen, 'Economic development and agriculture since 1945', p. 23; and W. P. Males, H. Davidson, P. Knopke, T. Loncar and M. J. Roarty, *Productivity Growth and Developments in Australia's Primary Industries*, Australian Bureau of Agricultural and Resource Economics, Discussion Paper 90.8, AGPS, Canberra, 1990.

36 Sloane, Cook and King Pty Ltd, *The Economic Impact of Pasture Weeds, Pests and Diseases on the Australian Wool Industry*, Australian Wool Corporation, Melbourne, 1988.

37 R. T. M. Pescott, 'Insect pests of subterranean clover', *Journal of Agriculture Victoria*, vol. 35, 1937, pp. 374–5.

38 R. L. Twentyman, 'Lime for pastures in the Gippsland region', *Journal of Agriculture Victoria*, vol. 53, 1955, p. 489.

39 The possibility of acidification was first reported by C. H. Williams and C. M. Donald, 'Changes in organic matter and pH in a podzolic soil as influenced by subterranean clover and superphosphate', *Australian Journal of Agricultural Research*, vol. 8, 1957, pp. 179–89.

40 K. R. Helyar, 'Nitrogen cycling and soil acidification', *Journal of the Australian Institute of Agricultural Science*, vol. 42, 1976, pp. 217–21.

41 G. S. P. Ritchie, 'The chemical behaviour of aluminium, hydrogen and manganese in acid soils' in A. D. Robson (ed.), *Soil Acidity and Plant Growth*, Academic Press, Sydney, 1989, p. 2.

42 B. Quin, 'Mineral phosphate for acid soils: will it be allowed to replace acidulated fertilisers in agriculture?', *Acres Australia*, No. 2, 1989, pp. 30–31.

43 J. G. Bath and I. H. Cameron, 'Rock phosphate for topdressing pasture: comparisons with superphosphate', *Journal of Agriculture Victoria*, vol. 58, 1960, pp. 459–67. The organic agriculture movement has promoted a return to rock phosphates. The opponents of superphosphate claim the early trials were made using relatively insoluble forms of rock phosphate from Christmas and Nauru Islands, arguing more soluble rock phosphates give better results. See Quin, *Acres Australia*, No. 2, 1989, pp. 30–31.

44 M. Boland and B. Gilkes, 'The poor performance of rock phosphate fertilisers in Western Australia', *Agricultural Science*, vol. 3, No. 1, 1990, pp. 43–8; and No. 2, pp. 44–7.

45 Long-term research plots at Rutherglen Research Station show a small increase in acidity on plots that have been continuously planted to wheat and fertilised with superphosphate for many years. Other plots that have been under sub clover pasture show much greater increases in acidity. Personal communication, T. Ellington, Rutherglen Research Institute.

46 K. R. Helyar, 'Soil acidity: the long term implications', Paper pre-

sented to the *Sustainable Agriculture: Farming For the Future* Conference, Benalla, 1988. (Proceedings available from Benalla Forum, Benalla, Vic.)

47   A. Ridley, 'Acid soils/perennial grass research program in the North East Region' in *Proceedings of Pasture Specialists Conference*, 16–18 August, 1989, Department of Agriculture and Rural Affairs, Melbourne, 1989, pp. 6–9.

Chapter 3: Symbolic trees and salinity

1   It is a matter of semantics as to whether the Grampians further to the west are a re-emergence of the Great Dividing Range or a separate range.

2   T. L. Mitchell, *Three Expeditions to the Interior of Eastern Australia* (vol. 2), T. & W. Boone, 1839 (facsimile edn, 1965), p. 168. (Diary entry near Bridgewater 9 July, 1836.)

3   Ibid., p. 171 (Diary entry 13 July, 1836.)

4   Ibid., p. 204.

5   Ibid., p. 211. Returning from Mount Gambier to the Glenelg at Casterton in 1857, James Bonwick 'exchanged sterility for fertility in a ramble through Wando Vale . . . The rich black soil was covered with luxuriant grass and a few trees'. James Bonwick, *Western Victoria: Its Geography, Geology, and Social Condition* (the narrative of an educational tour in 1857), Heinemann, Melbourne, 1969, p. 140.

6   Mitchell, *Three Expeditions to the Interior of Eastern Australia*, p. 207.

7   *The Journal of Granville William Chetwynd Stapylton*, 1836, Mitchell Library, Sydney. Entries for Western Victoria published in M. H. Douglas and L. O'Brien (eds), *The Natural History of Western Victoria*, Australian Institute of Agricultural Science, Horsham, 1971, pp. 85–115.

8   Mitchell, *Three Expeditions to the Interior of Eastern Australia*, pp. 215 & 236.

9   Glendenning run on the upper Glenelg River; T. F. Bride, *Letters from Victorian Pioneers*, pp. 290–91.

10 N. F. Learmonth, *The Portland Bay Settlement*, Historical Committee of Portland, McCarron Bird, Melbourne, 1934, p. 67.

11 Ibid., pp. 85–7.

12 Ibid., pp. 69–70; also 'Vegetation of western Victoria — a reconstruction' in Douglas and O'Brien, *The Natural History of Western Victoria*, p. 31.

13 The extent of Henty's downs country can be determined from maps in F. R. Gibbons and R. G. Downes, *A Study of the Land in South-Western Victoria*, Soil Conservation Authority, Victoria, 1964.

14 Mitchell, *Three Expeditions to the Interior of Eastern Australia*, p. 271. (Diary entry at Tatyoon, 21 September, 1836.)

15 Philip Brown, *The Challicum Sketch Book: 1842–53*, National Library

of Australia, Canberra, 1987. Challicum is west of Beaufort in the Western District of Victoria.

16 Mitchell, *Three Expeditions to the Interior of Eastern Australia*, p. 272. (Diary entry near Challicum, 22 September, 1836.)

17 Ibid., p. 274. (Diary entry two and a half miles south of Mount Cole, 23 September, 1836.)

18 M. Tipping, *An Artist on the Goldfields: The Diary of Eugene von Guérard*, Currey O'Neil, South Yarra, 1982.

19 Mitchell, *Three Expeditions to the Interior of Eastern Australia*, p. 274.

20 L. J. Peel, 'The first hundred years of agricultural development in western Victoria', in M. H. Douglas and L. O'Brien (eds), *The Natural History of Western Victoria*, p. 77. Between 1835 and 1851 pastoral licences were issued for all of the then Portland District to the exclusion only of the hinterland of the town of Portland and the reserves for the Aboriginal protectorate.

21 Tony Dingle, *The Victorians: Settling*, Fairfax, Syme & Weldon, McMahons Point, 1984, p. 41.

22 William Westgarth, *Victoria: Late Australia Felix, or Port Phillip District of New South Wales*, Oliver & Boyd, Edinburgh, 1853, p. 18 (appendix).

23 Ibid., p. 240.

24 Dingle, *The Victorians*, p. 44.

25 W. Howitt, *Land, Labour and Gold, or Two Years in Victoria*, Lowden Publishing, Kilmore, 1972 (Letter XI), p. 98. (First published Longmans, London, 1855.)

26 Howitt, *Land, Labour and Gold* (Letter XXV), p. 254.

27 See W. G. Tuddenham, *Landsat Imagery* in J. S. Duncan (ed.), *Atlas of Victoria*, Victorian Government Printing Office, Melbourne, 1982, p. 9.

28 Personal communication, Stephen Legg, Monash University Department of Geography in discussion of his forthcoming doctoral dissertation.

29 Joseph Jenkins, *Diary of a Welsh Swagman: 1869–1894*, ed. William Evans, Macmillan, Melbourne, 1975, p. 7.

30 N. Taylor, *Geological Survey of Learmonth*, Report of Progress, Geological Survey of Victoria, No. 4, 1876.

31 Jenkins, *Dairy of a Welsh Swagman*, pp. 128 & 166.

32 From the 1905 Report of the Sludge Abatement Board, cited in I. Forbes and L. R. East, *The Catchment of Cairn Curran Reservoir*, State Rivers and Water Supply Commission, Melbourne, 1950, p. 63.

33 Forbes and East, *The Catchment of Cairn Curran Reservoir*, p. 66.

34 C. Webb and J. Quinlan, *Greater Than Gold: A History of Agriculture in the Bendigo District 1935 to 1985*, Cambridge Press, Bendigo, 1985, pp. 104–5.

35 *The Eppalock Catchment Project*, Soil Conservation Authority, Melbourne, 1971; and G. T. Thompson, *A Brief History of Soil Conservation in Victoria: 1834–1961*, Soil Conservation Authority, Melbourne, 1971, pp. 54–5.

36 The improved pastures had a dramatic effect on the production of farms in the district. The wool yield for participating farmers increased, on average, from 10 to 18 kilograms per hectare. The Authority, perhaps optimistically, valued the *extra* income from this increase in wool production in 1969 at $45 600. See *The Eppalock Catchment Project*, p. 36.

37 *The Eppalock Catchment Project*, p. 32.

38 D. Brewin, An investigation of soil conservation programs and practices, MAgrSc Thesis, School of Agriculture and Forestry, University of Melbourne, 1978, p. 105.

39 F. Cope, *Catchment Salting in Victoria*, Soil Conservation Authority of Victoria, Government Printer, Melbourne, 1958, pp. 56-7.

40 Learmonth, one of the first selectors in the district, found Lake Burrumbeet to be intensely salty. See Bride, *Letters from Victorian Pioneers*, pp. 95 & 98. Mitchell observed salty lakes on the southwestern Victorian plains; see Mitchell, *Three Expeditions to the Interior of Eastern Australia*, p. 266. See also Charles Sturt, *Two Expeditions into the Interior of Southern Australia*, Boone, London, 1833.

41 Early miners knew the Loddon Valley deep lead was a prolific aquifer in the highlands. From upstream of Bridgewater the Loddon Valley deep lead pursues a course parallel to, and to the east of, the existing Loddon River. Early alluvial shaft mining was undertaken in the many upper tributaries of the Loddon Valley deep lead. Clunes, Creswick, Daylesford, Ballarat, Castlemaine, Maldon, Avoca, and Maryborough goldfields were all linked in this system. There was no mining in the lower main lead because of the presence of excessive water, which meant that it was all pumping and no mining. Some alluvial mines in the central Loddon Valley near Maryborough yielded up to 40 million litres of water per day.

42 Murray-Darling Basin Ministerial Council, Murray–Darling Basin Environmental Resources Study, State Pollution Control Commission, Sydney, 1987, p. 54.

43 J. Field, 'Research on trees as an agent for soil and land conservation', *Soil and Water Conservation Association of Australia Newsletter*, vol. 10, pp. 2–3.

44 *Draft Salinity Control Strategy: Ballarat Region*, Department of Conversation, Forests and Lands, Ballarat, 1988.

45 Phalaris is recommended for watertable control because of its high water-use capacity. Plant water use is only part of the water-use equation. Runoff or drainage is another important component. In recent measurements of water movement in phalaris pastures, water runoff was shown to be eliminated where phalaris was present. The decrease in runoff was greater than the increased water-use capacity of the plant. Despite increased plant water use there may be no net gain in controlling the watertable. (Research work by Jost Van Brower, Department of Conservation and Environment, Bendigo, Victoria.)

46 *Draft Salinity Control Strategy: Ballarat Region*.

47 Ibid., p. 18.

48 *Draft Regional Salinity Report: Loddon–Avoca Region*, Victorian Government Salinity Program, 1988.
49 *Draft Salinity Control Strategy: Ballarat Region*, p. 45.
50 Bride, *Letters from Victorian Pioneers*, p. 79.
51 L. Whitely, Community perceptions of dryland salinity in the Haddon area, unpublished report, Ballarat College of Advanced Education, 1988, p. 53.
52 Small areas of catchment salting have been present in adjacent areas of the Avoca catchment and in the Lexton area since before the 1950s. See F. Cope, *Catchment Salting in Victoria*, p. 11.
53 *Draft Salinity Control Strategy: Ballarat Region*, p. 20.
54 Ibid., p. 23. Many landholders are reluctant to undertake recharge works because they are not convinced such works will control salting.
55 Ibid., p. 24.
56 *Draft Regional Salinity Report: Loddon–Avoca Region*, p. 3.
57 Now part of the Department of Conservation and Environment.
58 *Soil and Water Notes: Burkes Flat Salinity Control Project*, Department of Conservation Forests and Lands, 1989.
59 *Soil and Water Notes: Burkes Flat Salinity Control Project*. The lack of substantive information is evident: 'It is hoped that as recharge is further reduced by tree and pasture growth on the mid and upper slopes, groundwater levels in lower areas and discharge sites will gradually be lowered and salinity brought under control. This is expected to occur over a number of years' (p.4). The idea of differential recharge via the upper rocky ridges has inherent logic, but as yet there is little empirical research evidence. See P. R. Dyson, 'Dryland salting and groundwater discharge in the Victorian uplands', *Proceedings of the Royal Society of Victoria*, vol. 95, No. 3, 1983, p. 115.
60 In a survey of 130 landholders in the upper Loddon–Avoca catchments nearly 90 per cent intended to plant trees in the future. In the previous five years, while 70 per cent of landholders had planted some trees, only ten farmers had planted over 1000 trees; the median planting was 100 trees. This is such a small rate of tree planting it would be unlikely to have a significant effect on salinity control. See F. M. Vanclay and J. W. Cary, *Farmers' Perceptions of Dryland Salinity*, School of Agriculture and Forestry, University of Melbourne, Parkville, 1989, p. 21.
61 Vanclay and Cary, *Farmers' Perceptions of Dryland Salinity*, pp. 13–14; and *Draft Salinity Control Strategy: Ballarat Region*, p. 18.
62 Whitely, Community perceptions of dryland salinity in the Haddon area.
63 The first signs of soil salting are reduced yields in crops, or reduced pasture production. The composition of pasture species changes, with the less salt-tolerant species such as subterranean clover, giving way to the more salt-tolerant species, such as strawberry clover. As the concentration of salt increases, other less palatable salt indicator species such as barley grass (*Hordeum leporinum*), sea barley grass(*Hordeum maritium*) take over. With higher concentrations of salt only more

salt-tolerant species, such as saltbush, grow. See Vanclay and Cary, *Farmers' Perceptions of Dryland Salinity*, p. 7.

64 Ibid., p. 15.
65 *Salinity Control in Northern Victoria*, Dwyer Leslie Pty Ltd, 1984, p. 41.
66 *Goulburn Dryland Draft Salinity Management Plan*, Goulburn Broken Region Salinity Pilot Program Advisory Council, August 1989, p. 95.
67 J. W. Cary, Unpublished research.
68 In unpublished University of Melbourne data, only 15 per cent of upper Loddon and Avoca catchment farmers planted trees on rocky hilltops, and only a third of these did so because of salinity. Only 13 per cent of farmers with preferential recharge sites on their property planted trees on hilltops.
69 Unpublished research, School of Agriculture and Forestry, University of Melbourne.
70 Vanclay and Cary, *Farmers' Perceptions of Dryland Salinity*, p. 22.
71 J. W. Cary, A. J. Beal and H. S. Hawkins, *Farmers' Attitudes to Land Management for Conservation*, School of Agriculture and Forestry, University of Melbourne, 1986, p. 18; and J. W. Cary, *Predicting the Effect of Communication Strategies on Attitudes and Beliefs Using Galileo Analysis*, School of Agriculture and Forestry, University of Melbourne, 1988, p. 41.
72 This is the map of the averaged beliefs of a random sample of 130 Loddon–Avoca landholders in late 1988; see Vanclay and Cary, *Farmers' Perceptions of Dryland Salinity*, pp. 16–19. Rather than draw the location of every person on the map, the locations of the average person and his beliefs are the centres of the circles. The size of the circle indicates the degree of assuredness of the correct location of the belief object. From a statistical viewpoint, for the sample we have used as representing the total population of Loddon–Avoca landholders, the circle depicts the area in which there is a 68 per cent likelihood that the belief object is actually located in that area for the total population.
73 'The economic value of plantations', *The Australasian*, 28 May, 1921.

Chapter 4: Visions of a timbered farmscape

1 W. A. W. de Beuzeville, *Australian Trees for Australian Planting*, Forestry Commission of New South Wales, 1947. This idea has been revived recently. The tree-planting organisation, Greening Australia, has proposed a 'ribbon of green' along the Great Eastern Highway from Perth to Kalgoorlie in Western Australia.
2 William Westgarth, *Victoria: Late Australia Felix, or Port Phillip District of New South Wales*, Oliver & Boyd, Edinburgh, 1853, p. 32.
3 J. M. Powell, 'Victoria's woodland cover in 1869: a bureaucratic venture in cartography', *New Zealand Geographer*, vol. 23, 1967, p. 113; Margaret Kiddle, *Men of Yesterday: A Social History of the Western*

District of Victoria 1834–1890. Melbourne University Press, Carlton, 1961, p. 317.

4   James Bonwick, *Western Victoria: It's Geography, Geology, and Social Condition* (the narrative of an educational tour in 1857), Heinemann, Melbourne, 1970, p. 160. One hundred years later the Victorian Department of Agriculture, noting that the clearing of trees 'is a natural accompaniment of land settlement and a necessary feature of rural development', provided advice on ringbarking and poisoning of trees. See K. V. M. Ferguson, 'Clearing of timber from farm lands', *Journal of Agriculture Victoria*, vol. 55, 1957, p. 137.

5   Margaret Kiddle, *Men of Yesterday*, p. 317.

6   A. W. Howitt, 'The Eucalypts of Gippsland', *Transactions of Royal Society of Victoria*, vol. 2, Part 1, 1890, pp. 116–20.

7   Eric Rolls, *A Million Wild Acres; 200 Years of Man and an Australian Forest*, Thomas Nelson, Sydney, 1981, Ch. 6.

8   The beginnings of the selection movement are discussed in Chapter 10.

9   *Hamilton Spectator*, 1 January 1870; cited in J. M. Powell, *The Public Lands of Australia Felix*, Oxford University Press, Melbourne, pp. 113 & 237; Powell, *New Zealand Geographer*, vol. 23, 1967, p. 116.

10 Joseph Jenkins, *Dairy of a Welsh Swagman*, ed. W. Evans, Macmillan, Melbourne, 1975, pp. 128 & 166.

11 W. A. Brodribb, *Recollections of an Australian Squatter*, John Ferguson, Sydney, 1978, p. 42 (first published 1883).

12 Ibid., p. 43.

13 W. H. Williams in *Land of the Lyre Bird*, Shire of Korumburra, 1966, p. 290 (first published 1920 by South Gippsland Pioneers' Association).

14 G. Matheson in *Land of the Lyre Bird*, pp. 280–81.

15 T. J. Cloverdale in *Land of the Lyre Bird*, p. 110.

16 G. Bolton, *Spoils and the Spoilers*, Allen and Unwin, Sydney, 1981, p. 43; I. Reeve, *A Squandered Land*, University of New England, Armidale, 1989.

17 Ferguson, *Journal of Agriculture Victoria*, vol. 55, 1957, pp. 137–40.

18 *Victorian Yearbook*, vol. 92, 1978, p. 360. Much of this area had not been successfully settled in the past because of soil trace element deficiencies.

19 Bob Twigg, 'What I've done to the land and what it has done to me' in Jenny Arnold (ed.), *The Bush Comes to the City: Fitzgerald Biosphere Project*, Fitzgerald Biosphere Project, 1989, pp. 16–17.

20 'Our Country Our Future', Statement on the Environment by the Prime Minister, AGPS, Canberra, July 1989, p. 48; and Section 75D, *Income Tax Assessment Act*.

21 Section 75(1) of the *Income Tax and Social Services Contribution Assessment Act* in 1947 provided for allowances in full, in the year of expenditure, of certain capital expenditure, such as the destruction and removal of timber, scrub or undergrowth indigenous to the land. See A. W. Hooke, 'Farm investment' in D. B. Williams (ed.), *Agricul-*

ture in the Australian Economy, 1st edn, Sydney University Press, Sydney, 1967, p. 205; and D. B. Williams, *Economic and Technical Problems of Australia's Rural Industries*, Melbourne University Press, Carlton, 1957, p. 67.

22   Rt Hon. J. McEwen, Address to Australian Agricultural Council, 1952, reprinted in *Report of the Committee of Economic Enquiry*, vol. 1, Commonwealth of Australia, May 1965, Ch. 8, para. 21.

23   For example, G. Bolton in *Spoils and Spoilers* entitles the chapter on agricultural development 'They hated trees'.

24   *Land of the Lyre Bird*, pp. 144, 213, 218, 262.

25   W. W. Johnstone in *Land of the Lyrebird*, pp. 213–14.

26   W. Gill, 'Forestry: notes on planting', *Journal of Agriculture and Industry of South Australia*, vol. 2, 1898, pp. 4–7; 'Robertstown Agricultural Society report', *Journal of Agriculture and Industry of South Australia*, vol. 2, 1898, p. 129; 'Amyton Agricultural Society report', *Journal of Agriculture and Industry of South Australia*, vol. 2, 1898, p. 150; G. Quinn, 'Orchard notes for August', *Journal of Agriculture and Industry of South Australia*, vol. 2, 1898, p. 31.

27   F. E. H. W. Krichauff, 'Chairman's address to the 1898 agricultural congress', *Journal of Agriculture and Industry of South Australia*, vol. 2, 1898, p. 188.

28   Margaret Kiddle, *Men of Yesterday*, p. 318.

29   *The Australasian*, 'Plantations at Lismore', 21 May, 1921.

30   Thomas Cherry, *Victorian Agriculture*, Oxford University Press, Melbourne, 1935, p. 65.

31   The first exhortation was by H. R. Gray, 'Farming and forestry', *Journal of Agriculture, Western Australia*, vol. 1, 1924, pp. 460–67.

32   J. G. O. Tepper, 'Trees and their role in nature', *Agricultural Gazette of New South Wales*, vol. 7, 1896, p. 40.

33   H. Pye, 'Trees on the farm', *Journal of Agriculture Victoria*, vol. 1, 1902, p. 847.

34   J. M. Reed, 'The importance of tree planting', *Journal of Agriculture Victoria*, vol. 6, 1907, pp. 562.

35   J. M. Reed, 'Victorian tree planting competition, 1912–15', *Journal of Agriculture Victoria*, vol. 9, 1911, pp. 721–4.

36   'Eucalypts for shelter and ornament', *The Australasian*, 28 May, 1921.

37   H. Mackay, 'Forestry and its relation to agriculture', *Journal of Agriculture Victoria*, vol. 12, 1914, p. 223; E. E. Pescott, 'The Australian flora from an ornamental aspect', *Journal of Agriculture Victoria*, vol. 17, 1919, p. 183; K. V. M. Ferguson, 'Tree windbreaks on the farm', *Journal of Agriculture Victoria*, vol. 38, 1940, pp. 46–67; and W. E. Hills, 'Tree shelter belts as a source of income', *Journal of Agriculture Victoria*, vol. 51, 1953, p. 493.

38   For example: 'Tree preservation is of national importance', *Agricultural Gazette of New South Wales*, vol. 52, 1941, p. 383.

39   G. H. Burvill, 'Control of clearing', *Journal of Agriculture, Western Australia*, Perth, vol. 28, 1951, p. 440.

40 The 1939 bushfires burned over 1.2 million hectares of Victorian forest.

41 'From ashes to disaster', *Victoria's Resources*, vol. 18, December 1976, pp. 2–3; and *Trees and Victoria's Resources*, vol. 26, September 1984, p.12. The league is a tree-growing and educational institution, which supplies trees to farmers, municipalities, government departments and community groups.

42 H. R. Dickinson and R. G. Downes, *The Westgate Farm Planning Project*, Soil Conservation Authority, Melbourne, 1953; and L. D. Garside, *The Westgate Story*, Soil Conservation Authority, Melbourne, undated.

43 P. A. Yeomans, *The Keyline Plan*, published by P. A. Yeomans, Wodonga, 1954.

44 P. A. Yeomans, *Water for Every Farm: Using The Keyline Plan*, Griffin Press, Adelaide, 1981, p. 58.

45 Ibid., pp. 58–9.

46 Ibid., p. 59.

47 A. Campbell, 'The Potter farmland plan: progress so far', *Trees and Natural Resources*, vol. 27, 1985, p. 30.

48 'Reversing rural tree decline', *Rural Research*, No. 146, 1990, p. 19.

49 J. A. Palmer, *William Moodie: A Pioneer of Western Victoria*, Hedges & Bell, Maryborough, 1973, p. 19.

50 Bruce Milne, Land degradation awareness — the Helm View experience, Unpublished paper prepared for people interested in arresting Australian land degradation, Coleraine, Victoria, 1989.

51 Bonwick, *Western Victoria: Its Geography, Geology, and Social Condition*, p. 150.

52 The Wando Dale run was subsequently made available for settlement under the *Closer Settlement Act* of 1898.

53 The Potter Foundation contributed $900 000 to the project of which $270 000 was spent on farm works. The participating farmers made a matching contribution, in cash, kind and labour, of $234 000.

54 J. W. Cary, A. J. Beal and H. S. Hawkins, *Farmers' Attitudes Towards Land Management For Conservation*, School of Agriculture and Forestry, University of Melbourne, Parkville, 1986, pp. 8 & 33. The survey data and the belief maps are drawn from interviews with the owners of fifteen Potter Farmland farms and a random sample of forty-one neighbourhood farmers in the Shire of Dundas.

55 The larger circles for the position of the locality farmers and for their beliefs about the Potter Farmland Plan indicate that there is less agreement, or more variance, about the location of these beliefs.

56 Cary, Beal and Hawkins, *Farmers' Attitudes Towards Land Management For Conservation*, pp. 23 & 56.

57 The Victorian LandCare program was first launched by the then Department of Conservation, Forests and Lands and the Victorian Farmers' Federation in November 1986, at Winjallok near St Arnaud in Central Victoria.

58 'Our Country Our Future', AGPS, Canberra, 1989.
59 Care will be needed to maintain the integrity of rural landscapes. Some farmer groups are interconnecting wildlife corridors and groves that transcend farm boundaries. The Dundas Black Range LandCare group in Western Victoria is providing a wildlife corridor between two separated forests. An interconnected set of plantations will weave its way across five farms to provide the corridor.
60 I. T. Loane, 'Economic evaluation of farm trees', Technical report for the Australian Special Rural Research Council, Project No. DAV 13A, Department of Agriculture and Rural Affairs, Melbourne, 1990, p. 1.
61 Ibid., pp. 1, 51–2.
62 *William Moodie: A Pioneer of Western Victoria*, p. 99. Some of those who had selected land near good wattle country made a good thing out of it by buying the bark in the paddocks on the trees, stripping it themselves and having their own teams cart it to the market (p. 100).
63 For example, in South Australia: 'Lucindale Agricultural Society report', *Journal of Agriculture and Industry of South Australia*, vol. 2, 1898, p. 287.
64 Loane, 'Economic evaluation of farm trees', pp. 104–7. Most of the recent Australian research has concerned species evaluation, planting density and management practices.
65 Edwin Adamson, *The Relationship Between Trees and Rural Productivity: A Bibliography*, Ministry of Planning and Environment, Victoria, 1988.
66 In many situations block plantings of native trees give real rates of return of less than 3 per cent and thus a negative net present value at normal discount rates of 4–8 per cent. Loane, 'Economic evaluation of farm trees', p. 1.
67 S. Gleeson, *Weekly Times*, April 25, 1990, p. 9.
68 Loane, 'Economic evaluation of farm trees', p. 106.
69 Pye, *Journal of Agriculture Victoria*, p. 847.
70 P. Woodgate and P. Black, *Forest Cover Changes in Victoria: 1869–1987*, Department of Conservation, Forests and Lands, Melbourne, 1989.

Chapter 5: Pastoralism in the rangelands

1 Geoffrey Dutton, *The Squatters*, Viking O'Neil, Ringwood, 1985, p. 14. The marginality of South Australia's northern country for cropping was clarified by the advance and retreat of an army of pioneer settlers beyond Goyder's Line. J. M. Powell, *An Historical Geography of Modern Australia: The Restive Fringe*, Cambridge University Press, Cambridge, 1988, p. 12.
2 John Pickard, 'Analysis of stocking records from 1884 to 1988 during the subdivision of *Momba*, the largest property in semi-arid New South Wales', *Proceedings of the Ecological Society of Australia*, vol. 16, 1990, pp. 245–53.

3   D. G. Yates, *The Sociology of Wool Production*, Department of Agricultural Economics and Business Management, University of New England, Armidale, 1974, p. 11; and M. D. Young, M. Gibbs, W. E. Holmes and D. M. D. Mills, 'Socio-economic influences on pastoral management' in G. N. Harrington, A. D. Wilson and M. D. Young (eds) *Management of Australia's Rangelands*, CSIRO Australia, 1984, p. 87.

4   Dutton, *The Squatters*, p. 112. Both were expert drovers and bushmen. In 1898 Tyson owned twenty stations — two million hectares — in Victoria, New South Wales and Queensland. Kidman owned chains of stations running from the Gulf of Carpentaria and the Victoria River in the Northern Territory down to the markets in Adelaide. In 1920 his stations covered more land than there is in Tasmania.

5   Ibid., p. 87.

6   Alan Barnard, *The Simple Fleece*, Melbourne University Press, Carlton, 1962, p. 436.

7   Yates, *The Sociology of Wool Production*, p. 10.

8   J. Macdonald Holmes, 'The erosion-pastoral problem of the Western Division of New South Wales', University of Sydney publications in Geography No. 2, University of Sydney, Australia, 1938.

9   J. W. Cary and W. E. Holmes, 'Relationships among farmers' goals and farm adjustment strategies: some empirics of a multidimensional approach' *Australian Journal of Agricultural Economics*, vol. 26, 1982, p. 116.

10 William Brodribb, *Recollections of an Early Squatter*. First published by John Woods & Co., Sydney, 1883, republished by John Ferguson Pty Ltd, Sydney, 1978, pp. 96 & 180. Brodribb summed up his experience from the early 1840s: 'We all paid dearly for experience; I had, as a colonist, paid dearly for mine; and for the future made up my mind to profit by it, if possible, and, at all events, not to speculate beyond my means', p. 54.

11 Ibid., p. 143.

12 J. Lester, 'Aboriginal lands', in J. Messer and G. Mosley (eds), *What Future for Australia's Arid Lands?* Australian Conservation Foundation, Melbourne, 1982, pp. 61–2.

13 Barnard, *The Simple Fleece*, p. 437. Short-term droughts of three to twelve months duration frequently occur in the rangelands. Most rangelands pastures will carry stock for twelve months into a drought, providing there is a reasonable component of perennials in the pasture. It is when stock numbers become excessively high during the unusual run of good seasons that greatest damage is done to pastures and soils during subsequent drought conditions.

14 R. W. Peacock, 'Our western lands. Their destruction and possible improvement', *Agricultural Gazette of New South Wales*, vol. 11, 1900, pp. 652–7.

15 G. N. Harrington and K. C. Hodgkinson, 'Shrub-grass dynamics in mulga communities of eastern Australia' in P. J. Joss, P. W. Lynch and O. B. Williams (eds), *Rangelands: A Resource Under Siege*, Austral-

ian Academy of Science, Canberra, 1986, p. 27; and R. W. Condon, 'Shrub invasion on semi-arid grazing lands in western New South Wales' in *Rangelands: A Resource Under Siege*, p. 6. Woody shrubs include green turkey bush (*Eremophila gilesii*), budda or false sandalwood (*Eremophila mitchellii*), turpentine (*E. sturtii*), hop bushes (*Dodonaea* spp.), *Cassia* spp., pine and poplar box (*E. populnea*). All of these are unpalatable and of no commercial value. The mulga tree (*Acacia aneura*) has value as a fodder source and for stabilising the ecosystems.

16 Harrington, Wilson and Young, *Management of Australia's Rangelands*, p. 194.

17 J. W. Freebairn, 'Drought assistance policy', *Australian Journal of Agricultural Economics*, vol. 27, 1983, pp. 185–99; and R. L. Heathcote, 'Drought perception' in J. V. Lovett (ed.), *The Environmental, Economic and Social Significance of Drought*, Angus and Robertson, Sydney, 1973, p. 35.

18 Alan Barnard, 'Aspects of the economic history of the arid land pastoral industry' in R. O. Slayter and R. A. Perry (eds), *Arid Lands of Australia*, Australian National University Press, Canberra, 1969, pp. 212 & 224.

19 D. McGhie, 'The Ord River Irrigation Area: success in the eighties', *Agricultural Science*, vol. 3, no. 1, 1990, p. 39.

20 Bruce Davidson, *The Northern Myth: Limits to Agricultural and Pastoral Development in Tropical Australia*, Melbourne University Press, Carlton, 1972.

21 J. J. Mott, J. C. Tothill and E. J. Weston, 'Animal production from the native woodlands and grasslands of northern Australia', *Journal of the Australian Institute of Agricultural Science*, vol. 47, 1981, p. 132; Victor Squires, 'Rangeland management', *Agricultural Science*, vol. 2, no. 4, 1989, p. 31; and B. R. Frank, Constraints to the adoption of management practices in North Queensland: a summary of sociological research in Dalrymple and Bowen Shires, unpublished paper, Department of Primary Industries, Queensland, 1990, p. 6.

22 *South Australian Pastoral Act*, 1936–1974, Schedule A. At the time of writing a new Pastoral Land Management and Conservation Bill was proposed in South Australia.

23 M. D. Young, 'Rangeland administration', in Harrington, Wilson and Young (eds), *Management of Australia's Rangelands*, p. 165.

24 K. O. Campbell 'Problems of adoption of pastoral businesses in the arid zone.' *Australian Journal of Agricultural Economics*, vol. 10, 1966, pp. 14–26.

25 See R. B. Hacker, 'The development of rangeland monitoring in Western Australia', *Proceedings of the Fifth Australian Soil Conservation Conference*, Perth, Western Australia, 1990; and compare Brendan Lay 'Monitoring under a new Pastoral Lands Act in South Australia, practical necessity versus scientific excellence', *Proceedings of the Fifth Australian Soil Conservation Conference*. Lay considers some of the difficulties of reconciling scientific approaches and administrative

requirements in a monitoring system. There is no consideration of whether the cost of such a system will be covered by the revenue collected. See also M. D. Young, 'Pastoral land tenure options in Australia', *Australian Rangelands Journal*, vol. 7, 1985, pp. 43–6.

26   D. G. Wilcox, 'Fair use and a fair go', *Australian Rangelands Journal*, vol. 10, 1988, pp. 77–8.

27   W. E. Holmes, 'Property build-up in a semi-arid grazing area', MAgrSc Thesis, University of Melbourne, 1980.

28   Cary and Holmes, *Australian Journal of Agricultural Economics*, vol. 26, 1982, p. 118.

29   See J. R. Mills and D. M. D. Mills, 'The pastoral industry in south-west Queensland 1960–2000', in Joss, Lynch and Williams, *Rangelands: A Resource Under Siege*, p. 191; B. R. Crouch and G. Payne 'Value orientation of pastoralists in the arid zone of Queensland and its relation to adoption of sheep management practices' *Proceedings of Symposium on Social Sciences in the Arid Zone*, CSIRO, Deniliquin, NSW, 1983; and Frank, 'Constraints to the adoption of management practices in North Queensland'.

30   Adapted from J. R. Childs, An economic and psychological assessment of managerial ability, MAgrSc Thesis, University of Melbourne, 1977.

31   Ibid., pp. 171–2; J. W. Cary and R. E. Weston, *Social Stress in Agriculture: The Implications of Rapid Economic Change*, School of Agriculture and Forestry, University of Melbourne, Parkville, 1978; and M. D. Young 'Influencing land use in pastoral Australia', *Journal of Arid Environments*, vol. 2, 1979, pp. 284.

32   C. Harris, 'Pastoralism: an industry of erosion and subsidies', *Conservation News*, Newsletter of the Australian Conservation Foundation, vol. 22, no. 5, June 1990, pp. 8–9.

33   CSIRO, *A Policy for The Future of Australia's Rangelands*, CSIRO Division of Wildlife and Ecology, Canberra, 1990, p. 15.

34   Frank, 'Constraints to the adoption of management practices in North Queensland'.

35   N. D. MacLeod and B. G. Johnston, 'An economic framework for evaluating rangeland restoration options', presented to the Annual Conference of the Australian Agricultural Economics Society, University of Queensland, Brisbane, 1990, p. 7.

36   The pro development line is best exemplified in Bill Burrows, 'Prospects for increased production in the north east Australian beef industry through pasture management', *Agricultural Science*, vol. 3, no. 1, 1990, pp. 19–24. Burrows cites evidence that clearing the woodland will double cattle-carrying capacity and liveweight gains. P. Gillard, J. Williams and R. Monypenny in 'Clearing trees from Australia's semi arid tropics', *Agricultural Science*, vol. 2, no. 2, 1989, pp. 34–9, cite evidence that pasture improvement can be just as profitable if the trees are left where they are. One can only conclude that there is no hard and fast rule for all woodland.

37   C. J. Gardener, J. G. McIvor and J. Williams, 'Dry tropical rangelands:

solving one problem and creating another', *Proceedings of the Ecological Society of Australia*, vol. 16, 1990, p. 286.

38 Queensland's Department of Primary Industries, which has displayed a somewhat unrestrained enthusiasm for northern development, has issued staff guidelines for woodland development to prevent excessive clearing of trees. See Burrows, *Agricultural Science*, vol. 3, no. 1, 1990, p. 21.

39 Frank, 'Constraints to the adoption of management practices in North Queensland'.

40 CSIRO, *A Policy for The Future of Australia's Rangelands*; also Gardener, McIvor and Williams, *Proceedings of the Ecological Society of Australia*, vol. 16, 1990, p. 284.

41 See Harris, *Conservation News*, p. 8.

Chapter 6: Sustaining the wheat crop

1 B. R. Davidson, *European Farming in Australia*, Elsevier, Amsterdam, 1981, Ch. 3; A. G. L. Shaw, 'History and Development of Australian Agriculture', in D. B. Williams (ed.), *Agriculture in the Australian Economy*, Sydney University Press, 1967, pp. 1–3; and E. Dunsdorfs, *The Australian Wheat-Growing Industry*, Melbourne University Press, Carlton, 1956, pp. 39–50.

2 William Howitt, *Land, Labour and Gold*, Longmans, London 1855, Letter XV.

3 Alfred Joyce, *A Homestead History: Being the Reminiscences of Alfred Joyce of Plaistow and Norwood, Port Phillip 1843 to 1864*, G. F. James (ed.), Oxford University Press, Melbourne, 3rd edn, pp. 57–8.

4 J. W. Bull, *Early Recollections and Experiences of Colonial Life*, cited in A. R. Callaghan and A. J. Millington, *The Wheat Industry in Australia*, Angus and Robertson, Sydney, 1956, pp. 337–8.

5 For a description of domestic wheat production by the squatters, see Joyce, *A Homestead History*, Chs 4 & 5.

6 Howitt, *Land, Labour and Gold*, Letter XVII.

7 Joyce, *A Homestead History*, p. 155.

8 N. F. Barr and E. F. Almond, *The Third Way: Case Studies in Commercial Part Time Farming*, School of Agriculture and Forestry, University of Melbourne, 1981, Ch. 4; and E. F. Almond and N. F. Barr, *The Agricultural Work Force in Victoria: Case Studies in the Division of Labour on Farms*, School of Agriculture and Forestry, University of Melbourne, 1981, Ch. 3.

9 Davidson, *European Farming in Australia*, pp. 144–7.

10 J. M. Powell, *Yeomen and Bureaucrats: The Victorian Crown Lands Commission 1878–79*, Oxford University Press, Melbourne, 1973, p. 53.

11 Notable exceptions are the rich soils of northern Tasmania and the black soils of the Darling Downs, northern New South Wales and the

Wimmera. G. W. Leeper, in *The Australian Environment*, Melbourne University Press, Carlton, 1970, pp. 22–3.

12 Powell, *Yeomen and Bureaucrats*, pp. 48–50, 53, 136–7. In Victoria the farmers of Bellarine and Barrabool seem to have been an exception, running mixed wheat and grazing farms and sowing pastures of Dutch clover.

13 E. Dunsdorfs, *The Wheat Growing Industry in Australia, 1788–1948*, Melbourne University Press, Carlton, 1956, p. 142.

14 Evidence to the Crown Lands Commission of Inquiry, Victorian Parliamentary Papers, September 1879.

15 Joseph Jenkins, *Dairy of a Welsh Swagman: 1869–1884*, (ed. W. Evans), Macmillan, Melbourne, 1975, pp. 13, 70, 106, 110, 132, 149.

16 D. W. Meinig, *On the Margins of the Good Earth*, Rigby, Adelaide, 1962.

17 *The Farmers' Weekly Messenger*, May 22, 1874, cited in Meinig, *On the Margins of the Good Earth*, p. 55.

18 South Australian Parliamentary Papers No. 23, 1870–71, p. 8, cited in Meinig, *On the Margins of the Good Earth*, pp.117–18.

19 For example; 'Dawson Agricultural Society report', *Journal of Agriculture and Industry of South Australia*, vol. 2, 1898, p. 130.

20 H. Pye, 'The improvement of cereals: some suggestions for farmers', *Journal of Agriculture Victoria*, vol. 9, 1911, p. 256; and C. C. Brittlebank, 'Green manurial crops and take all', *Journal of Agriculture Victoria*, vol. 17, 1919, p. 171.

21 R. L. Pudney, 'Advantages of bare fallow', *Agricultural Gazette of New South Wales*, vol. 1, 1890, pp. 28–32.

22 'The growing of wheat', *Journal of Agriculture Victoria*, vol. 3, 1905, p. 182.

23 Davidson, *European Farming in Australia*, p. 189; and Callaghan and Millington, *The Wheat Industry in Australia*, p. 275.

24 T. A. J. Smith, 'Nhill Agricultural Society annual crop and fallow competitions, 1916', *Journal of Agriculture Victoria*, vol. 15, 1917, pp. 169–75.

25 Ibid., p. 173.

26 H. A. Mullett, 'Crop competitions in the Wimmera', *Journal of Agriculture Victoria*, vol. 18, 1920, p. 251.

27 H. A. Mullet, 'Horsham crop competition', *Journal of Agriculture Victoria*, vol. 19, 1921, p. 356.

28 J. G. O. Tepper, 'Take all and its remedies', *Agricultural Gazette of New South Wales*, vol. 3, 1892, pp. 69–70; J. G. O. Tepper, 'Remarks on weeds: the soil and its fertility', *Agricultural Gazette of New South Wales*, vol. 6, 1895, pp. 385–90.

29 'Mundoora Agricultural Society report', *Journal of Agriculture and Industry of South Australia*, vol. 2, 1898, p. 133; 'Calca Agricultural Society report', *Journal of Agriculture and Industry of South Australia*, vol. 2, 1898, p. 370.

30 E. S. Clayton, 'The control of soil erosion on wheatlands', *Agricultural Gazette of New South Wales*, vol. 17, 1931, p. 826.

31 C. M. Donald, 'Innovation in Agriculture' in Williams, *Agriculture in the Australian Economy*, p. 72.

32 A. J. Holt, *Wheat Farms of Victoria*, School of Agriculture, University of Melbourne, 1947.

33 H. L. Hore, 'Loss of topsoil: effect on yield and quality of wheat', *Journal of Agriculture Victoria*, vol. 52, 1954, p. 241.

34 A. E. V. Richardson, 'Wheat and its cultivation', *Journal of Agriculture Victoria*, vol. 11, 1913, pp. 418–19.

35 Ibid., p. 419.

36 'Changing agriculture: an urgent national need', *Agricultural Gazette of New South Wales*, vol. 52, 1941, p. 1; J. N. Whittet, 'Lucerne: the urgent need for its cultivation', *Agricultural Gazette of New South Wales*, vol. 52, 1941, pp. 3–4; 'Soil erosion: its control by systems of cropping', *Agricultural Gazette of New South Wales*, vol. 52, 1941, pp. 506–16; G. H. Burvill, 'Changing ideas in soil cultivation', *Journal of Agriculture Western Australia*, vol. 22, 1945, pp. 3–10.

37 J. A. Morrow and R. H. Hayman, 'Clover ley farming for mixed farming areas of moderate rainfall', *Journal of Agriculture Victoria*, vol. 38, 1940, p. 205.

38 R. H. Hayman, 'The maintenance of soil fertility and production', *Journal of Agriculture Victoria*, vol. 41, 1943, pp. 331–5.

39 'Red Hill Agricultural Society report', *Journal of Agriculture and Industry of South Australia*, vol. 2, 1898, p. 131.

40 Davidson, *European Agriculture in Australia*, p. 353.

41 H. L. Hore and H. J. Sims, 'The Wimmera championships', *Journal of Agriculture Victoria*, vol. 50, 1952, p. 194; and H. J. Sims and J. M. McCann, 'Wheat crop championships', *Journal of Agriculture Victoria*, vol. 51, 1953, p. 149.

42 D. R. Rooney and C. T. Patton, 'The Wimmera wheat crop championship', *Journal of Agriculture Victoria*, vol. 59, 1961, p. 191.

43 In wetter areas some farmers eliminated fallow in a pasture–pasture–wheat–wheat–oats rotation. Fallow was eliminated, but repeated ploughing remained. On these farms wheat paddocks were burnt in April and cultivated four or five times to control weeds and prepare a seedbed prior to planting.

44 Sims and McCann, *Journal of Agriculture Victoria*, vol. 51, 1953, p. 149.

45 A. Morgan, 'The absorption and translocation of herbicides', *Journal of Agriculture Victoria*, vol. 33, 1935, pp. 200–208.

46 There are several disadvantages associated with conservation cropping. In some soils direct drilling can create a hardened soil layer just beneath the depth reached by the seeding drill. Decomposing stubbles compete with wheat plants for nitrogen and standing stubbles can release chemicals into the soil that reduce the germination and growth of following crops. Stubbles can carryover fungal diseases from one crop to the next. Generations of farmers have burnt stubbles to control disease, against the long standing advice of soil scientists. (See H. Pye, 'Bush fires and stubble burning', *Journal of Agriculture Victoria*, vol. 2, 1904, p. 428). Standing stubbles make weed control

harder to achieve by hindering the application of herbicides and block machinery, making sowing difficult.

47 A. Rovira, 'Tillage and soil-borne root diseases of winter cereals' in P. S. Cornish and J. E. Pratley (eds), *Tillage: New Directions in Australian Agriculture*, Inkata Press, Melbourne, 1987, p. 335.

48 T. Cherry, 'Present day problems in agriculture', *Journal of Agriculture Victoria*, vol. 5, 1907, p. 480; J. R. Goldstein, 'Cow-peas', *Journal of Agriculture Victoria*, vol. 6, 1908, p. 652; 'The benefit of legumes', *Journal of Agriculture Victoria*, vol. 11, 1913, p. 677; A. E. V. Richardson, 'Results of experiments, 1915', *Journal of Agriculture Victoria*, vol. 14, 1916, p. 147; 'Green manuring', *Agricultural Gazette of New South Wales*, vol. 2, 1891, pp. 687–8; 'General notes', *Agricultural Gazette of New South Wales*, vol. 2, 1891, p. 711.

49 B. Bardsley, *Farmers' Assessment of Information and Its Sources*, School of Agriculture and Forestry, University of Melbourne, 1982, p. 89.

50 A. D. Rovira, D. K. Roget, S. M. Neate, K. R. J. Smetterton, R. W. Fitzpatrick, N. J. MacKenzie, J. C. Buckerfield and A. Simon, 'The effect of conservation tillage practices on soil properties, earthworms, cereal root diseases and wheat yields', *Proceedings of the Fifth Australian Soil Conservation Conference*, Perth, 1990.

51 The claims of declining protein content of Australian wheat may be partly exaggerated by changing classification standards for the highest quality wheats.

52 A major recommendation of 'alternative agriculture' advocates in the United States is the use of legume rotations in the wheat industry; see *Alternative Agriculture*, Committee on the role of Alternative Farming Practices in Modern Production Agriculture, National Academic Press, Washington D. C., 1989, pp. 138–9.

53 Rock phosphate is often water insoluble. When spread over the soil, the phosphate is not in a readily available form. It is concentrated in areas around the comparatively large grains of rock phosphate. Superphosphate is a water soluble compound. When applied to soil it is widely dispersed in its soluble form before being converted to insoluble forms of phosphate. A given amount of phosphorus applied to soil as superphosphate will be more widely dispersed in smaller particles than an equivalent amount of rock phosphate. Debate on the merits of rock phosphate and superphosphate has waged since the turn of the century. Except in the occasional location, superphosphate has consistently outperformed rock phosphate as a fertiliser when judged by pasture yields.

54 Drawn from a paper presented by Alan Druce, farmer from Temora NSW, at the 'Sustainable Agriculture: Farming For the Future' Conference, Benalla, 1988. See also Els Wynen and Geoff Edwards, 'Towards a comparison of chemical-free and conventional farming in Australia', *Australian Journal of Agricultural Economics*, vol. 34, 1990, p. 46.

55 Wynen and Edwards, *Australian Journal of Agricultural Economics*, vol. 34, 1990, p. 48.

56 B. J. Goddard and P. Nash, 'Farmers' attitudes and intentions towards

conservation cropping practices', Paper presented to the Fifth Australian Soil Conservation Conference, Stable Cropping Systems Workshop, Perth, 1990.

57 The most commonly used contact herbicide, paraquat, is claimed to leave no harmful residues in soil. (*Paraquat: Its Fate and Effects in the Soil*, Agrochemical Monograph, ICI, 1984.) But residues of the supposedly residue-free systemic herbicide glyphosate have been shown to persist in some soils for up to 120 days, reducing the vigour of pasture plants. P. L. Eberbach & L. A. Douglas, 'Persistence of glyphosate in a sandy loam soil', *Soil Biology and Biochemistry*, vol. 15, 1983, pp. 485–7.

58 In the 1986–87 National Residue Survey, 5 per cent of wheat samples contained cadmium at levels above the proposed 0.05 parts per million food standard. In Victoria, 19 per cent of samples were above the proposed limit. In tracebacks of contaminated samples, some soils were found to have cadmium at levels of 5 to 10 parts per million.

Chapter 7: Selling sustainable cropping

1   S. S. Cameron, 'Lime in Victoria', *Journal of Agriculture Victoria*, vol. 10, 1912, p. 585.
2   Ibid., p. 586.
3   Ibid., p. 587.
4   A. E. V. Richardson, 'Rutherglen Experimental Farm', *Journal of Agriculture Victoria*, vol. 12, 1914, p. 241.
5   A. E. V. Richardson, 'The value of experimental research in agriculture', *Journal of Agriculture Victoria*, vol. 12, 1914, pp. 349–50.
6   A. E. V. Richardson, 'Rutherglen Experimental Farm: report on permanent experimental field season 1913', *Journal of Agriculture Victoria*, vol. 12, 1914, p. 142.
7   A. E. V. Richardson, 'Rutherglen Experimental Farm: second annual farmers' field day', *Journal of Agriculture Victoria*, vol. 12, 1914, p. 4.
8   H. A. Mullet, 'Shepparton crop and fallow competition', *Journal of Agriculture Victoria*, vol. 21, 1923, p. 82.
9   H. A. Mullett, 'Numurkah crop and fallow competition 1924', *Journal of Agriculture Victoria*, vol. 22, 1924, p. 350.
10 See Chapter 2.
11 H. A. Mullet, 'Dookie crop competition 1923', *Journal of Agriculture Victoria*, vol. 22, 1924, p. 158.
12 Mullett, *Journal of Agriculture Victoria*, vol. 22, 1924, pp. 160–61.
13 Mullet, *Journal of Agriculture Victoria*, vol. 21, 1923, p. 83.
14 Mullet, *Journal of Agriculture Victoria*, vol. 22, 1924, p. 86.
15 R. G. Thomas, 'Numurkah crop and fallow competition', *Journal of Agriculture Victoria*, vol. 23, 1925, p. 339.
16 R. H. Billing, 'Farm competitions in extension', Unpublished Report, School of Agriculture and Forestry, University of Melbourne, 1968.

17 J. A. Morrow and R. H. Hayman, 'Clover ley farming for mixed farming areas of moderate rainfall', *Journal of Agriculture Victoria*, vol. 38, 1940, p. 205.
18 Ibid., p. 206.
19 R. H. Hayman, 'The maintenance of soil fertility and production', *Journal of Agriculture Victoria*, vol. 41, 1943, pp. 331–5.
20 Ibid., p. 335.
21 J. A. Morrow, 'Rutherglen crop competition', *Journal of Agriculture Victoria*, vol. 40, 1942, pp. 131–3.
22 Morrow and Hayman, *Journal of Agriculture Victoria*, vol. 38, 1940, pp. 205–10.
23 R. G. Thomas and W. G. Andrew, 'Soil conservation competition in the Goulburn catchment area', *Journal of Agriculture Victoria*, vol. 46, 1948, p. 355.
24 W. J. Tame, 'Soil conservation competition in the Goulburn catchment 1947', *Journal of Agriculture Victoria*, vol. 45, 1947, p. 254.
25 In the early 1950s a small group of farmers at Goorambat organised themselves into a cooperative erosion control group. Earlier generations had cleared trees on the hillsides above the Goorambat cropping flats. The increased flood flows had created large gullies across paddocks. These needed to be removed and structures needed to be erected; but the whole gully required treatment. Today driving through parts of the Goorambat district one can see large concrete drop structures at the heads of shallow sloped watercourses. These were once deep canyons of erosion, which the group battered in. In some paddocks you can see where the slopes turn to old gullies again, monuments to old animosities and lack of cooperation. The rules of the Hanslow Cup did not award points for the intangible work of getting on with your neighbours and planning together. See various reports of the Hanslow Cup Competition judges in the *Journal of Agriculture Victoria* between 1945 and 1955.
26 Morrow and Hayman, *Journal of Agriculture Victoria*, vol. 38, 1940, p. 205.
27 If a paddock was sown to pasture for 5 years, wild oats would be almost eliminated, and the farmer could plough in autumn before planting and grow a crop, confident wild oats would not be a problem. Where a paddock was in pasture for only a couple of years, the only way of controlling wild oats was by fallowing for a summer before sowing.
28 J. A. Morrow, 'Northern district championship', *Journal of Agriculture Victoria*, vol. 46, 1948, p. 157.
29 The availability of earlier maturing sub clover strains was being promoted for the northern parts of the district in the early 1950s. H. C. H. Watson, 'The northern district championship', *Journal of Agriculture Victoria*, vol. 51, 1953, p. 186. Between 1948 and 1958 the Dookie Competition was consistently won by crops grown on long fallow. Adoption of ley pastures accelerated between 1955 and 1965. It was not till 1966 that the Yarrawonga area was to win the champi-

onship with a crop grown on ley ground. J. B. McPherson, 'The northern district championship', *Journal of Agriculture Victoria*, vol. 57, 1959, p. 268. R. D. Whitaker, 'The northern district championship', *Journal of Agriculture Victoria*, vol. 65, 1967, p. 184.

30 Morrow, *Journal of Agriculture Victoria*, vol. 40, 1942, p. 133.

31 Morrow, *Journal of Agriculture Victoria*, vol. 46, 1948, p. 157.

32 J. G. Bath, 'The northern district championship', *Journal of Agriculture Victoria*, vol. 53, 1955, p. 191.

33 Hayman, *Journal of Agriculture Victoria*, vol. 41, 1943, p. 335.

34 D. R. Coventry and H. D. Brooke, 'Development of a minimum tillage system: Rutherglen experience', *Proceedings of the 5th Australian Agronomy Conference*, 1989, pp. 245–51.

35 Ibid., p. 246.

36 The difficulties of running experimental plots on farms before the arrival of motor transport are well described in H. J. Sims and C. J. Webb, *Mallee Sand To Gold: The Mallee Research Station Walpeup 1932–82*, Department of Agriculture Victoria, 1982, pp. 89–91. Travelling by train to farms with bags of wheat under the arm limited what could be achieved.

37 Coventry and Brooke, *Proceedings of the 5th Australian Agronomy Conference*, 1989, pp. 246–7.

38 Ibid., p. 247; T. G. Reeves and I. S. Smith, 'Wheat without cultivation', *Journal of Agriculture Victoria*, vol. 71, 1973, pp. 74–5; and A. Ellington and T. Reeves, 'Minimum cultivation saves soil, time and energy', *Journal of Agriculture Victoria*, vol. 76, 1978, pp. 150–56.

39 P. J. Haines and T. G. Reeves, cited in Coventry and Brooke, *Proceedings of the 5th Australian Agronomy Conference*, 1989, p. 248.

40 Craig Ewers, Innovation in response to environmental degradation, MAgrSc Thesis, School of Agriculture and Forestry, University of Melbourne, 1990, p. 94.

41 Ibid., p. 95.

42 Ibid., p. 97.

43 J. W. Cary, R. L. Wilkinson and C. R. Ewers, *Caring for the Soil on Cropping Lands*, School of Agriculture and Forestry, University of Melbourne, 1989, p. 15.

44 Coventry and Brooke, *Proceedings of the 5th Australian Agronomy Conference*, 1989, p. 248.

45 R. D. Whitaker, Farmers' attitudes to minimum tillage in northern Victoria, unpublished Report, School of Agriculture and Forestry, University of Melbourne, 1977, p. 16.

46 Coventry and Brooke, *Proceedings of the 5th Australian Agronomy Conference*, 1989, p. 249; and J. C. Avery, P. Ockenden and C. Lindsay, 'Cropland SoilCare in north east Victoria — an integrated extension research package', Working paper delivered to the Group Extension Workshop, Australian Soil Conservation Conference, Perth, 1990.

47 Cary, Wilkinson and Ewers, *Caring for the Soil on Cropping Lands*, p. 15; and J. T. Harvey and F. T. Hurley, *Cropping and Conservation*,

Regional Studies Unit, Ballarat University College, Ballarat, 1990, p. 18.

48 Ewers, 'Innovation in response to environmental degradation', p. 108.

49 Cary, Wilkinson and Ewers, *Caring for the Soil on Cropping Lands*, pp. 15–16.

50 Ibid., p. 5.

51 Harvey and Hurley, *Cropping and Conservation*, p. 51; and reanalysis of the Cary, Wilkinson and Ewers study data.

52 Based on a re-analysis of the data reported in Cary, Wilkinson and Ewers; and Harvey and Hurley, *Cropping and Conservation*, p. 60.

53 Ewers 'Innovation in response to environmental degradation', p. 108; Cary, Wilkinson and Ewers, *Caring for the Soil on Cropping Lands*, p. 28; Harvey and Hurley, *Cropping and Conservation*, p. 19. This complaint is supported by some research work. R. A. Fischer, I. B. Mason, and G. N. Howe, 'Tillage practices and the growth and yield of wheat in southern New South Wales: Yanco, in a 425 mm rainfall region', *Australian Journal of Experimental Agriculture*, 1988, vol. 28, pp. 223–36; I. B. Mason and R. A. Fischer, 'Tillage practices and the growth of wheat in southern New South Wales: Lockhart, in a 450 mm rainfall region', *Australian Journal of Experimental Agriculture*, 1986, vol. 26, pp. 457–68.

54 Derived from a reanalysis of the Cary, Wilkinson and Ewers study data.

55 Harvey and Hurley, *Cropping and Conservation*, p. 66. In this study of farmers in north-eastern, north-central and north-western Victoria there was no statistical relationship between perception of erosion or concern about erosion and use of direct drilling or post crop cultivation. See also J. A. Sinden and D. A. King, 'Who adopts conservation practices', *Australian Journal of Soil and Water Conservation*, vol. 1, no. 1, November, 1988, pp. 32–6. This study of wheat producers in the Manilla Shire of Northern NSW found 'adherence to the conservation ethic does not significantly differ between the adopters and non adopters of conservation cropping'; F. Vanclay, Socio-economic characteristics of adoption of soil conservation, MEnvS Thesis, Griffith University, Queensland, 1988, found farmers adhering to a conservation ethic were less likely to adopt conservation farming.

56 Harvey and Hurley, *Cropping and Conservation*, p. 61.

57 F. T. Hurley, B. C. Fitzgerald, J. T. Harvey and P. P. Oppenheim, *Cropping and Conservation: A Survey of Conservation Practices in Victorian Grain Growing Areas'*, Ballarat College of Advanced Education, Ballarat, 1985, p. 38; and Harvey and Hurley, *Conservation and Cropping*, p. 24.

58 Based on a stratified random sample survey of 146 cropping farmers in north-eastern Victoria. See Cary, Wilkinson and Ewers, *Caring for the Soil on Cropping Lands*, pp. 11–14.

59 Harvey and Hurley, *Conservation and Cropping*, pp. 22, 56, 60.

60 Cary, Wilkinson and Ewers, *Caring for the Soil on Cropping Lands*, p. 15; and Hurley and Harvey, *Conservation and Cropping*.

61 Cary, Wilkinson and Ewers, *Caring for the Soil on Cropping Lands*, p. 15; and Harvey and Hurley, *Cropping and Conservation*, p. 56.

62 Reanalysis of the Cary, Wilkinson and Ewers data.

63 Harvey and Hurley, *Cropping and Conservation*, pp. 30 & 44.

64 Hurley, Fitzgerald, Harvey and Oppenheim, *Cropping and Conservation*, p. 43. Some forms of trash clearance machinery, such as the Ryan disc, have been more successful in wet conditions and with stubbles from higher yielding crops.

65 The diseases striped rust and yellow spot are encouraged by stubble retention; Harvey and Hurley, *Cropping and Conservation*, p. 46.

66 D. R. Coventry, T. G. Reeves, H. D. Brooke, A. Ellington and W. J. Slattery, 'Increased wheat yields in north east Victoria by liming and deep ripping', *Australian Journal of Experimental Agriculture*, vol. 27, 1987, pp. 679–85; A. Ellington, *Soil Acidity — Effects and Remedies*, Department of Agriculture and Rural Affairs Technical Report Series no. 102, 1985; and A. Ellington, 'Effects of deep ripping, direct drilling, gypsum and lime on soils, wheat growth and yield', *Soil Tillage Research*, vol. 8, 1986, pp. 29–49.

67 Cary, Wilkinson and Ewers, *Caring for the Soil on Cropping Lands*, p. 16.

68 J. P. Brennan, 'Potential impact of soil acidification on farm productivity and survival', Paper presented to the Soil Acidification Workshop, Rutherglen Research Institute, June 1990.

Chapter 8: Potatoes and processors

1 C. E. Oldaker, 'Potatoes', *Tasmanian Journal of Agriculture*, vol. 1, 1930, p. 171.

2 C. E. Oldaker, 'Blight of potatoes in Tasmania', *Tasmanian Journal of Agriculture*, vol. 18, 1947, pp. 137–40.

3 W. J. Dowson, 'The prevention of late Irish blight of potatoes', *Tasmanian Journal of Agriculture*, vol. 2, 1931, p. 211.

4 J. O. Henrick, 'A survey of farming in north western Tasmania', *Tasmanian Journal of Agriculture*, vol. 1, 1930, p. 181.

5 G. C. Wade, 'Potato fire blight', *Tasmanian Journal of Agriculture*, vol. 21, 1950, p. 211.

6 'Comparisons of potato yields in Tasmania and New Zealand', *Tasmanian Journal of Agriculture*, vol. 17, 1946, p. 228.

7 C. A. Holland, 'Reconstruction in post-war agriculture', *Tasmanian Journal of Agriculture*, vol. 17, 1945, p. 126.

8 Ibid., pp. 126–9.

9 I. G. Inglis, 'Trends in the Tasmanian potato industry since 1915', *Tasmanian Journal of Agriculture*, vol. 24, 1953, pp. 260–64.

10 'Pesticide residues in dairy produce', *Tasmanian Journal of Agriculture*, vol. 38, 1967, p. 318.

11   C. R. Ewers, H. S. Hawkins, A. W. Kennelly and J. W. Cary, *Onion Growers' Perceptions of Soil Management Problems in Northern Tasmania*, University of Melbourne, School of Agriculture and Forestry, 1989.

12   A. G. Volum, A. J. Myers and R. Biasi, *A Survey of the Soil Management Practices of Gippsland Potato Growers*, Research Report Series No. 76, Department of Agriculture and Rural Affairs, Melbourne, 1988, p. 22.

13   C. R. Ewers, 'Implications of corporate involvement in agriculture', Paper presented to the Rural Policy Workshop, Annual Conference of Australian Institute of Geographers, Adelaide, February 1989.

14   Volum, Myers and Biasi, *A Survey of the Soil Management Practices of Gippsland Potato Growers*, p. 33.

15   Ewers, 'Implications of corporate involvement in agriculture'.

16   Based on a survey of the population of 91 onion growers in the Wynyard and Devonport areas of Northern Tasmania. See Ewers, Hawkins, Kennelly and Cary, *Onion Growers' Perceptions of Soil Management Problems in Northern Tasmania*, p. 17.

17   In 1990 CIG won a Tasmanian Landcare award.

Chapter 9: Beyond the silent spring

1   Parliament of Australia, *Senate Select Enquiry into Agricultural and Veterinary Chemicals*, Australian Government Printer, Canberra, 1990, p. 228.

2   Australian Science and Technology Council, *Health, Politics, Trade: Controlling Chemical Residues in Agricultural Products*, Australian Government Publishing Service, 1989, pp. 56–64.

3   N. F. Learmonth, *The Portland Bay Settlement*, Historical Committee of Portland, McCarron Bird, Melbourne, 1934, p. 93.

4   William Howitt, *Land Labour and Gold, or Two Years in Victoria*, Longman, London, 1855, Letter XLII.

5   C. French 'Economic entomology and ornithology', *Journal of Agriculture Victoria*, vol. 1, 1902, p. 59.

6   A. S. Olliff, 'Codlin moth', *Agricultural Gazette of New South Wales*, vol. 1, 1890, p. 4.

7   'Notes and comments', *Journal of Agriculture and Industry of South Australia*, vol. 2, 1898, p. 174; and G. Quinn, 'Orchard notes for August', *Journal of Agriculture and Industry of South Australia*, vol. 2, 1898, pp. 31–2.

8   Advisory officers attempted to revive the practice in the 1920s. W. C. Rigg, 'Poultry in the orchard', *Journal of Agriculture Victoria*, vol. 22, 1924, p. 186.

9   'Norton Summit Agricultural Society report', *Journal of Agriculture and Industry of South Australia*, vol. 2, 1898, p. 131; and G. R. Laffer, 'Experience with codlin moth 1898', *Journal of Agriculture and Industry of South Australia*, vol. 2, 1898, p. 190–93.

10   L. J. Newman, 'Codlin moth: warning', *Journal of Agriculture Western*

Australia, vol. 1, 1924, p. 483; L. J. Newman, 'Codlin moth', *Journal of Agriculture Western Australia*, vol. 2, 1925, pp. 340–45; L. J. Newman, 'Codlin moth', *Journal of Agriculture Western Australia*, vol. 3, 1926, pp. 531–34; G. Wickens, 'Codlin moth control', *Journal of Agriculture, Western Australia*, vol. 4, 1927, p. 13.

11 D. McAlpine, 'Black spot experiments', *Journal of Agriculture Victoria*, vol. 2, 1904, p. 761.

12 G. R. Laffer, *Journal of Agriculture and Industry of South Australia*, vol. 2, 1898, pp. 190–3.

13 C. B. Luffman, 'The vegetable foods of different races', *Journal of Agriculture Victoria*, vol. 3, 1905, p. 156.

14 G. R. Laffer, 'Experience with codlin moth 1898', *Journal of Agriculture and Industry of South Australia*, vol. 2, 1898, pp. 190–93.

15 'Catching moths', *Journal of Agriculture and Industry of South Australia*, vol. 2, 1898, pp. 331–2.

16 Transcript of the 1898 agricultural congress, *Journal of Agriculture and Industry of South Australia*, vol. 2, 1898, p. 199.

17 P. O. Hutchinson, 'Controlling fruit pests', *Journal of Agriculture and Industry of South Australia*, vol. 2, 1898, pp. 193–201.

18 'Notes and comments', *Journal of Agriculture and Industry of South Australia*, vol. 2, 1899, p. 470.

19 'Balaklava and Naracoorte Agricultural Society reports', *Journal of Agriculture and Industry of South Australia*, vol. 2, 1898, pp. 150–51; F. E. H. W. Krichauff, 'Chairman's address to the 1898 agricultural congress', *Journal of Agriculture and Industry of South Australia*, vol. 2, 1898, p. 186; A. S. Olliff, 'Entomological notes', *Agricultural Gazette of New South Wales*, vol. 2, 1891, pp. 385–6.

20 G. Quinn, 'Use of Bordeaux mixture in South Australia', *Journal of Agriculture and Industry of South Australia*, vol. 2, 1898, p. 33.

21 A. S. Olliff, 'Insect friend and foes', *Agricultural Gazette of New South Wales*, vol. 2, 1891, pp. 63–6; A. S. Olliff, 'Entomological notes', *Agricultural Gazette of New South Wales*, vol. 2, 1891, p. 158; G. Quinn, 'Orchard notes for August', *Journal of Agriculture and Industry of South Australia*, vol. 2, 1898, p. 31–2.

22 A. S. Olliff, 'Economic entomology in Victoria', *Agricultural Gazette of New South Wales*, vol. 2, 1891, pp. 489–91.

23 D. McAlpine, 'A fungus parasite on codlin moth', *Journal of Agriculture Victoria*, vol. 2, 1904, pp. 468–71.

24 F. E. H. W. Krichauff, 'Chairman's address to the 1898 agricultural congress', *Journal of Agriculture and Industry of South Australia*, vol. 2, 1898, p. 186; Transcript of the 1898 agricultural congress, *Journal of Agriculture and Industry of South Australia*, vol. 2, 1898, p. 198.

25 McAlpine, *Journal of Agriculture Victoria*, vol. 2, 1904, p. 469.

26 E. E. Pescott, 'Orchard and garden notes', *Journal of Agriculture Victoria*, vol. 8, 1910, p. 478.

27 G. Quinn, 'Use of Bordeaux mixture in South Australia', *Journal of Agriculture and Industry of South Australia*, vol. 2, 1898, p. 33.

28 A. S. Olliff, 'Entomological notes', *Agricultural Gazette of New South*

Wales, vol. 2, 1891, p. 72; A. H. Benson, 'The use of Paris green as an insecticide', *Agricultural Gazette of New South Wales*, vol. 3, 1892, pp. 835–6.

29 Transcript of the 1898 agricultural congress, *Journal of Agriculture and Industry of South Australia*, vol. 2, 1898, p. 200.
30 The only time the codlin grub was vulnerable to chemical attack was during the short period between hatching and burrowing into the apple. Lead arsenate was a particularly persistent chemical. When sprayed on the leaves it was ingested by the passing parade of caterpillars. 'Codlin moth: the calyx spray', *Agricultural Gazette of New South Wales*, vol. 52, 1941, pp. 531–3.
31 Instructions on this dangerous procedure can be found in C. French, 'Report of the entomologist', *Journal of Agriculture Victoria*, vol. 1, 1902, p. 795.
32 J. Lang, 'The codlin moth', *Journal of Agriculture Victoria*, vol. 2, 1903, p. 59.
33 G. Quinn, 'Use of Bordeaux mixture in South Australia', *Journal of Agriculture and Industry of South Australia*, vol. 2, 1898, pp. 34–5.
34 McAlpine, *Journal of Agriculture Victoria*, vol. 2, 1904, p. 765.
35 A. G. McCalman, 'Peach growing in Victoria', *Journal of Agriculture Victoria*, vol. 19, 1921, pp. 7–16.
36 J. M. Ward, 'The apple industry', *Journal of Agriculture Victoria*, vol. 25, 1927, p. 464.
37 A. A. Hammond, 'Fumigation for destruction of scale insects', *Journal of Agriculture Victoria*, vol. 10, 1912, pp. 366–74.
38 E. E. Pescott, 'Orchard and garden notes', *Journal of Agriculture Victoria*, vol. 8, 1910, p. 478; and R. T. Pescott, 'Codling moth control: experiments at Harcourt', *Journal of Agriculture Victoria*, vol. 29, 1931, p. 538.
39 R. T. Pescott, 'Codling moth control: experiments at Harcourt', *Journal of Agriculture Victoria*, vol. 31, 1933, pp. 484–9.
40 Ibid., p. 487.
41 L. Newman, 'Protection of useful insects, birds and animals', *Journal of Agriculture Western Australia*, vol. 1, 1924, pp. 45–7.
42 K. M. Ward, 'Insect pests of stone fruit: control measures recommended', *Journal of Agriculture Victoria*, vol. 30, 1932, p. 344; and H. W. Davey, 'Current orchard work', *Journal of Agriculture Victoria*, vol. 27, 1929, p. 719.
43 Davey, *Journal of Agriculture Victoria*, vol. 27, 1929, p. 719.
44 A. A. Hammond, 'Common orchard diseases: control measures recommended', *Journal of Agriculture Victoria*, vol. 30, 1932, p. 261.
45 '*Report on the Research Work on Brown Rot of Stone Fruit Conducted Under the Auspices of the Brown Rot Research Committee*', Victorian Government Printer, Melbourne, 1963.
46 A. G. McCalman, 'Peach growing in Victoria', *Journal of Agriculture Victoria*, vol. 19, 1921, p. 16.
47 G. W. Gayford, 'Cultivation and no cultivation in orchards', *Journal of Agriculture Victoria*, vol. 48, 1950, p. 512.

48	G. W. Gayford, 'In the orchard: soil erosion', *Journal of Agriculture Victoria*, vol. 58, 1960, p. 125.

49	G. T. Thompson, *A Brief History of Soil Conservation in Victoria*: 1834–1961, Soil Conservation Authority, Melbourne.

50	H. A. Presser, '*Did They Go Willingly Or Were They Pushed?*', School of Agriculture and Forestry, University of Melbourne, 1974.

51	C. J. R. Johnston, 'Red and other scale pests of citrus', *Journal of Agriculture Victoria*, vol. 58, 1960, p. 169.

52	Chemists turned to organic tin compounds to control the pests. Cyhexatin (Plictran) gave the best control of any available chemical, until it was banned.

53	H. B. Wilson, 'Poisonous organic phosphorus insecticides: care needed in their use', *Journal of Agriculture Victoria*, vol. 58, 1960, p. 453. See also R. J. Nancarrow, Practices and precautions with pesticides, unpublished Report, School of Agriculture and Forestry, University of Melbourne, 1977; and R. J. Nancarrow, Applying adult education principles to agricultural extension programs: A case study of pesticide extension, MAgrSc Thesis, School of Agriculture and Forestry, University of Melbourne, 1979.

54	*Annual Report*, Commission of Public Health, Victoria, 1976.

55	H. A. Presser, '*The Red Hill Survey*', School of Agriculture and Forestry, University of Melbourne, 1969.

56	A. Allen, The predatory mite in The Goulburn Valley: A survey of adopters and non-adopters, unpublished Report, School of Agriculture and Forestry, University of Melbourne, 1979.

57	These controls may include cultural practices to disrupt the life cycle of an insect pest; interplanting crops of different maturity, or the rotation of annual crops, to reduce the advantage a crop monoculture gives to pests; and the introduction of predators and disease and natural resistance.

58	R. Carson, *Silent Spring*, Houghton Mifflin, New York, 1962.

59	Carson, *Silent Spring*, Penguin, London, 1982, p. 1.

60	R. Clarke, *A Survey of Organic Retailers*, Department of Agriculture and Rural Affairs Research Report, Melbourne, 1989; and P. Morey, 'Marketing of Organics', Paper presented to the 'Sustainable Agriculture Conference', Benalla, 1988.

61	A. Podolinski, *Bio-dynamic Agriculture Lectures: vol 1*, Gavemer Foundation Publishing, Sydney, 1985, pp. 1788–9.

62	One aspect of bio-dynamic farm management is the use of the bio-dynamic fertilisers, the most important being called formulae 500 and 501. Formula 500 is manufactured from cow manure stored in cows' horns, which are buried and subsequently used at a time determined by phases of the moon. The application rate is extremely low, approximately 100 to 150 grams to the hectare. It is claimed this acts not as a fertiliser, but as a catalyst to unlock the organic nutrients in a bio-dynamic soil. Formula 501 is made of ground quartz and sprayed onto growing pasture to concentrate light onto plants to stimulate the 'life forces'. More mystical is an interest in planting by the moon and the influence of planets on the growth habits of various plant species. See A. Podolinski, *Bio-dynamic Agriculture Lectures: Volume 1*; H. H. Koepf, 'The principle and practice of bio-dynamic agriculture', in B. Stonehouse (ed.), *Biological Husbandry: A Scientific Approach to Organic Farming*, Butterworths, London, 1981, p. 237; and R. C. Oelhaf, *Organic Agriculture: Economic and Ecological Comparisons with Conventional Methods*, Wiley, New York, 1978, p. 118.

63   A. Podolinski, 'Agri-culture 1990', *Acres Australia*, No. 3, 1990, p. 24.
64   The term 'eco-agriculture' has also been claimed by others in the United States promoting conservation cropping using chemicals.
65   B. Mollison and D. Holmgren, *Permaculture One: A Perennial Agriculture for Human Settlements*, Corgi, Melbourne, 1978, p. 1.
66   B. Mollison, 'Appropriate scales for food production', *Acres Australia*, No. 3, 1990, pp. 28–9.
67   For example, see the description of the marketing of herbicides in C. Ewers, Innovation in response to environmental problems, MAgrSc. Thesis, School of Agriculture and Forestry, University of Melbourne, Parkville, 1990.
68   An example of chemical company promotion of IPM can be found in 'Research money pays off', *The Weekly Times*, No. 6197, 27 Feb., 1991, p. 28.
69   Parliament of Australia, *Senate Select Enquiry into Agricultural and Veterinary Chemicals*, Australian Government Printer, Canberra, 1990, pp. 190–91.
70   Ibid., pp. 5–6.
71   D. Crawford, A. Worsley and M. Peters, 'Is food a health hazard? Australian's beliefs about the quality of food', *Food Technology in Australia*, vol. 36, 1984, p. 9. H. Thompson, J. Ashley, C. Stopes and L. Woodward, 'Consumer Survey', Research Notes No. 6, Elm Farm Research Centre, Kintbury, 1988.
72   K. Love, 'Victorian produce monitoring: results of residue testing 1989–1990', Research Report Series No. 105, Department of Agriculture, Melbourne, May 1991.
73   *Pesticides: Issues and Options For New Zealand*, Ministry of Environment, Wellington, 1989, p. 9.
74   World Health Organisation, *Public Health Impact of Pesticides Used in Agriculture*, WHO, Geneva, 1990, p. 48.
75   Keith Schneider, 'IBT–guilty: how many studies are no good?', *Amicus Journal*, vol. 5, no. 2, 1983, pp. 4–7.
76   See Bruce Ames, 'Dietary carcinogens and anticarcinogens', *Science*, vol. 221, no. 4617, 1983, pp. 1256–64.
77   Ibid., p. 1258.
78   See B. Ames, M. Profet and L. S. Gold, 'Nature's chemicals and synthetic chemicals: comparative toxicology', '*Proceedings of the National Academy of Sciences of the United States*, vol. 87, Oct. 1990, pp. 7782–86.

79   The Australian Government has seen the existing ban on apple and pear imports as a trade barrier and has expressed willingness to negotiate district quarantining within New Zealand.

80   World Health Organisation, *Public Health Impact of Pesticides Used in Agriculture*, WHO, Geneva, 1990, pp. 86–9.

81   Allen, The predatory mite in the Goulburn Valley: A survey of adopters and non-adopters, unpublished Report, School of Agriculture and Forestry, University of Melbourne, 1979.

82   In a survey of Victorian farmers, 88 per cent of horticulturalists described themselves as 'edgy' about the long-term chemical effects of the sprays they used. Only a third believed all chemicals released were safe. Despite this, in the same survey 85 per cent of horticulturalists had used herbicides in the past 12 months, 85 per cent had used insecticides and 90 per cent had used fungicides. *Attitudes and Behaviour of Users and Retailers towards Rural Chemicals in Victoria*, Agrimark Consultants, Collingwood, 1988.

83   Parliament of Australia, *Senate Select Enquiry into Agricultural and Veterinary Chemicals*, Australian Government Printer, Canberra, 1990.

84   Results of an as yet unpublished survey of orchardists' attitudes to Integrated Pest Management, V. Bates, Tatura Institute for Sustainable Agriculture, pers. comm..

85   Hassell and Associates, 'The market for Australian produced organic food', Report prepared for the Rural Industries Development Corporation, Canberra, 1990.

86   Ibid., pp. 9–16.

87   Els Wynen and Geoff Edwards, 'Towards a comparison of chemical free and conventional farming in Australia', *Australian Journal of Agricultural Economics*, vol. 34, 1990, pp. 39–55.

88   Hassell and Associates, 'The market for Australian produced organic food', p. 37.

89   J. Cameron et al., *Recovering Ground*, Australian Conservation Foundation, Melbourne, 1991.

90   *Attitudes and Behaviour of Users and Retailers towards Rural Chemicals in Victoria*, Agrimark Consultants, Collingwood, 1988.

Chapter 10: The thirst for water

1   W. R. D. Sewell, D. I. Smith and J. W. Handmer, *Water Planning in Australia: From Myths to Reality*, Centre for Resource and Environmental Studies, Australian National University, Canberra, 1985.

2   Irrigated horticulture and orcharding had failed extensively in the Murray Valley because of rising watertables and salinity. Examples were the Swan Hill area, Mystic Park and Tresco near Kerang, the west of Echuca at Ballendella, Tongala and Shepparton; irrigated pasture failed in a wide area around Kerang and Swan Hill.

3   William Howitt, *Land, Labour and Gold*, Lowden Publishing, Kilmore,

1972, (Letter XI), pp. 76, 130, 150, 324, 406 (first published Longmans, London, 1855).

4   The money from Crown land sales was to be used to subsidise immigration to ensure a continual supply of fresh labour. S. H. Roberts, *History of Australian Land Settlement*, Macmillan, Melbourne, 1968.

5   E. Curr, *Recollections of An Australian Squatter*, Melbourne University Press, Carlton, 1965, pp. 157–8.

6   Howitt, *Land, Labour and Gold*, Letter XXXIX.

7   W. Brodribb, *Recollections of An Early Squatter*, first published by John Woods & Co., Sydney, 1883, republished by John Ferguson Pty Ltd, Sydney, 1978, p. 18.

8   J. Jenkins, *Diary of a Welsh Swagman* (ed. W. Evans), Macmillan, Melbourne, 1975, p. 86.

9   J. H. McColl 'Hugh McColl and the water question in northern Victoria' *Victorian Historical Magazine*, 1917.

10 *Report of the Water Conservancy Board Report on Irrigation*, 1882, Section 15.

11 'H. McColl memorial: surface irrigation canals', 12 September, 1883, *Victorian Parliamentary Papers*, 1883, Second Session, i, p. 653; C. S. Martin, *Irrigation and Closer Settlement in the Shepparton District 1836–1906*, Melbourne University Press, Carlton, 1955, p. 33.

12 J. M. Powell, *Watering the Garden State*, Allen & Unwin, Sydney, 1989, p. 104.

13 McColl resigned from Parliament, declaring he had achieved all he had fought for. In the following year he died. Ironically, in the same year his protagonist, Gordon, was destroyed politically by the lobby for irrigation development.

14 *Second Report of the Water Conservancy Board on Irrigation*, 1884.

15 In the same year, the embarrassing failure of the Pine Lodge Weir on the Broken River within one month of construction had further undermined the credibility of Gordon and Black. The position of the weir had been recommended by the Water Conservancy Board. See Martin, *Irrigation and Closer Settlement in The Shepparton District*, pp. 41–4.

16 Martin, *Irrigation and Closer Settlement in The Shepparton District*.

17 M. Sharland, *These Verdant Plains: A History of the East Loddon Shire*, The Hawthorn Press, Melbourne, 1971.

18 A. S. Kenyon, 'Stuart Murray and irrigation in Victoria', *Victorian Historical Magazine*, vol. 10, no. 3, 1925, p. 118.

19 Henry Gyles Turner, *A History of the Colony of Victoria*, Longmans, Green and Co., London, 1904, vol. 2, p. 256.

20 B. R. Davidson, *European Farming in Australia: An Economic History*, Elsevier Publishing, Amsterdam, 1981, pp. 163–4.

21 E. Mead, 'Irrigated agriculture in the Goulburn Valley', *Journal of Agriculture Victoria*, vol. 6, 1908, p. 263.

22 Tony Dingle, *Settling*, Fairfax, Syme and Wheldon, McMahons Point, 1984, p. 127.

23 Royal Commission relating to the working of the Closer Settlement

Acts, 'Progress report', *Victorian Parliamentary Papers*, vol. 2, no. 21, 1915; and 'Final report', *Victorian Parliamentary Papers*, vol. 2, no. 29, 1916.

24   *Victorian Parliamentary Debates: Legislative Assembly*, vol. 143, 27 July, 1916, p. 399.

25   For a discussion of soldier settlement schemes, see Marilyn Lake, *The Limits of Hope: Soldier Settlement in Victoria 1915–38*, Oxford University Press, Melbourne, 1987, Chs 2 & 3.

26   H. W. Forster, *Waranga*, Cheshire, Melbourne 1965, p. 105.

27   Ibid., pp. 105 & 113.

28   A. S. Kenyon, 'Drainage and irrigation', *Journal of Agriculture Victoria*, vol. 5, 1907, p. 207.

Chapter 11: A bitter legacy

1   A. B. Paterson, *A Vision Splendid*, Angus and Robertson, Sydney, 1990, p. 121.

2   When rain falls on irrigated land in spring, summer or autumn, the soil profile is already wet from irrigation. There is no dry soil buffer to absorb the rainfall, so the rain water enters an already wet soil profile and quickly percolates to the watertable. Variations due to rainfall tend to mask both the steady rise in watertables and the contribution of irrigation to this rise.

3   See A. S. Kenyon, 'Drainage and irrigation', *Journal of Agriculture Victoria*, vol. 5, 1907, p. 208; and G. H. Tolley, 'Irrigation', *Journal of Agriculture Victoria*, vol. 9, 1911, p. 223.

4   B. A. Santamaria, *The Earth Our Mother*, Araluen Publishing, Melbourne, 1945, pp. 7 & 13.

5   Ibid., pp. 61–2.

6   Ibid., pp. 52 & 55.

7   It was generally estimated that there needed to be 1 metre of drainage for each metre of supply channel. When Eildon was enlarged there was 0.4 metres of drainage for each metre of channel for the Goulburn Valley irrigation areas and only 0.2 metres of drainage for the Rodney District. See 'Land and water in the Goulburn Valley', *Aqua*, December 1965, p. 91.

8   'Reclamation in Gippsland', *Aqua*, Feb 1961, p. 132.

9   Farmers lucky enough to have exploitable shallow aquifers under their farm can recycle their groundwater for a time to avoid salinity problems. This recycling cannot be continued forever; it will inevitably increase the salinity of the watertable as more salt arrives in rainfall or in irrigation water.

10   A. J. Peck, J. F. Thomas and D. R. Williamson, *Salinity Issues: Effects of Man on Salinity in Australia*, Water 2000; Consultant Report No. 8, AGPS, Canberra, 1983.

11   D. J. Blackmore, 'Water — past developments, future consequences',

Proceedings of the Symposium on Murray–Darling Basin: A Resource to be Managed, Albury, NSW, 1989, Australian Academy of Technology, Science, and Engineering, Melbourne.

12   N. Mackay, T. Hillman and J. Rolls, 'Water quality of the Murray River: review of monitoring 1978 to 1986', Murray–Darling Basin Commission, Water Quality Report No. 1, 1988; and R. Morton and R. B. Cunningham, 'Longitudinal profile of trends in salinity in the River Murray', *Australian Journal of Soil Research*, vol. 23, 1985, pp. 1–13.

13   G. B. Alison, P. G. Cook, S. R. Barnett, G. R. Walker, I. D. Jolly and M. W. Hughes, 'Land clearance and river salinisation in the western Murray basin, Australia', *Journal of Hydrology*, vol. 119, 1990, p. 18.

14   T. Blake and P. Cock, *Community Awareness for Salinity Control: An Evaluation of Community Response and Future Strategies in the Goulburn Broken Region*, Graduate School of Environmental Science, Monash University, 1988, pp. 59–60.

15   N. F. Barr and J. W. Cary, *Farmer Perceptions of Soil Salting: Appraisal of an Insidious Hazard*, School of Agriculture and Forestry, University of Melbourne, 1984.

16   The reason for surprise is because dairy farmers rarely retire on their farms. There is a constant turnover in irrigation property ownership. Those who take over farms often do not have a knowledge of the history of the area to which they have moved.

17   Barr and Cary, *Farmer Perceptions of Soil Salting*.

18   J. Watson, *Riverine Plain Groundwater Usage Survey: Community Attitudes Towards Groundwater Use*, Rural Water Commission Investigations Branch Report No. 1988/28.

19   Recognising early salt damage means being able to detect decreased pasture growth. This is not a simple task. The farmer must recognise an initial change in pasture production. There are often other possible explanations for poor growth.

20   Farmers who did see salt damage on their own farm were much more likely to see high levels of salt damage in the district. Farmers who did not see salt on their farms saw comparatively little salt damage in the district.

21   Local reaction to the basin proposal is recounted in J. Sawtell and J. Bottomley, *The Social Impact of Salinity in the Shire of Deakin and Waranga*, Urban Ministry Incorporated, 1989, pp. 8–12.

22   H. G. Turner, *A History of the Colony of Victoria*, Longmans, Green and Co., London, vol. 2, 1904, p. 256.

23   B. R. Davidson, *Australia Wet or Dry?*, Melbourne University Press, Carlton, 1969.

24   Murray–Darling Basin Ministerial Council, *Salinity and Drainage Strategy: Background Paper*, Background Paper No. 87/1, December 1987, pp. 39–69.

25   Murray–Darling Basin Ministerial Council, *Salinity and Drainage Strategy: Background Paper*.

26   *Shepparton Land and Water Salinity Management Plan*, Draft Salinity

Management Plan, Goulburn Broken Region Salinity Pilot Program Advisory Council, Shepparton, 1989, p. 40.

27 The regional consequence was estimated, pessimistically, to be a loss of 3500 jobs, $48 million in wages and $120 million in agricultural output. *Shepparton Land and Water Salinity Management Plan*, pp. 51–5.

28 R. L. Wilkinson and N. F. Barr, *Evaluation of community consultation in the Victorian salinity program*, Department of Agriculture, Melbourne, 1992.

29 *Shepparton Land and Water Salinity Management Plan*, pp. 70–91.

30 Research at Tatura had shown that trees in irrigation areas lower the watertable immediately under the tree, but have little effect beyond the tree's drip zone. Urban views of trees as ameliorators of salinity were excessively optimistic. Shepparton townspeople overwhelmingly believed re-afforestation was the best way to control local salinity. They were unaware of the scale of planting required to achieve watertable control on the irrigation farms. See T. Blake and P. Cock, *Community Awareness for Salinity Control: An evaluation of Community Response and Future Strategies*, Department of Environmental Science, Monash University, 1988.

31 *Victorian Government Support for Salinity Plans*, Government of Victoria, Department of Premier and Cabinet, Melbourne, 1990, p. 2.

32 Ibid., p. 28.

33 Watson, *Riverine Plain Groundwater Usage Survey: Community Attitudes Towards Groundwater Use.*

34 A. Heuperman, 'Battling salt — a success story', *Rural Quarterly*, vol. 2, no. 3, Spring 1989, pp. 32–3.

35 J. W. Cary, unpublished research. The comparison was made by contacting the same irrigators at the beginning and end of the two-year period.

36 *Campaspe West: Planning for its Future*, Draft Salinity Management Plan, Campaspe West Sub-Regional Working Group, Echuca, 1989, p. 9.

37 C. M. Alaouze, M. Read and N. H. Sturgess, 'A statistical analysis of a pilot survey of salt affected dairy farms', *Australian Journal of Agricultural Economics*, vol. 29, 1985, pp. 49–62.

38 Barr and Cary, *Farmer Perceptions of Soil Salting.*

39 The government was to bear installation costs and the irrigation community bear the cost of operation and the cost of disposal of saline water into the drain as required under the Murray–Darling Basin Agreement. See *Cost Sharing Guidelines for the Development of Salinity Management Plans*, Salinity Bureau, Department of Premier and Cabinet, Melbourne, 1988; and *Shepparton Land and Water Salinity Management Plan*, p. 140.

40 Background Papers, Campaspe West Salinity Management Plan, Department of Agriculture and Rural Affairs, Melbourne, 1990.

41 For an American example see W. E. Martin, H. Ingram and N. K.

Laney, 'A willingness to pay: analysis of water resources development', *Western Journal of Resource Economics*, July 1982, pp. 133–9.
42 John Dainton, 'Community approaches to water salinity and management', Paper presented to the Shepparton Regional Outlook Conference, 3 October, 1990.

Chapter 12: Historic misjudgments

1 T. L. Mitchell, *Three Expeditions to the Interior of Eastern Australia*, vol. 2, T & W. Boone, London, 1838, pp. 271–2.
2 Joseph Hawdon, *The Journal of a Journey from New South Wales to Adelaide, Performed in 1838*, Georgian House, Melbourne, 1952, pp. 19 & 21.
3 T. F. Bride, *Letters from Victorian Pioneers*, Heinemann, Melbourne, 1969, p. 60.
4 Ibid., p. 328.
5 J. A. Kent, *The Major's Vision and its Fulfilment: A History of Pyramid Hill and District*, Cambridge Press, Bendigo, February, 1974, p. 173.
6 E. M. Curr, *Recollections of Squatting in Victoria*, 2nd edn, Melbourne University Press, Carlton, 1965, p. 169.
7 L. C. Bartels, 'Irrigation methods', *Journal of Agriculture Victoria*, vol. 26, 1928, p. 594.
8 L. C. Bartels, 'Pyramid Hill irrigated pastures competition', *Journal of Agriculture Victoria*, vol. 36, 1938, pp. 599–601.
9 K. R. Garland, 'Macorna land use survey: part 1', *Journal of Agriculture Victoria*, vol. 50, 1952, pp. 493–6.
10 J. A. Aird, 'Pyramid Hill competition: the efficient use and effective control of water', *Journal of Agriculture Victoria*, vol. 42, 1944, pp. 249–52.
11 Hawdon, *The Journal of a Journey from New South Wales to Adelaide*, pp. 19 & 21.
12 Curr, *Recollections of Squatting in Victoria*, p. 86.
13 O. B. Williams, 'The Riverina and its pastoral industry, 1860–1896' in A. Barnard (ed.), *The Simple Fleece*, Melbourne University Press, 1962, pp. 412–13.
14 L. F. Bartels, 'Flood irrigation of pastures: development of border check irrigation in Victoria', *Journal of Agriculture Victoria*, vol. 61, 1963, pp. 369–73.
15 L. C. Bartels, 'Irrigation practices', *Journal of Agriculture Victoria*, vol. 30, 1932, pp. 565–8.
16 Bartels, *Journal of Agriculture Victoria*, vol. 36, 1938, pp. 599–601.
17 H. Hanslow, 'Pyramid Hill competitions: control of water–summer fodders', *Journal of Agriculture Victoria*, vol. 41, 1943, pp. 295–8.
18 A. Morgan, 'Pasture improvement in alkali soils: results in Kerang district', *Journal of Agriculture Victoria*, vol. 40, 1942, pp. 171–7.

19   A. Morgan, 'Pasture investigations: observations of salt reclamation trials', *Journal of Agriculture Victoria*, vol. 41, 1943, pp. 349–53.

20   K. R. Garland, 'Macorna land use survey: part 3', *Journal of Agriculture Victoria*, vol. 51, 1953, pp. 29–33.

21   A. Morgan, 'Reclamation of salt affected land in the Kerang district', *Journal of Agriculture Victoria*, vol. 45, 1947, pp. 111–15.

22   Morgan, *Journal of Agriculture Victoria*, vol. 41, 1943, pp. 349–53.

23   Morgan, *Journal of Agriculture Victoria*, vol. 45, 1947, pp. 111–15.

24   K. R. Garland, 'Macorna land use survey: part 2', *Journal of Agriculture Victoria*, vol. 50, 1952, p. 512.

25   Bartels, *Journal of Agriculture Victoria*, vol. 36, 1938, pp. 599–601; H. Hanslow and L. C. Bartels, 'Pyramid Hill pasture competition, 1942', *Journal of Agriculture Victoria*, vol. 41, 1943, pp. 113–17; Aird, *Journal of Agriculture Victoria*, vol. 42, 1944, pp. 249–52; L. C. Bartels and M. Harvey, 'Pyramid farm competition', *Journal of Agriculture Victoria*, vol. 29, 1931, pp. 283–6.

26   A. Morgan, 'Pyramid Hill district pasture competition', *Journal of Agriculture Victoria*, vol. 39, 1931, pp. 93–6.

27   Hanslow and Bartels, *Journal of Agriculture Victoria*, vol. 41, 1943, pp. 113–17.

28   Hanslow and Bartels, *Journal of Agriculture Victoria*, vol. 41, 1943, pp. 113–17; Hanslow, *Journal of Agriculture Victoria*, vol. 41, 1943, pp. 295–8; and Aird, *Journal of Agriculture Victoria*, vol. 42, 1944, pp. 249–52.

29   Bartels, *Journal of Agriculture Victoria*, vol. 36, 1938, pp. 599–601; and G. B. Rayner, 'Pyramid Hill irrigated pasture competition 1946', *Journal of Agriculture Victoria*, vol. 45, 1947, pp. 107–110.

30   A. Morgan, 'Irrigated pastures in Victoria', *Journal of Agriculture Victoria*, vol. 48, 1950, pp. 363–4.

31   Morgan, *Journal of Agriculture Victoria*, vol. 45, 1947, pp. 111–15.

32   K. R. Garland, *Journal of Agriculture Victoria*, vol. 50, 1952, pp. 493–6.

33   J. K. M. Skene, *Soils and Land Use in the Mid-Loddon Valley, Victoria*, Technical Bulletin No. 22, Department of Agriculture Victoria, Melbourne, 1971, pp. 12–14.

34   Morgan, *Journal of Agriculture Victoria*, vol. 41, 1943, pp. 349–53.

35   K. R. Garland and G. O. Jones, 'Reclamation of salt affected land — test at Kerang', *Journal of Agriculture Victoria*, vol. 59, 1961, pp. 327–9.

36   C. Nicholson and K. Heslop, *A Survey of Current Irrigation and Pasture Management Practices in the Pyramid Hill Area*, Dept of Agriculture and Rural Affairs, Victoria, 1990.

37   R. Standen, *Survey of Current Farming Practices in the Tragowel Plains: Winter 1982*, Department of Agriculture Victoria, Research Series.

38   J. O. Ferguson, A. W. Smith, and P. J. Taylor, *Economic Aspects of the Use of Water in the Kerang Region*, Institute of Applied Economic and Social Research, University of Melbourne, 1979.

39 Standen, *Survey of Current Farming Practices in the Tragowel Plains.*
40 N. F. Barr, R. E. Weston and G. Wyatt, *The Tragowel Plains: A Study of an Agricultural Community*, Report prepared for the Department of Agriculture and Rural Affairs, Victoria, 1988.
41 S. N. Stone, Rural Communities: Facility Development, Attitudes and Self Determination in the Gordon Shire, unpublished MA Thesis, La Trobe University, Bundoora, 1977.
42 C. R. Ewers, *Laser Grading in the Goulburn–Murray Irrigation District: An Innovation Diffusion Study*, Monash Publications in Geography No. 35, Monash University, Clayton, 1988.
43 C. Norman, C. Lyle, A. Heuperman and D. Poulton, *Pyramid Hill Irrigation Area Soil Salinity Survey May–June 1988*, Institute for Irrigation and Salinity Research, Department of Agriculture and Rural Affairs, Tatura.
44 ACIL Australia, 'Agricultural conditions on the Tragowel Plains', Tragowel Plains Area Salinity Management Plan, Part B, vol. 2, Draft Report.
45 'The Tragowel Plains Area Salinity Management Plan', The Tragowel Plains Sub-Regional Working Group and Inter Department Planning Support Group, Preliminary Draft, 1989.
46 The rapid spread of laser grading is documented in Ewers, *Laser Grading in the Goulburn–Murray Irrigation District.*
47 N. F. Barr and J. W. Cary, *Farmers' Perceptions of Soil Salting: Appraisal of an Insidious Hazard*, University of Melbourne, School of Agriculture and Forestry, 1984.
48 P. M. Patto, 'Laser landforming/grading and salinity', A position paper for the State Salinity Research and Investigation Working Group, Irrigation Services Branch, Rural Water Commission, June, 1988, p. 18.
49 Ewers, *Laser Grading in the Goulburn–Murray Irrigation District.*
50 P. Sayer, 'Laser grading survey of cropping farms 1985', Rural Water Commission of Victoria, 1986.
51 Paper presented to Kerang Salinity Update Seminar, May 1986, by Don Naunton, Principal Consultant with ACIL Australia, acting at the time as a counsellor with the farm debt mediation scheme.
52 Barr, Weston and Wyatt, *The Tragowel Plains: A Study of an Agricultural Community.*
53 Norman, Lyle, Heuperman and Poulton, *Pyramid Hill Irrigation Area Soil Salinity Survey.*
54 G. H. Adcock, 'Native fodder plants: value of saltbush in arid and saline regions', *Journal of Agriculture Victoria*, vol. 3, 1905, p. 117.
55 In the period 1971 to 1986 in the Australian Bureau of Statistics Collector District around Swan Hill the population declined approximately 25 per cent. The population of Pyramid Hill fell by 15 per cent. Source: Australian Bureau of Statistics Survey of Population and Housing.
56 Barr and Cary, *Farmers' Perceptions of Soil Salting.*

Chapter 13: A tale of two settlements

1   One of the earliest squatters in the Swan Hill district, Andrew
    Beveridge, and several shepherds were killed by local Aborigines. As
    a result there were savage white reprisals against the blacks. See
    Michael Cannon, *Who Killed the Koories?*, William Heinemann, Mel-
    bourne, 1990, pp. 221–5.

2   J. W. Cary and N. F. Barr, 'Social profile of the Kerang Lakes region',
    Paper produced for the Kerang Lakes Area Working Group, 1988.

3   J. E. Robertson, *The Progress of Swan Hill and District*, Melbourne,
    1912.

4   L. Scholes, *A History of the Swan Hill Shire; Public Land, Private Profit
    and Settlement*, Shire of Swan Hill, 1989, p. 119.

5   The law determined that drainage works would only be built where
    there was 80 per cent local approval.

6   Royal Commission on Soldier Settlement, Majority Report and Mi-
    nority Report, *Victorian Parliamentary Papers*, vol. 2, 1925.

7   Narrow, hand dug channels would have been more efficient and
    decreased loss of water to the watertable but wide channels were
    cheaper to build. F. M. Read, 'Soil alkali investigations at Tresco',
    Journal of Agriculture Victoria, vol. 28, 1930, p. 66.

8   There were a number of reasons for the more successful promotion of
    Tresco. There were less onerous residential conditions that allowed
    owners to reside in nearby Lake Boga and commute to their block and
    allowed city speculators to buy a block and not farm it, but await the
    expected capital gains. Several prominent politicians were associated
    with the scheme, seemingly assuring it of favoured treatment. Glow-
    ing letters and articles of recommendation appeared in newspapers
    written by shareholders purporting to be independent evaluators of
    the project.

9   Read, *Journal of Agriculture Victoria*, vol. 28, 1930, p. 69.

10 J. K. Taylor, F. Penman, T. J. Marshall and G. W. Leeper, *A soil
    survey of the Nyah, Tresco, Tresco West, Kangaroo Lake and Goodnight
    Settlements*, Council for Scientific and Industrial Research, Bulletin
    No. 73, Melbourne, 1933.

11 A. S. Kenyon, 'Drainage and irrigation', *Journal of Agriculture Victoria*,
    vol. 5, 1907, p. 206.

12 Answers to Correspondents, *The Australasian*, November 1921.

13 Some claim this settlement was developed to help re-establish dis-
    placed Tresco settlers. *Report of the Committee of Inquiry Into the Mid-
    Murray Dried Vine Fruits Area*, Commonwealth Government, 1956,
    p. 54.

14 Reminiscences of Winnifred Trinham in Scholes, *A History of the
    Swan Hill Shire*; p. 144.

15 Taylor, Penman, Marshall and Leeper, *A Soil Survey of the Nyah,
    Tresco, Tresco West, Kangaroo Lake and Goodnight Settlements*, pp.
    18–21.

16 Read, *Journal of Agriculture*, vol. 28, 1930, pp. 65–90; F. M. Read, 'An investigation of the non-thrifty condition of citrus trees at Tresco', *Journal of Agriculture*, vol. 29, 1931, pp. 551–71; and J. E. Thomas, *An Investigation of the Problems of Salt Accumulation on a Mallee Soil in the Murray Valley Irrigation Area*, Council for Scientific and Industrial Research, Bulletin, No. 128, 1939.

17 A. J. McClatchie, 'Crops under irrigation', *Journal of Agriculture Victoria*, vol. 2, 1903, p. 15 reprinted from Arizona Agricultural Experimental Station Bulletin 41.

18 Because there were only a few V-shaped furrows in the middle of each row, water seeped from the centre towards the trees, talking salt with it. When the water evaporated near the tree, the salt around the roots of the trees remained there until it was leached by winter rains.

19 F. M. Read, 'Drainage investigations on the experimental citrus grove, Tresco, 1926–31', *Journal of Agriculture Victoria*, vol. 31, 1933, pp. 140–5.

20 Taylor, Penman, Marshall and Leeper, *A Soil Survey of the Nyah, Tresco, Tresco West, Kangaroo Lake and Goodnight Settlements*, pp. 15–16.

21 A. J. McIntyre, *Sunraysia: A Social Survey of a Dried Fruits Area*, Melbourne University Press, Carlton, 1948, p. 44.

22 *Report of the Committee of Inquiry into the Mid Murray Dried Vine Fruits Area.*

23 A farmer needed 20 to 25 acres (8 to 10 hectares) to survive with an acceptable income. In Nyah, 80 per cent of blocks were smaller than 20 acres. Only 29 of the 161 blocks were judged to be viable. In Tresco only 20 of the total 90 blocks were viable. *Dried Vine Fruit Industry: Management Practices in Victoria and New South Wales*, Bureau of Agricultural Economics, Canberra, 1956, p. 21.

24 *The Dried Vine Fruit Industry: An Economic Survey, 1965–66 to 1967–68*, Bureau of Agricultural Economics, Canberra, 1971, p. 13.

25 Scholes, *A History of the Swan Hill Shire*, p. 173.

26 Between 1954 and 1970 the number of Italian-owned farms in Tresco increased from four to twenty-five of the 90 settlement blocks.

27 A. J. Heslop, The Italian community: aspects of and barriers to communication in the Woorinen area of the Victorian Mid-Murray Horticultural District, unpublished report, University of Melbourne, School of Agriculture and Forestry, 1973.

28 R. T. Sloane, A survey of Tresco Irrigation District: a case study in the use of irrigation water of high salinity for horticultural purposes, unpublished report, University of Melbourne, School of Agriculture and Forestry, November 1971.

29 Sloane, A survey of Tresco Irrigation District, p. 31; and H. A. Presser and J. B. Cornish, *Channels of Information and Farmers' Goals in Relation to the Adoption of Recommended Practices*, Bulletin No. 1, Rural Sociology Department, University of Melbourne, 1968, p. 9.

30 Sloane, A survey of Tresco Irrigation District, pp. 27–8; and R. T. J.

Webber, 'Importance of irrigation layout and method', *Water Talk*, Summer 1962, p. 10.

31 N. Chamberlain, *Factors Affecting Some Farming Decisions In the Districts of Woorinen, Nyah, Tresco and Tyntynder: A Report to the Rural Water Commission*, Department of Political Science, University of Melbourne, 1988, p. 127; and Sloane, 'A survey of Tresco Irrigation District', p. 31.

32 Sloane, 'A survey of Tresco Irrigation District', pp. 27–8.

33 Ibid., pp. 29–30, 33.

34 Ibid., pp. 27–40.

35 *Dried Vine Fruit Industry: Management Practices in Victoria and New South Wales*, p. 22.

36 In 1963 the Bureau of Agricultural Economics reported that severe income problems continued, especially in the Mid-Murray where farm incomes were 30 per cent below sultana farm incomes at Mildura, *The Australian Dried Vine Fruits Industry: An Economic Survey*, The Bureau of Agricultural Economics, Canberra, 1966.

37 H. A. Presser, Preliminary report to the Nyah Development Committee, unpublished report, School of Agriculture and Forestry, University of Melbourne, 1971.

38 The scheme merely transferred the low income problem from the responsibility of one government department to another.

39 Chamberlain, *Factors Affecting Some Farming Decisions In the Districts of Woorinen, Nyah, Tresco and Tyntynder*, p. 21; N. Barr and K. McDougall, Social profile of the Nyah Settlement, unpublished report, Department of Agriculture and Rural Affairs, East Melbourne, 1990.

40 K. C. Hawke, *The Ideas and Perceptions of Year Ten Students in the Swan Hill Area on Community Salinity Mitigation Practices and Ground Water Tables*, Shire of Swan Hill, 1988, p. 16.

41 Chamberlain, *Factors Affecting Some Farming Decisions In the Districts of Woorinen, Nyah, Tresco and Tyntynder*, p. 78; and B. J. Menzies, and P. N. Gray, *Technological Change: Salinity and Irrigation Systems — The Response of Fruit Growers in the South Australian Riverland*, Department of Agriculture South Australia, Adelaide, Technical Paper No. 8, 1984, p. 240.

42 A. J. Heslop, Agricultural extension among Australian fruit growers of Italiana descent, MAgrSc Thesis, School of Agriculture and Forestry, University of Melbourne, 1974, pp. 12–24.

43 Cary and Barr, 'Social profile of the Kerang Lakes region', p. 23.

44 In spring each year, pumps on the Murray River pump fresh water across Pental Island into the Little Murray Weir to dilute the salt levels of the water flowing to Woorinen.

45 Loddon flood water is salty and if it is diverted down the overflow, fruit trees at Tresco and Woorinen are damaged. The flood water instead is sent down the Loddon River. Rural Water Commission and Camp Scott and Furphy Pty Ltd, 'Kerang Lakes flood study: draft options for distribution of floodwaters in the Kerang Lakes', May 1990.

46 *The Conservation Value of Wetlands in the Kerang Lakes Area*, Department of Conservation, Forests and Lands, Victoria, October 1989, p. 57.

47 *The Conservation Value of Wetlands in the Kerang Lakes Area*, p. 140; 'Hydrologic and environmental impact of the sill on third marsh: summary report to the Kerang Lakes working group', Kerang lakes Planning Support Group, 1988.

48 *The Conservation Value of Wetlands in the Kerang Lakes Area*, p. 108.

49 Ibid., p. 82; and 'Kerang Lakes Area management plan: draft issues and options', Kerang Lakes Area Working Group, July, 1988, p. 17–20.

50 *The Conservation Value of Wetlands in the Kerang Lakes Area*, p. 92.

51 Lake Tutchewop was the first of a planned series of evaporation basins, culminating in a pipeline to Lake Tyrrell in the Mallee. Landholder opposition to the Mineral Reserves Basin near Lake Tutchewop culminated in a class action against the Rural Water Commission. The Commission won the class action suit, but then announced it was cancelling the project for economic reasons.

52 In 1988 the Government established the Kerang Lakes Area Working Group, a group of local people supported by government scientists. The group held more than eighty meetings over four years and, at the time of writing, was still grappling with developing a management plan. They were presented with over thirty major technical reports, and many minor reports, by engineers, economists, agricultural scientists and biologists.

Conclusion

1 C. M. Donald, 'The progress of Australian agriculture and the role of pastures in environmental change', *Australian Journal of Science*, vol. 27, 1965, p. 188.

2 C. M. Donald, 'Innovation in Australian Agriculture' in D. B. Williams (ed.) *Agriculture in the Australian Economy*, Sydney University Press, Sydney, 1967.

3 Karl Popper, *Objective Knowledge: An Evolutionary Approach*, Oxford, London, 1979.

4 R. G. Dumsday and G. Edwards, 'Recent uncertainty about land degradation policies', Paper presented to the 5th Australian Soil Conservation Conference, Perth, 1990; and Geoff Edwards 'Regreening Australia: what rationale and what price?' *Agricultural Science*, vol. 3, no. 2, 1990, pp. 40–43.

5 L. E. Woods, *Land Degradation in Australia*, Australian Government Publishing Service, Canberra, 1983.

6 L. Zarsky, *Sustainable Development: Challenges for Australia*, Occasional Paper, The Commission for the Future, Canberra, 1990, p. 10.

7 R. Eckersley, *Regreening Australia: The Environmental, Economic and*

Social Benefits of Reforestation, Occasional paper No. 3, CSIRO, Canberra, June 1989, p. 8.

8   Tom Griffiths, 'History and natural history: conservation movements in conflict?' in D. J. Mulvaney (ed.) *The Humanities and the Australian Environment*, The Australian Academy of the Humanities, Canberra, 1991.

9   K. Boulding, *Evolutionary Economics*, Sage Publications, Beverly Hills, 1981, p. 18.

10 Jerry Ravetz, 'Knowledge in an uncertain world', *New Scientist*, 22 September 1990.

11 Primo Levi, *If This is a Man* and *The Truce*, Abacus, London, 1987, p. 396.

Index